Insecticides:
Mechanism of Action
and Resistance

Insecticides:
Mechanism of Action
and Resistance

Edited by

D. OTTO
B. WEBER

Intercept

Andover

Published by Intercept Limited, P.O. Box 716, Andover, Hants SP10 1YG, England

British Library CIP data available
ISBN 0 946707 38 3

Typeset by Bookman, Bristol

Printed in Great Britain by Athenaeum Press Ltd,
Newcastle upon Tyne

Preface

We present here the papers given in Reinhardsbrunn, Germany, during the Second Symposium on Insecticides, their Mechanisms of Action and Resistance. The Castle of Reinhardsbrunn has a history of holding specialist meetings in pesticide basic research: since 1974 our colleagues from fungicides research have held eight symposia there, organized by Professor Lyr. The response to the First Symposium on insecticides in 1988 encouraged us to hold the Second Symposium, once again inviting specialists in insecticide and resistance research.

The topics of this second meeting are of current interest. In future, agricultural production will be more intensive and economical as farmers consider financial restraints and the pressing demands for food around the world. There is no doubt that the chemical plant protection agents will continue to be needed in integrated pest management programmes. Plant protection, however, has to take into account the requirements of a healthy environment and the quality of pesticides must be very high. Pesticides have to be highly effective with high safety and environmental profiles. Thus, the increasing demands and costs for the development of new pesticides combine to promote the 'biorational ways'.

Many efforts have been made to find new insecticidal key structures and processes in naturally occurring products and to identify new targets in insects suitable for selective attacks by safe chemical compounds. The physiology and biochemistry of insect developmental processes are of special importance in this respect and this is reflected in the contributions at the Symposium. There is a noticeable concentration of papers describing the hormonal regulation of insect moulting and developmental processes as a target for insecticidal action. The search for new insecticides among natural compounds and their derivatives, as well as for new insecticidal principles of action, is also strongly emphasized in these proceedings. Neurophysiology, formerly the principal domain of research into insecticide mechanisms is represented by only a few but interesting contributions. This, however, may have been due to other important meetings on neurophysiology and neurotoxicology taking place elsewhere and concurrently with this Symposium.

The problems of insecticide resistance are also of immediate concern. As the number of available active compounds is restricted as a consequence of increasing ecotoxicological demands, farmers as well as manufacturers are interested in ensuring a high effectiveness of insecticides which guarantee environmental safety. The resistance of insects to such compounds is central to this matter. Emphasis is laid more and more on understanding the genetics of insect resistance to insecticides. I believe much remains to be done to turn the theoretical insights into practical tools for resistance management.

Well-known specialists from the various areas of insecticide research attended this Symposium. It showed that the need exists for collaboration among experts from basic research, applied entomology and the pesticide-developing companies to discuss results, concepts and the different approaches into pesticide and resistance research. We are very grateful to the sponsors, whose generous support made it possible to bring together specialists from East and West, from overseas and from the so-called Third World. The Symposium was supported by the Deutsche Forschungsgemeinschaft and by the companies BASF, Bayer, Ciba-Geigy, Hoechst AG and Schering AG. It was organized by staff members of the Biological Research Centre Berlin in Kleinmachnow, Germany.

We were pleased to welcome 60 participants from 15 countries including specialists from universities and other scientific institutions of basic research, from institutes of plant protection and medicinal epidemiology and from the agrochemical industry.

Dieter Otto
Bernd Weber
Kleinmachnow[1]
April 1991

---

[1]Present address: KAI e.V., Projectgroup, Natural Compounds at the University of Potsdam, D-O-1531 Kleinmachnow, Stahnsdorfer Damm 81, Germany

Acknowledgements

The Second International Symposium on Insecticide Research held in Reinhardsbrunn was kindly sponsored by the **Deutsche Forschungsgemeinschaft (DFG)** as well as by several companies of the chemical industry (listed in alphabetical order):

BASF AG, Ludwigshafen (Germany)
BAYER AG, Leverkusen (Germany)
CIBA-GEIGY Ltd, Basel (Switzerland)
HOECHST AG, Frankfurt a.M. (Germany)
SCHERING AG, Berlin (Germany)

The editors wish to express their gratitude to the above mentioned sponsors on behalf of all symposium participants. Furthermore, the editors are grateful to Dr A. Butterfield (Birstall, Batley, UK) and to Miss S.V. Norsinski (BBA Kleinmachnow) for carefully styling, correcting and retyping numerous manuscripts.

Contributors

A.M. ABDEL-LATIEF *Faculty of Agriculture at Moshtohor, Tukh, Egypt*

M. AGARWAL *Université de Pau et des Pays de l'Adour, Pau, France*

S.M. AHMED *Faculty of Agriculture at Moshtohor, Tukh, Egypt*

C.S. AVEYARD *Schering AG, Plant Protection, Berlin, Germany*

L. BANASIAK *Biological Research Centre Berlin, Kleinmachnow, Germany*

H. BARTOSZ-BECHOWSKI *University Wrocklaw, Poland*

J.A. BENSON *Ciba-Geigy Ltd, Agricultural Division, Basel, Switzerland*

J. BRIDE *INRA, Laboratoire de biologie de invertebrés, Antibes, France*

G.T. BROOKS *University of Reading, Reading, UK*

K. CAMPBELL *Université de Pau et des Pay de l'Adour, Pau, France*

G. CASPERSON *Biological Research Centre Berlin, Kleinmachnow, Germany*

I. DENHOLM *AFRC Institute of Arable Crops Research, Rothamsted Experimental Station, Harpenden, Herts, UK*

A.L. DEVONSHIRE *AFRC Institute of Arable Crops Research, Rothamsted Experimental Station, Harpenden, Herts, UK*

M. DOMBROWSKI *University Potsdam, Potsdam, Germany*

B. DOROBEK *Biological Research Centre Berlin, Kleinmachnow, Germany*

A.Y.M. EL-LAITHY *National Research Centre, Cairo, Egypt*

F.A. EL-LAKWAH *Faculty of Agriculture at Moshtohor, Tukh, Egypt*

M. ETO *Kyushu University, Fukuoka, Japan*

N.A. FILIPPOV *Institute of Biological Methods for Plant Protection, Kishinev, Moldova*

D. FOURNIER *INRA, Laboratoire de biologie des invertebrés, Antibes, France*

I. GELBIČ *Czechoslowak Academy of Science, Institute of Entomology, Česke Budějovice, Czechoslowakia*

G.P. GEORGHIOU *University of California, Riverside, USA*

K.G. GORMAN *AFRC Institute of Arable Crops Research, Rothamsted Experimental Station, Harpenden, Herts, UK*

M. GUNDEL *Friedrich Schiller University, Jena, Germany*

M.S. HAMED *University of Missouri, Columbia, USA*

M. HIRANO   *Sumitomo Chemical Ltd, Takarazuki Research Centre, Takarazuka-Shi, Japan*
T. HIROYOSHI   *Osaka University, Osaka, Japan*
I. ISHAAYA   *Agricultural Research Centre, Volcani Centre, Bet Dagan, Israel*
T.N. KARELINA   *Institute of Biological Methods for Plant Protection, Kishinev, Moldova*
H. KAYSER   *Ciba-Geigy Ltd, Agricultural Division, Basel, Switzerland*
H.C. von KEYSERLINGK   *Schering Ag, Agrochemical Research, Berlin, Germany*
M.M. KHATTAB   *Faculty of Agriculture at Moshtohor, Tukh, Egypt*
H. KLEEBERG   *Trifolio-M GmbH, Lahnau, Germany*
W. KNAUF   *Höchst Ag, Agricultural Division, Biological Research, Frankfurt a. Main, Germany*
Ch. O. KNOWLES   *University of Missouri, Columbia, USA*
A.C. KOTZE   *NSW Agriculture and Fisheries, Rydalmere, Australia*
B.G. KOVALOV   *Institute of Biological Methods for Plant Protection, Kishinev, Moldova*
D. KONOPINSKA   *University of Wroclaw, Poland*
K. KUWANO   *Kyushu University, Fukuoka, Japan*
H. LEISNER   *Medical Academy Erfurt, Erfurt, Germany*
D. MARTIN   *Biological Research Centre Berlin, Kleinmachnow, Germany*
G. MATOLSCY   *Plant Protection Institute, Hungarian Academy of Science, Budapest, Hungary*
E. MOLL   *Biological Research Centre Berlin, Kleinmachnow, Germany*
G.D. MOORES   *AFRC Institute of Arable Crops Research, Rothamsted Experimental Station, Harpenden, Herts, UK*
G. MOUCHÉS   *Université de Pau et des Pays de l'Adour, Pau, France*
A. MUELLER   *Biological Research Centre Berlin, Kleinmachnow, Germany*
A. MUTERO   *INRA Laboratoire de biologie des invertebrés, Antibes, France*
I. NAWROT   *Institute of Plant Protection, Poznan, Poland*
S.V. NEDELKINA   *Institute of Cytology and Genetics Nowosibirsk, Siberian branch of the USSR Academy of Science, Nowosibirsk, USSR*
D. OTTO   *Biological Research Centre Berlin, Kleinmachnow, Germany*

H. PENZLIN   *Friedrich Schiller University, Jena, Germany*
A. PLECH   *Silesian Academy of Medicine, Zabrze, Poland*
M. PRALVAVORIO   *INRA Laboratoire de biologie de invertebrés, Antibes, France*
T.A. PUDOVA   *A.N. Nesmeyanow Institute of Organo-element Compounds, Moscow, Russia*
N.A. PUHALSKYA   *Institute of Biological Methods for Plant Protection, Kishinev, Moldova*
H. REMBOLD   *Max Planck Institute for Biochemistry, Martinsried, Germany*
S.E. REYNOLDS   *School of Biological Sciences, University of Bath, Bath, UK*
K. RICHTER   *Saxon Academy of Sciences Leipzig, Research Group Jena, Jena, Germany*
P. RICHTER   *Biological Research Centre Berlin, Kleinmachnow, Germany*
G.A. ROSENTHAL   *University of Kentucky, Lexington, USA*
G. ROSINSKI   *A. Mickiewicz University Poznan, Poznan, Poland*
F.J. RUDER   *Ciba-Geigy Ltd, Agricultural Division, Basel, Switzerland*
R.I. SALGANIK   *Institute of Cytology and Genetics Nowosibirsk, Siberian Branch of the USSR Academy of Science, Nowosibirsk, USSR*
I. SCHEFFLER   *University Potsdam, Potsdam, Germany*
H. SCHMUTTERER   *Justus Liebig University Giessen, Giessen, Germany*
U. SCHOKNECHT   *Biological Research Centre Berlin, Kleinmachnow, Germany*
W. SOBOTKA   *Institute of Industrial Organic Chemistry, Warsaw, Poland*
U. STARK   *Biological Research Centre Berlin, Kleinmachnow, Germany*
G. STERK   *Research Station of Gorsem, Sint Truiden, Belgium*
A. STOLTNER   *Max Planck Institute for Biochemistry, Martinsried, Germany*
B. STYCZYNSKA   *State Institute of Hygiene, Warsaw, Poland*
G. SUKHOROCHENKO   *All-Union Institute of Plant Protection, Leningrad, USSR*
J. ŠULA   *Czechoslowak Academy of Science, Institute of Entomology Česke Budéjovice, Czechoslowakia*
Y. TAKADA   *Sumitomo Chemical Ltd, Takarazuka Research Centre, Takarazuka-Shi, Japan*
Z. TANG   *Shanghai Institute of Entomolgy, Academia Sinica,*

Shanghai, China

I. UJVARY   *Plant Protection Institute, Hungarian Academy of Science, Budapest, Hungary*

L. VARJAS   *Plant Protection Institute, Hungarian Academy of Science, Budapest, Hungary*

L.A. VICHREVA   *A.N. Nesmeyanov Institute of Organo-element Compounds, Moscow, USSR*

B. WEBER   *Biological Research Centre Berlin, Kleinmachnow, Germany*

R.J. WILLIS   *Schering Agrochemicals Ltd, Essex, UK*

U. WUTTIG   *Friedrich Schiller University Jena, Jena, Germany*

M. ZARCHADA   *Czechoslowak Academy of Science, Institute of Entomology, Česke Budéjovice, Czechoslowakia*

Contents

PART 2   MOULTING AND DEVELOPING PROCESSES AND THEIR HORMONAL REGULATION AS TARGETS FOR INSECTICIDE ACTIONS

Part 1

Natural Compounds and Their Derivatives as a Source of New Insecticides and Insecticidal Principles

1
Higher Plants as Sources of Novel Pesticides

H. SCHMUTTERER

Institut für Phytopathologie und Angewandte Zoologie, Ludwigstr. 23, Giessen, Germany

Introduction

Simple crude extracts from plants such as the common yew tree, *Taxus baccata*, have been used for centuries but have never played a major role in pest control. This is due not only to, *inter alia*, lack of knowledge but also to the relatively laborious production of such natural pesticides. During the twentieth century pyrethrum, rotenone, sabadilla, quassin and nicotine have been used on a limited scale, and some of these are still used (Jacobson, 1989).

Pyrethrum is the best known botanical. It is a mixture of various pyrethrins, jasmolins and cinerins, extracted at low temperatures with hexane from young flowerheads of *Chrysanthemum* spp. (Asteraceae). Rotenone is obtained from various parts of plants from different families.

Extracts containing the original active compounds can be used directly, but these can also form the basis for synthesizing products with similar or even better insecticidal properties. The natural pyrethrins, for instance, are the precursors of the synthetic pyrethroids that play an important role in modern chemical crop protection. The synthetic carbamates derive from physostigmine, a natural carbamate from *Physostigma venenosa*.

The intensive search over the past twenty years for plants with novel pesticidal constituents has had some success. There are

Insecticides: Mechanism of Action and Resistance
© 1992 Intercept Ltd, P.O. Box 716, Andover, Hants SP10 1YG, UK

numerous such plant species in the families Meliaceae, Rutaceae, Asteraceae (Compositae), Labiatae, Aristolochiaceae and Annonaceae among others. The active principles of some selected species of these families and – as far as are known – their modes of action are discussed in the present chapter.

Screening of Botanical Pesticides

Screening methods are of paramount importance in obtaining promising insecticides from plants. Many researchers focus on tropical plants as more active ingredients are expected from them. This is because they are exposed to a greater selection pressure by phytophagous animals under constantly warm climatical conditions. Plants in temperate climates should not be neglected however, as they may also yield interesting compounds. There are a number of problems in screening extracts of many plant species. The extract's activity may depend on the plant part used, the test organism, the method and time of collection, the method of extraction and the solvent used. In 1984, for instance, *Calystegia sepium* (Convolvulaceae) plants were collected near Giessen (Germany) and their methanolic extracts tested against fourth instar larvae of the Mexican bean beetle, *Epilachna varivestis*. The results were convincing, as the growth of the test insects was retarded and many of them died during moults, thus indicating that the active ingredient(s) had a growth regulating (IGR) effect. In 1985, however, extracts from the same plant species collected near Giessen were not effective. The summer climate conditions of the two years were entirely different: 1984 was warm and dry; 1985 was cool and humid.

Plant ecotypes in different geographical regions, too, may differ in their active ingredients. Consideration must also be given to the fact that some active principles show a very low stability and are lost during the extraction and purification process. In general, plant species of the same or closely related genera contain the same or similar active principles. This applies, for instance, to *Azadirachta* (Meliaceae) and *Ajuga* (Labiatae). Consequently, a screening system which was successful with one species, should also be attempted with closely related species.

In screening systems emphasizing plant constituents with IGR properties, very sensitive test insects are essential. To shorten relatively long tests the last instar larvae should be used in holometabolous insects. In our laboratories in Giessen the Mexican bean beetle, *E. varivestis*, tested on treated bean (*Phaseolous vulgaris*) plants, proved very suitable. A standardized biotest system, developed by Steets in 1975, gave reliable results in studies using extracts of *Azadirachta* spp., *Melia* spp., *Ajuga* spp., *Asarum europaeum*, *Taxus baccata fastigiata*, and *Gaillardia* sp. This phytophagous coccinellid does not react on β–ecdyson.

Plants/Plant Constituents with Insectical Properties

Meliaceae

The literature on ingredients of Meliaceae has grown rapidly during the last ten years. Many phytochemists have concentrated mainly on the constituents of the neem tree, *Azadirachta indica*. The most important properties of the main active principle from the seed kernels, the tetranortriterpenoid (limnoid) azadirachtin, were described in a contribution to the first symposium in Reinhardsbrunn (Schmutterer, 1989a). Since then endocrinologists have 'fleshed out' the description of azadirachtin's unique mode of action. It is a phagodeterrent to some insects, for instance *Schistocerca gregaria*; in others an antifeedant. Its main effect from the viewpoint of pest control is on the endocrine system: it leads to a moulting disturbance which is often lethal. Also the fecundity of many insect species is reduced. These effects are mainly caused by the reduction of ecdysteroid titres in nymphs/larvae; in adults (locusts) also of juvenile hormone (Rembold *et al.*, 1984). The main reason for the reduction in ecdysteroid titre seems to be that the release of the prothoracicotropic hormone from the insect's brain is prevented by azadirachtin (Barnby and Klocke, 1990). According to Rembold (1989), the corpus cardiaca is the main target of azadirachtin. In Table 1 some effects of azadirachtin or azadirachtin-containing extracts are listed.

Some of the effects on the fitness (vigour) of adult insects may be important for future pest management programmes. The reduction of fecundity and flying ability, in particular, could be utilized. The influence of neem products, especially neem oil, on the phases of the desert locust, *S. gregaria*, may also be of practical importance (Nicol and Schmutterer, 1991). Gregarious locusts are harmful, solitary ones are not.

After the procedures for producing enriched neem insecticides based on alcoholic neem seed kernel extracts were first published (Feuerhake and Schmutterer, 1982) the first commercial product Margosan-O was registered by the Environmental Protection Agency (EPA) for ornamentals in the United States in 1985. In 1990, another US firm applied for the registration of Azatin, which mainly contains hydrogenated azadirachtin as the active principle, and shows a much greater stability than azadirachtin itself. Margosan-O is used in the United States in greenhouses against whiteflies, such as *Bemisia tabaci* and *Trialeurodes vaporariorum*, apparently with success.

Table 1   Effects of azadirachtin and of extracts from neem seed kernels on nymphs/larvae and adults of insects.

Nymphs/larvae	Adults
Repellent effect	Repellent effect
Antifeedant effect	Antifeedant effect
Disturbance or prevention of moults	Antiovipositional effect
Prolongation of larval/nymphal development	Attraction of egglaying females (*Plutella xylostella*)[*]
Morphogenetic defects	Morphogenetic defects
Reduction of walking and spinning ability	Reduction of fecundity
Change of gregarious to transient and solitary forms (locusts)	Egg sterility (moderate and rare)
	Impotence of males (Heteroptera)
	Reduction of flying and walking ability
	Reduction of ability to recognize the male pheromone (Tephritidae)
	Prolongation of lifespan (females of Coleoptera)

[*]Klemm, personal communication

Some attempts to synthesize molecules simpler than azadirachtin have been successful (Nishikimi *et al.*, 1989; Anderson and Ley, 1990). However, it must be determined, especially under field conditions, whether the synthetic compounds can compete with the natural product(s) which show, as already indicated, a great variety of effects on insects. The mixtures of active principles in neem pesticides are, as a rule, significantly more effective than azadirachtin alone.

Water extracts, seed oil and seed powder (ground seed kernels) gave convincing results against a wide spectrum of pests in developing countries where farmers can produce them easily with simple methods and equipment. Although preparing such natural pesticides, from the harvest of fruit to the extraction of ground seed or seed kernels, is more laborious than the use of a ready-to-use synthetic pesticide, the acceptance by farmers in parts of Latin America is surprisingly good, especially in areas where they have had bad experiences with highly toxic synthetics. Furthermore, the problem of pest resistance to conventional pesticides requires new, expensive, unaffordable products. In India, at least eight neem products are on the market, but their quality is definitely poor because they are not standardized. It is as essential to standardize botanicals as it is any pesticide and to formulate them in a way that guarantees a satisfactory shelf life; otherwise they will fail. Some plant species related to neem probably contain identical or similar active principles. For example, in extracts from the seed of the chinaberry tree, *Melia azedarach*, there are ingredients with antifeedant and IGR properties. This plant, also of Asian origin, has a wide distribution including the Mediterranean, and it is particularly common in North Africa and Greece. Like neem, locusts do not feed on it (Ascher, 1987). Our working group found that extracts from young fruit and leaves are clearly more effective than those from old ones (Zhu and Schmutterer, in press). Based on the *E. varivestis* test for IGR properties, a highly active pure substance has recently been isolated from the leaves and awaits characterization (Zhu and Ermel, 1991). *Melia toosandan* is another interesting species from southern China. It contains antifeedant and IGR phytochemicals (Chiu Shin-Foon, 1989).

The disadvantage of *M. azadarach* is the high toxicity of its fruit to warm-blooded organisms. But its leaves seem to be much less toxic. In Australia toxic and non-toxic ecotypes occur. The recently 'rediscovered' species *Azadirachta excelsa* (= *A. integrifoliola*), the marrango or Philippine neem tree, is an endangered species, at least in parts of its distribution range in South-East Asia. Its seed kernels contain, besides azadirachtin, a closely related substance which will be characterized soon (Ermel, personal communication). Water extracts, methanolic extracts and oil from the seed kernels of *A. excelsa* gave better results in biotests with *E. varivestis* and against *S. gregaria* than the corresponding extracts from neem (Schmutterer, 1989b; Doll and Schmutterer, in press). Salannin, a major component in extracts from *A. indica*, is practically non-existent in *A. excelsa* extracts. This compound has only antifeedant properties. In a comparison of methanolic extracts from the leaves of various Meliaceae, using first instar larvae of *Peridroma saucia* (Noctuidae) as test insects, extracts from *Aglaia odorata* and *Turrea holstii* proved almost as active as those of *A. indica* (Champagne *et al.*, 1989).

Rutaceae

There are numerous publications on the ingredients of Rutaceae and their effects on the feeding behaviour of insects. Many of them describe limonoids from *Citrus* spp., for instance nomilin and limonin. In a feeding assay with *Heliothis virescens* as test insect nomilin showed the same activity as azadirachtin, but limonin was ten times less effective. Enormous differences in effectiveness were found in tests with different insect species. Limonoids could be obtained from citrus peels in large quantities. Unfortunately, their short residual effect under field conditions precluded their application. However, it should be kept in mind that the use of antifeedants in crop protection if they possess no additional other properties (systemic action, etc.) could be problematic. Adult insects will fly to neighbouring crops and nymphs/larvae with high mobility may also leave treated fields and reach untreated ones. Interestingly, limonoid bitters produced as a by-product of the citrus fruit industry has been proposed as a cheap starting material for the synthesis of azadirachtin.

Various furanocoumarins including bergapten and isopiminellin were extracted from *Thamnosa montana* (Klocke *et al.*, 1989). Bergapten showed a phototoxic effect on *Heliothis virescens* larvae, isopimpinellin did not. The diet ingestion and weight gain of the test insects was greatly reduced. Apart from the Rutaceae tropical Asteraceae synthesize the widest range of phototoxic phytochemicals (Downum, 1985). Other Rutaceae are sources of isobutylamides of unsaturated C_{10}–C_{18} acids. They cause knockdown effects on flying insects and are toxic to cockroaches and some other insects with chewing mouthparts (Miyakado *et al.*, 1989).

Asteraceae (Compositae)

Chrysanthemum spp. are, as already mentioned, sources of pyrethrins, which are a group of closely related esters.

Encecalin, an acetylchromene, was obtained from the desert sunflower *Encelia* (Isman, 1989). This chemical is toxic to a number of insect species but its decomposition under field conditions is relatively quick. Consequently the prospects for its use in pest control are nil. Precocene I and II (7-methoxy-chromenes), isolated more than ten years ago from the flower heads of *Ageratum houstonianum*, attracted much attention because of their mode of action and relatively simple structure. They act as anti-juvenile hormones in Heteroptera, in particular, because of their cytotoxic action on the tissue of the corpora allata. Encecalin, which differs in structure only slightly from precocene II, has no anti-juvenile hormone activity (Isman, 1989). The narrow spectrum of efficacy in insects and cytotoxic effects on liver cells of rats precluded further studies on practical use of precocenes.

The sesquiterpene lactone alantolactone, which has antifeedant properties, was obtained from extracts of *Inula helenium* (Jacobson, 1989). Other Asteraceae contain a range of repellent and insecticidal ingredients, such as flavonoids, alkaloids, sesquiterpenes and acetylene derivatives (Bell *et al.*, 1990).

Labiatae

Some *Ajuga* spp. from Africa (*A. remota*) and Europe (*A. reptans*) have been studied during the last ten years. First the ajugarins were identified (Kubo *et al.*, 1976); they are clerodane diterpenes. Later, a number of phytoecdysoids affecting moults and fecundity of insects were isolated (Schmutterer and Tervooren, 1980; Kubo *et al.*, 1981). Ajugarins were also isolated from *Plectranthus* and *Teucrium*, but *Ocimum* and *Mentha* contain volatile monoterpenes. Basil and mint have been used for pest deterrence in the home for centuries. Crude extracts from *Ocimum sanctum* caused a high mortality in the aphid *Myzus persicae* (Stein and Klingauf, 1990). In extracts of *O. basilicum* the clerodane juvocimenes I and II were identified, which cause IGR effects in the large milkweed bug, *Oncopeltus fasciatus* after topical application. The larvicidal effect of basil oil is at least partly attributed to the two juvocimenes but there are also some other active compounds, including linalool and some sesquiterpenes.

Some isomeric ajugarins have been synthesized. However, the application of these compounds is doubtful as their biological activity is relatively low.

Aristolochiaceae

Aristolochia spp. contain aristolochic acid and asarone. The former is toxic to insects, the latter is regarded as a sterilant. During recent years a substance was found in *Asarum europaeum*, which causes IGR effects in larvae and sterilization in adults of both sexes of *E. varivestis*. This compound has not been characterized but it differs from aristolochic acid, α- and β-asarone (Schmutterer and Kleffner, 1988). Aristolochic acid and asarone are suspected of carcenogenicity, so further studies with them in conjunction with crop protection are not promising.

Annonaceae

Annona spp. are rich in alkaloids. Extracts from the seeds were used in the past in South-East Asia for pestcontrol, but severe cases of human poisoning were reported. Toxic effects of ethereal extracts from twigs and IGR effects of ethanolic extracts on *Diabrotica* beetles were attributed to kaurane and 16-kaurane diterpenes. The papaw tree, *Asimina triloba*, contains some potent acetogenins in its bark. One of them, asimicin, has been characterized. It has antifeedant effects and causes mortality in a number of pest species. A synthesis of asimicin would be difficult because of a great number of isomers, but the bark of annonaceous trees may be utilized as a source of effective crude extracts (Alkofahi *et al.*, 1989). Annonin, another acetogenin, was isolated from the seeds of *Annona squamosa*. This compound causes high mortality in a wide range of insects but is also very toxic to warm-blooded organisms.

Other plant families

Several sesquiterpene dialdehydes were detected in extracts from East African trees, for instance *Warburgia ugandensis* (Canellaceae). In the laboratory they showed strong antifeedant properties against some noctuid larvae, e.g. *Helicoverpa zea*, *Heliothis virescens* and *Spodoptera exigua* (Hassanali and Lwande, 1989). In addition they proved effective against aphid colonization and virus transmission (Gibson *et al.*, 1982).

Polygodial was identified in extracts of the herb *Polygonum hydropiper* (Polygonaceae). If it was sprayed on to barley plants infested by aphids the yield was significantly increased (Pickett *et al.*, 1987).

Some members of the family Fabaceae (Leguminosae) have been utilized for a long time as fish poisons and insecticides. *Lonchocarpus* spp., *Derris* spp. and *Tephrosia* spp. yield rotenones and related isoflavonoids in their roots or seeds. They show a strong antifeedant effect but also toxicity to a wide range of insects. The synthesis of rotenoids has not been attractive commercially until now, but some local interest now exists to use the root extracts of *Lonchocarpus* spp., for instance in southern China, where several hundred hectares are grown (Chiu Shin-Foon, personal communication). Some polyhydroxy-alkaloids, extracted from *Lonchocarpus* and *Derris*, are potent glycosidase inhibitors in insects (Bell *et al.*, 1990).

The rather toxic alkaloid nicotine from *Nicotiana* spp. (Solanaceae) and the related anabasine from *Anabasis* spp. (Chenopodiaceae) are well known and still used in some countries. Petuniasterones, which are steroidal ketones, occur in the common garden plant *Petunia hybrida*. They inhibit the growth of *Helicoverpa zea* larvae. Another solanaceous plant, *Physalis peruviana*, contains several withanolides, among which highly glycosilated derivatives are the most active (Elliger and Waiss, 1989). Some species of the Apocynaceae (*Thevetia* spp., *Nerium* spp.) contain cardiac glycosides with a high toxicity to insects and mammals. Extracts from seeds of *Thevetia peruviana* are used in South America (Ecuador) for seed treatment against birds.

Conclusions

Higher plants are doubtless an extremely rich source of chemicals that influence the behaviour, development, reproduction and fitness of insects in various ways. Other compounds of plant origin exhibit a high toxicity in a wide range of arthropods and warm-blooded organisms. The use of these in crop protection should be avoided if possible. By improving screening methods and further studies on structure–activity relationships and modes of action a considerable number of promising botanicals could be characterized over the next few years. But though there are an enormous number of defensive compounds in plants, only some will meet the requirements of modern crop protection – a substantial reduction of harmful pest populations below the economic threshold level – azadirachtin is one of them.

Normally the synthesis of natural compounds is a prerequisite for their commercial production, but in some cases the extraction from the raw material may also be feasible. This, however, depends on the availability of the raw material, the obtainable yield of active principles, the costs of the extraction and the efficacy of the active principles. In other cases a simple water extraction may be an alternative for countless small farmers in developing countries, putting them in a position to produce their own effective pesticides of plant origin. In organic farming, such simple extraction procedures may also be satisfactory if only small quantities of sprays are needed.

Plant constituents which resist insects may be of great importance for future plant breeding. The transfer of the genetic characteristics by intergenic hybridization into economically important plant species could be a solution for a number of pest problems. Many of these important plant species are already endangered, so the studies on plant ingredients with insecticidal and other properties should be intensified. Otherwise many valuable ingredients offered to us by nature will be lost forever.

References

ALKOFAHI, A., RUPPRECHT, J.K., ANDERSON, J.E., MCLAUGHLIN, J.L., MIKOLAJCZAK, K.L. and SCOTT, B.A. (1989). Search for new pesticides from higher plants. In *Insecticides of Plant Origin* (J.T. Arnason, B.J.R. Philogène and P. Morand, eds), pp. 25–43. American Chemical Society Symposium 387, Washington DC.

ANDERSON, J.C., and LEY, S.V. (1990). Chemistry of insect antifeedants from *Azadirachta indica* (part 6): Synthesis of an optically pure acetal intermediate for potential use in the synthesis of azadirachtin and novel antifeedants. *Tetrahedron Letters*, **31**, 431–432.

ASCHER, K.R.S. (1987). Notes on Indian and Persian lilac pesticide research in Israel. In *Proceedings of the 3rd International Neem Conference*, Nairobi 1986 (H. Schmutterer and K.R.S. Ascher, eds), pp. 45–53. GTZ Eschborn, Germany.

BARNBY, M.A. and KLOCKE, J.A. (1990). Effects of azadirachtin on levels of ecdysteroids and prothoracicotropic hormone-like activity in *Heliothis virescens* (Fabr.) larvae. *Journal of Insect Physiology*, **36**, 125–131.

BELL, E.A., FELLOWS, L.E. and SIMMONDS, M.S.J. (1990). Natural products from plants for the control of insect pests. In *Safer Insecticides* (E. Hodgson and J.R. Kuhr, eds), pp. 337–350. Marcel Dekker, Inc., New York and Basel.

BOWERS, W.S. (1976). Discovery of insectantiallotropins. In *The Juvenile Hormones* (L.I. Gilbert, ed). Plenum Press, New York and London.

CHAMPAGNE, D.E., ISMAN, M.B. and TOWERS, G.H.N. (1989). Insecticidal activity of phytochemicals and extracts of the Meliaceae. In *Insecticides of Plant Origin* (J.T. Arnason, B.J.R. Philogène and P. Morand, eds), pp. 95–109. American Chemical Society Symposium 387, Washington DC.

CHIU SHIN-FOON (1989). Recent advances in research on botanical insecticides in China. In *Insecticides of Plant Origin* (J.T. Arnason, B.J.R. Philogène and P. Morand, eds), pp. 69–77. American Chemical Society Symposium 387, Washington DC.

DAWSON, G.W., GRIFFITHS, D.C., MERRIT, L.A., MUDD, A., PICKETT, J.A., WADHAMS, L.J. and WOODCOCK, C.M. (1990). Aphid semiochemicals – a review, and recent advances on the sex pheromone. *Journal of Chemical Ecology*, **16**, 3019–3030.

DOLL, M. and SCHMUTTERER, H. (in press). Vergleich der

Wirkung von Samenkernextrakten und Öl von *Azadirachta excelsa* Jack (= *A. integrifoliola* Merr.) und *Azadirachta indica* (A. Juss.) beim Mexikanischen Bohnenkäfer *Epilachna varivestis* (Muls.) *Mitteilungen der Deutschen Gesellschaft für Allgemeine und Angewandte Entomologie*, **8**.

DOWNUM, K.R. (1985). Photoactivated biocides from higher plants. In *Natural Resistance of Plants to Pests* (M.B. Green and P.A. Hedin, eds), pp. 197–205. American Chemical Society Symposium 296, Washington DC.

ELLIGER, C.A. and WAISS JR. A.C. (1989). Insect growth inhibitors from *Petunia* and other solanaceous plants. In *Insecticides of Plant Origin* (J.T. Arnason, B.J.R. Philogène and P. Morand, eds), pp. 188–205. American Chemical Society Symposium 387, Washington DC.

FEUERHAKE, K.J. and SCHMUTTERER, H. (1982). Einfache Verfahren zur Gewinnung und Formulierung von Niemsamenextrakten und deren Wirkung auf verschiedene Schadinsekten. *Zeitschrift für Pflanzenkrankheiten und Pflanzenschutz*, **89**, 737–747.

GIBSON, R.W., RICE, A.D., PICKETT, J.A., SMITH, M.C. and SAWICKI, R.M. (1982). The effects of the repellents dodecanoic acid and polygodial on the acquisition of non-, semi- and persistent plant viruses by the aphid *Myzus persicae*. *Annals of Applied Biology*, **100**, 55–59.

HASSANALI, A. and LWANDE, W. (1989). Antipests secondary metabolites from African plants. In *Insecticides of Plant Origin* (J.T. Arnason, B.J.R. Philogène and P. Morand, eds), pp. 78–94. American Chemical Society Symposium 387, Washington DC.

ISMAN, M.B. (1989). Toxicity and fate of acetylchromenes in pest insects. In *Insecticides of Plant Origin* (J.T. Arnason, B.J.R. Philogène and P. Morand, eds), pp. 44–58. American Chemical Society Symposium 387, Washington DC.

JACOBSON, M. (1989). Botanical insecticides: past, present and future. In *Insecticides of Plant Origin* (J.T. Arnason, B.J.R. Philogène and P. Morand, eds), pp. 1–10. American Chemical Society Symposium 387, Washington DC.

KLOCKE, J.A., BALANDRIN, M.F., BARNBY, M.A. and YAMASAKI, R.B. (1989). Limonoids, phenolics, and furanocoumarins as insect antifeedants, repellents, and growth inhibitory compounds. In *Insecticides of Plant Origin* (J.T. Arnason, B.J.R. Philogène and P. Morand, eds), pp. 136–149. American Chemical Society Symposium 387, Washington DC.

KUBO, I., LEE, Y.W., BALOGH-NAIR, V., NAKANISHI, R.

and CHAPYA, A. (1976). Structure of ajugarins. *Journal of Chemical Society*, Chemical Communication **22**, 949–950.

KUBO. I., KLOCKE, J.A. and ASANO, S. (1981). Insect ecdysis inhibitors from the East African medicinal plant *Ajuga remota* (Labiatae). *Agricultural Biology and Chemistry*, **45**, 1925–1927.

MIYAKADO, M., NAKAYAMA, I. and OHNO, N. (1989). Insecticidal unsaturated isobutylamides. From natural products to agrochemical leads. In *Insecticides of Plant Origin* (J.T. Arnason, B.J.R. Philogène and P. Morand, eds), pp. 173–187. American Chemical Society Symposium 387, Washington DC.

NICOL, C.M.Y. and SCHMUTTERER, H. (1991). Kontaktwirkungen des Samenöls des Niembaumes *Azadirachta indica* (A. Juss.) bei gregären Larven der Wüstenheuschrecke *Schistocerca gregaria* (Forskål). *Journal of Applied Entomology*, **111**, 197–205.

NISHIKIMI, Y., IIMORI, T., SODEOKA, M. and SHIBASAKI, M. (1989). Synthetic studies of azadirachtin. Synthesis of the cyclic acetal intermediate in the naturally occurring form. *Journal of Organic Chemistry*, **54**, 3354–3359.

PICKETT, J.A., DAWSON, W.G., GRIFFITHS, D.C., HASSANALI, A., MERRITT, L.A., MUDD, A., SMITH, M.C., WADHAMS, L.J., WOODCOCK, C.M. and ZHANG, Z.-N. (1987). Development of plant-derived antifeedants for crop protection. In *Science and Biotechnology* (R. Greenhalgh and T.R. Roberts, eds), pp. 125–128. Blackwell Scientific Publications, Oxford.

REMBOLD, H. (1989). Azadirachtins: their structure and mode of action. In *Insecticides of Plant Origin* (J.T. Arnason, B.J.R. Philogène and P. Morand, eds), pp. 150–163. American Chemical Society Symposium 387, Washington DC.

REMBOLD, H., FORSTER, M., CZOPPELT, CH., RAO, P.S. and SIEBER, K.-P. (1984). The azadirachtins, a group of growth regulators from the neem tree. In *Proceedings of the 2nd International Neem Conference* Rauischholzhausen 1983 (H. Schmutterer and K.R.S. Ascher, eds), pp. 153–162. GTZ Eschborn, Germany.

SCHMUTTERER, H. (1989a). Environmentally sound pest control by application of neem (*Azadirachta indica*) based natural pesticides. *Tagungsberichte der Akademie der Landwirtschaftswissenschaften der DDR, Berlin*, **274**, 135–144.

SCHMUTTERER, H. (1989b). Erste Untersuchungen zur insektiziden Wirkung methanolischer Blatt- und Rindenextrakte des philippinischen Niembaumes *Azadirachta integrifoliola* Merr. im Vergleich zu Extrakten aus *Azadirachta indica* (A. Juss) und *Melia azedarach* L. *Journal of Economic Entomology*, **108**, 483–489.

SCHMUTTERER, H. (1990). Properties and potential of natural pesticides from the neem tree, *Azadirachta indica*. *Annual Review of Entomology*, **35**, 271–297.

SCHMUTTERER, H. and KLEFFNER, L. (1988). Zur metamorphosestörenden, fekunditäsmindernden und eisterilisierenden Wirkung von Rohextrakten aus *Asarum europaeum* L. *Mitteilungen der Deutschen Gesellschaft für Allgemeine und Angewandte Entomologie*, **6**, 296–301.

SCHMUTTERER, H. and TERVOOREN, G. (1980). Die Wirkung von Rohpreßäften und Rohextrakten aus *Ajuga*-Arten auf Fraßaktivität und Metamorphose von *Epilachna varivestis*. *Journal of Applied Entomology*, **89**, 470–478.

STEETS (1975). Die Wirkung von Rohextrakten aus den Meliaceen *Azadirachta indica* und *Melia azedarach* auf verschiedene Insektenarten. *Journal of Applied Entomology*, **77**, 306–312.

STEIN, U. and KLINGAUF, F. (1990). Insecticidal effect of plant extracts from tropical and subtropical species. *Journal of Economic Entomology*, **110**, 160–66.

ZHU, J. and ERMEL, K. (1991). Isolierung einer auf den Mexikanischen Bohnenkäfer *Epilachna varivestis* Muls. (Col., Coccinellidae) metamorphosestörend wirkenden Substanz aus Blättern von *Melia azedarach* L. *Zeitschrift für Pflanzenkrankheiten und Pflanzenschutz*, **98**.

ZHU, J. and SCHMUTTERER, H. (in press). Wirkung von Extrakten aus *Melia azedarach*- Früchten und -Blättern unterschiedlichen Alters auf die Metamorphose und Mortalität *Epilachna varivestis*. *Mitteilungen der Deutschen Gesellschaft für Allgemeine und Angewandte Entomologie*, **8**.

2
Isolation and Structural Elucidation of Three New Insect Growth Inhibitors, Prunellins A–C, from *Prunella vulgaris*

HEINZ REMBOLD and ANTON STOLTNER

Max-Planck-Institut für Biochemie, 8033 Martinsried, Germany

Introduction

The great number of insect species – recent estimates indicate the existence of as many as 30 million species (Stork and Gaston, 1990) – together with their tremendous capability to reproduce and their long evolutionary history enable them to occupy almost every environmental niche. By virtue of this adaptive power, some insects have become vectors of infectious diseases, while others are important agricultural pests. A rapidly increasing resistance against every type of insecticide along with an acute off-target neurotoxicity makes their further application an impending threat to ecology and human health. For this reason, more target-specific methods of insect control are the goal of world-wide scientific efforts. Manipulation of the endocrine control of growth, development and reproduction appears a promising alternative for insect control in the future. The development of such a strategy is a tremendous challenge for basic research.

It was established quite early that regulation of growth and development of insects works in a hierarchic manner (Highnam, 1979). This means that, in close analogy to the hypthalamo-hypophysial system of mammals, neuropeptides from the central nervous system control the synthesis of steroid (moulting) and sesquiterpenoid (juvenile) hormones in peripheric glands. Within this

Insecticides: Mechanism of Action and Resistance
© 1992   Intercept Ltd, P.O. Box 716, Andover, Hants SP10 1YG, UK

framework, a question of major concern is how such a network of hormonal control can be influenced by external chemical factors. Promising candidates are secondary metabolites of plant origin which are naturally occurring insect growth inhibitors. Such compounds may provide ideas and models for new chemical structures which are able to interfere with insect reproduction in a target-specific way.

A prerequisite for the isolation of such botanical insect growth inhibitors is a sensitive bioassay which discriminates between the ubiquitous feeding repellents and growth inhibitors, both of which usually are present in plant extracts. Starvation due to feeding inhibition may also induce endocrine disturbance to some extent with the consequence of a pseudo-growth-inhibiting effect. Selection of a well suited test organism can therefore become vital when screening plant material for insect growth inhibitors. Another important factor is a remarkable diversity in sensitivity and tolerance of the different insect species to botanicals which is primarily due to their evolutionary history. It is well known from research on the group of azadirachtins, the main insect growth inhibitors from *Azadirachta indica* seed, that for some insects these triterpenoids are extremely strong feeding inhibitors, for others growth inhibitors, whereas other insects even tolerate this group of natural compounds (Rembold, 1989).

With the present investigation we have selected an indigenous medicinal plant, *Prunella vulgaris*, which is widely distributed and has, e.g. long been used in traditional Japanese medicine as a diuretic (Kojima and Ogura, 1986). In China it is applied as a medicinal herb against tuberculosis, icterus, bacterial infections and cancer (Chung, 1979). Tabba *et al.* (1989) reported on the isolation of an anti-HIV active component, after Chang and Yeung (1988) had described an *in vitro* activity of an aqueous *P. vulgaris* extract against the AIDS virus. An insecticidal activity of *P. vulgaris* seed extract on *Culex* and *Aedes* larvae has been demonstrated by Sharma and Wattal (1982). Based on this scarce information, and after crude methanol extracts from dried whole plants had shown promising growth inhibitory effects in the standard *E. varivestis* assay (Rembold *et al.*, 1980), we have isolated three new insect growth inhibitors, Prunellins A–C, and elucidated their structures as derivatives of α- (Prunellin A and B) and β–amyrin (Prunellin C).

Material and Methods

Plant material

P. vulgaris, 40.7 kg of shade, air-dried leaves, stalks and flowers, which had been harvested from an insecticide-free field of H. Mack, Illertissen, FRG, were crushed in a mill and sieved through 0.5 cm mesh. The flour was used in several batches for extraction.

Chemicals

All chemicals were of analytical and, resp., HPLC purity (Merck). For open column chromatography either silica gel KG-60, 0.063–0.2 mm (Merck 7754), or RP-8 silica gel Lichroprep Si-60, 0.15–0.25 mm (Merck 9336) was used. The following TLC-plates were used: Si 50 000 F_{254}S (Merck 15132), RP-18 F_{254}S (Merck 13724), both 0.2 mm for analysis; for preparative TLC, 0.5 mm unmodified silica gel 20×20 cm (Merck 5744) and 0.25 mm (Merck 5721). Spray reagent anisaldehyde–sulphuric acid was applied for detection of substances.

Instruments

For analytical and semi-preparative HPLC the following equipment was applied: LC pump Perkin-Elmer, series 1; Philips Pye-Unicam PV 4020 detector; Spectra Physics SP 4270 integrator; Rheodyne type 7105 injector system; 500×8 and 230×4 mm columns with Polygosil 60-5, CH 8091 (Macherey & Nagel).

Ei-ms spectra were measured on a Varian CH 7A, ^{1}H-nmr spectra on a Bruker AM-500 instrument. Each 4.2 mmol of substance was dissolved in 0.5 ml $CdCl_3$ for nmr spectroscopy.

Bioassays

The methods as described for the *E. varivestis* cage test by Rembold *et al.* (1980) were followed. Under our standard conditions, the developmental periods shown in Table 1 were found.

A colony of *Culex pipiens molestus* originated from the Biology Department, University of Mosul, Iraq and was kindly supplied by Prof. Z. Al-Sharook. The culture was maintained and the assay performed as described already (Al-Sharook *et al.*, 1991). Development of *C. pipiens* under our standard conditions is given in Table 2.

The compounds to be tested were added, dissolved in 400 µl ethanol, or the same volume of alcohol for blank, to 40 ml 0.08% NaCl in a beaker. To each solution 20 fourth-instar larvae (24 h old) were added and kept at 27 ± 1°C. Evaporated water was replaced regularly. Dead larvae or pupae were removed every 24 hours. The assay was terminated after 10 days.

For statistical calculation of the MC_{50} values for *E. varivestis* and *C. pipiens*, the probit analysis after Noack and Reichmuth (1978) was applied.

Table 1.   Life cycle of the Mexican bean beetle, *E. varivestis*, under the standard cage test conditions as described by Rembold *et al.* (1980).

Stage		Period (days)
Egg		5
Larva	L1	5
	L2	5
	L3	3–4
	L4	5–6
Pharate pupa	PP	2
Pupa	P	5–8
Adult	A	4–6 weeks

Table 2.   Life cycle of *C. pipiens* under standard laboratory conditions as described by Al-Sharook *et al.* (1991).

Stage		Period (days)
from	to	
Egg	Larva L1	2–3
Li	L2	2
L2	L3	2
L3	L3	2
L4	Pupa	5–7
Pupa	Adult	2.5–3
Adult		50–70

Results and Discussion

Bioassay and isolation of Prunellins A–C

The individual steps as used for the isolation of the three bioactive compounds from *P. vulgaris* is summarized in the flow chart of Figure 1.

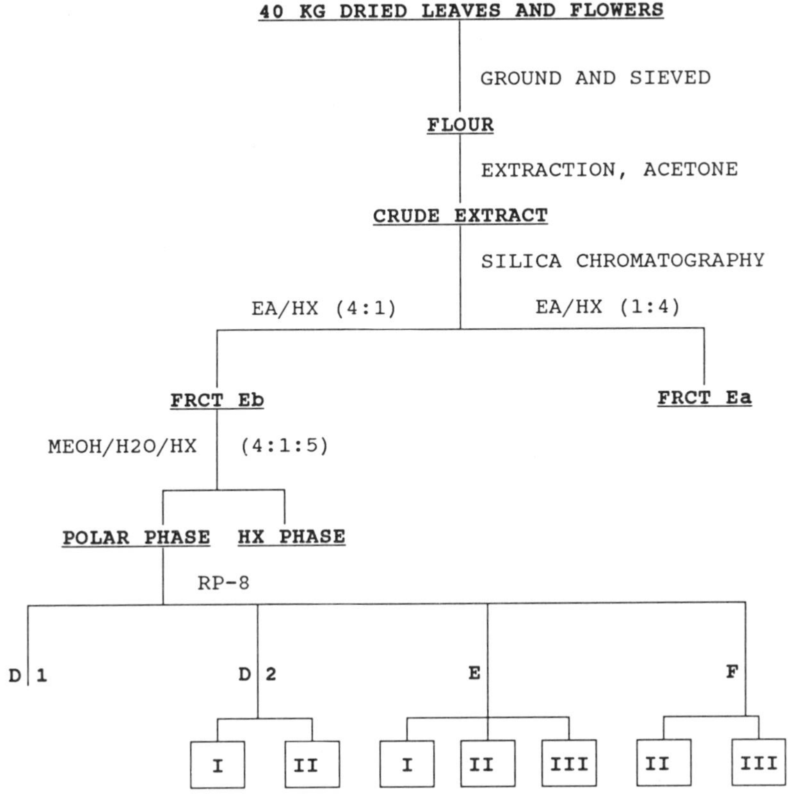

Figure 1. Flow chart for the isolation of the insect growth inhibitors I–III from *P. vulgaris*.

The ground and sieved plant powder was extracted with several solvents and the crude extracts were tested in the *E. varivestis* bioassay. These crude extracts significantly differ in their biological activities, but all of them exert a strong feeding repellent effect as reflected by the reduced growth rate of the treated fourth-instar larvae (Table 3). In addition, the alcoholic extracts caused a very hard cover on the sprayed bean leaves which made feeding almost impossible for the larvae. The hexane extracts were dissolved and sprayed in pentane. Methanol and hexane extracts caused highest and acetone extract lowest feeding inhibition in relation to growth inhibition and was therefore used for further purification of the active fractions.

Table 3.   Effect of *P. vulgaris* crude extracts on 20 L4 each in the *E. varivestis* cage test.

Extract (%)	L4*	PP*	P*	A*	NA	wtg (%)	NA (%)
Acetone							
0	0	0	0	0	20	146.4	100
5	2	0	1	3	14	61.6	70
10	0	0	1	8	11	30.4	55
Methanol							
0	0	0	0	0	20	109.8	100
5	1	1	1	5	12	27.7	60
10	4	2	0	2	12	18.1	60
Ethanol							
0	1	0	0	2	17	116.5	85
5	1	1	1	5	12	54.0	60
10	0	0	0	7	13	52.0	65
Hexane							
0	0	1	0	0	19	111.8	95
5	2	0	1	3	14	32.9	70
10	1	1	1	5	12	8.2	60

L4, fourth instar larva; PP, pharate pupa; P, pupa; A, adult; *, dead animals; NA, normal adults; wtg, weight gain after 48 h of test, expressed as percentage.

Chromatography of the crude acetone extract on an open silica gel column yield two fractions, Ea and Eb, both of which interfered with growth and development of *E. varivestis* larvae (Table 4). Fraction Ea exhibits feeding inhibition without remarkable effects on development, whereas EB clearly demonstrates an inhibition of adult morphogenesis. Subsequent distribution of this fraction between the solvents methanol/water/hexane yielded a lipophilic fraction HX and a polar phase PP. The bioassay (Table 4) indicates no growth but strong feeding inhibition for HX and a clear growth inhibition in PP. This fraction was therefore chromatographed on an open RP-8 column by stepwise elution with ethyl acetate/hexane mixtures, yielding the four fractions D1, D2, E and F (Figure 1). Results from the *E. varivestis* bioassays with these fractions are also collated in Table 4.

Table 4.   Biological activity in the *E. varivestis* cage test (20 L4 each) of chromatographic fractions from a crude *P. vulgaris* acetone extract.

Fraction (%)		L4*	PP*	P*	A*	NA	wtg (%)	NA (%)
	0	0	0	0	1	19	101.8	95
Ea	2.5	1	0	1	2	16	56.4	80
Eb	2.5	0	0	1	8	11	30.6	55
	0	0	0	1	1	18	118.8	90
HX	5	2	0	0	1	17	32.0	85
PP	2.5	0	1	0	7	12	56.5	60
D1	0	0	0	0	2	18	64.4	90
	1	0	0	0	3	17	52.3	85
	2	0	0	0	11	9	69.2	45
D2	0	0	0	0	0	20	89.7	100
	1	20	0	0	0	0	31.7	0
	2	20	0	0	0	0	50.7	0
E	0	0	0	0	0	20	135.2	100
	0.1	0	0	1	8	11	109.1	55
F	0	0	0	0	0	20	135.2	100
	0.1	0	0	1	8	11	109.1	55

* All symbols as for Table 3 and Figure 1.

Fractions D1 and D2 are eluted from RP-8 with an 1:1 mixture of ethyl acetate and hexane. Both are active in the *E. varivestis* assay, however D2 exhibits a very strong activity on larval development ending up in a complete arrest of further growth and development.

The larvae finally died due to starvation. This fraction was selected for further purification. The more polar 3:2 mixture of ethyl acetate and hexane elutes fraction E with high biological activity at 0.1% already. With a final 4:1 mixture fraction F was eluted with similar growth disrupting activity like fraction E.

Both these fractions as well as D2 were further purified by use of analytical and preparative TLC, yielding three pure compounds I–III. Their MC_{50} values are given for the *E. varivestis* cage test in Table 5 and for *C. pipiens* in Table 6. In a typical isolation procedure, these substances were contained in the crude acetone extract in a proportion of 4 mg I, 1.5 mg II and 2.5 mg III.

Table 5.   Inhibition of metamorphosis by insect growth inhibitors from *P. vulgaris.*

Triterpenoid	n	MC_{50} (ppm)[*]
Prunellin A I	16	3139.0 ± 14.0
Prunellin B II	10	1327.8 ± 15.0
Prunellin C III	9	433.0 ± 11.4

[*] MC_{50} values for compounds I–III in the *E. varivestis* bioassay. n, number of independent tests.

Table 6.   Inhibition of metamorphosis by insect growth inhibitors from *P. vulgaris.*

Triterpenoid	n	MC_{50} (ppm)[*]
Prunellin A I	11	44.90 ± 1.12
Prunellin B II	9	45.46 ± 15.0
Prunellin C III	9	56.49 ± 1.09

[*] MC_{50} values for compounds I–III in the *C. pipiens* bioassay. n, number of independent tests.

Structural elucidation of Prunellins A–C

Pentacyclic triterpenoids of the ursolic and oleanoic acid type have been isolated from *P. vulgaris* by several authors (Shimano *et al.*, 1956; Kojima and Ogura, 1986; Kojima *et al.*, 1987; Lee *et al.*, 1988).

One of the three compounds, 3β–hydroxyurs-12-en*28-oic acid (I), has already been isolated and structurally identified from *Ilex rotunda* (Nakatani *et al.*, 1989). Other ursolic and oleanoic acid derivatives have been isolated and identified from *Salvia argentea* by Bruno *et al.* (1987). However, none of these compounds has been identified as insect growth inhibitor. By comparison of the spectroscopic data from the published data with our own ones as collated in Table 7, the structures of the three pentacyclic triterpenoid insect growth inhibitors could be elucidated as shown in Figure 2.

Table 7.   Spectroscopic data for the prunellins I – III.

3β–hydroxyurs–12–en–28–oic acid (I). ms:$C_{30}H_{48}O_3$ (M^+ 456). ^{1}H–nmr: H–le 1,0; H–la 1,0; H–2a 1,63; H–2e 1,58; H–3α 3,23 (dd,4,9/11,2); H–9 1,51 (t); 2 H–11 1,93 (dd); H–12 5,26 (t, 3,8/3,8); H–18a 2,20 (d, 11,5); H–19a 1,34 (m); Me 0,78; H–20 1,03 (m) ; H–16a 1,89 (dt, 4,5/13,6/13,6); H–15a 2,03 (dt, 4,5/13,6/13,6); H–16e 1,10; H–15e 1,69; Me–19 0,88 (6,3); Me–20 0,95 (6,3); Me 0,81; Me 0,97; Me 1,01; Me 1,09.

2α,3α–dihydroxyurs–12–en–28–oic acid (II). ms: $C_{30}H_{48}O_4$ (M^+ 472). ^{1}H–nmr: H–le 1,67 (dd, 4/14); H–1'a 1,28 (dd, 11/14); H–2β 3,98 (dt, 3,5/3,5/11,8); H–3β 3,48 (d, 2,8); H–9 1,65 (dt); H–11e 1,98; H–11a 2,05; H–12 5,28 (t, 3,7/3,7); H–15a 2,02 (dt, 4,3/13,4/ 13,4); H–15e 1,69 (m); H–16a 1,88 (dt, 4,5/13,4/13,4); H–16e 1,10 (m); H–18a 2,20 (d, 11,2); H–19a 1,34 (m); H–20 1,01; Me–19 0,87 (7,0); Me–20 0,96 (6,6); Me 1,11; Me 0,98; Me 1,04; Me 0,85; Me 0,80; H–21 2,29; H–21' 2,33; H–22 2,22; H–22' 1,63.

2α,3β–dihydroxolean–12–en–28–oic acid (III). ms: $C_{30}H_{48}O_4$ (M^+ 472). ^{1}H–nmr: H–le 2,0 (dd); H–la 0,97; H–2α 3,70 (m, 4,5/9,6/11,0) H–3β 3,02 (d, 9,6); H–9 1,62; H–11 1,94 (m, 4,2/13,6/14,0); H–11' 1,90 (m); H–12 5,26 (t, 4,4/4,4); H–15a 1,98; H–15e 1,63; H–16a 1,72; H–16e 1,08; H–18a 2,83 (dd, 4,1/13,0); H–19a 1,64 (13/15); H–19e 1,16 (dd, 4/ 15); H–21e 1,21; H–21a 1,36; H–22e 1,58; H–22a 1,78; Me 0,78; Me 0,82; Me 0,90; Me 0,93; Me 0,98; Me 1,03; Me 1,15.

Prunellin A (I) is a 3β–mono-, Prunellin B (II) a 2α, 3α-dihydroxyursenoic acid which again is an oxidation product of α-amyrin (Figure 3). Prunellin C (III) is a 2α, 3β–dihydroxyoleanenoic acid which is derived from β–amyrin. The two amyrins differ in their E-ring substitution, where the two methyl groups occupy an *ortho* position in the α and a *geminal* position in the β) isomer (Figure 3). Consequently, due to their two hydroxyl groups, Prunellin B (II) and C (III) have the same molecular weight.

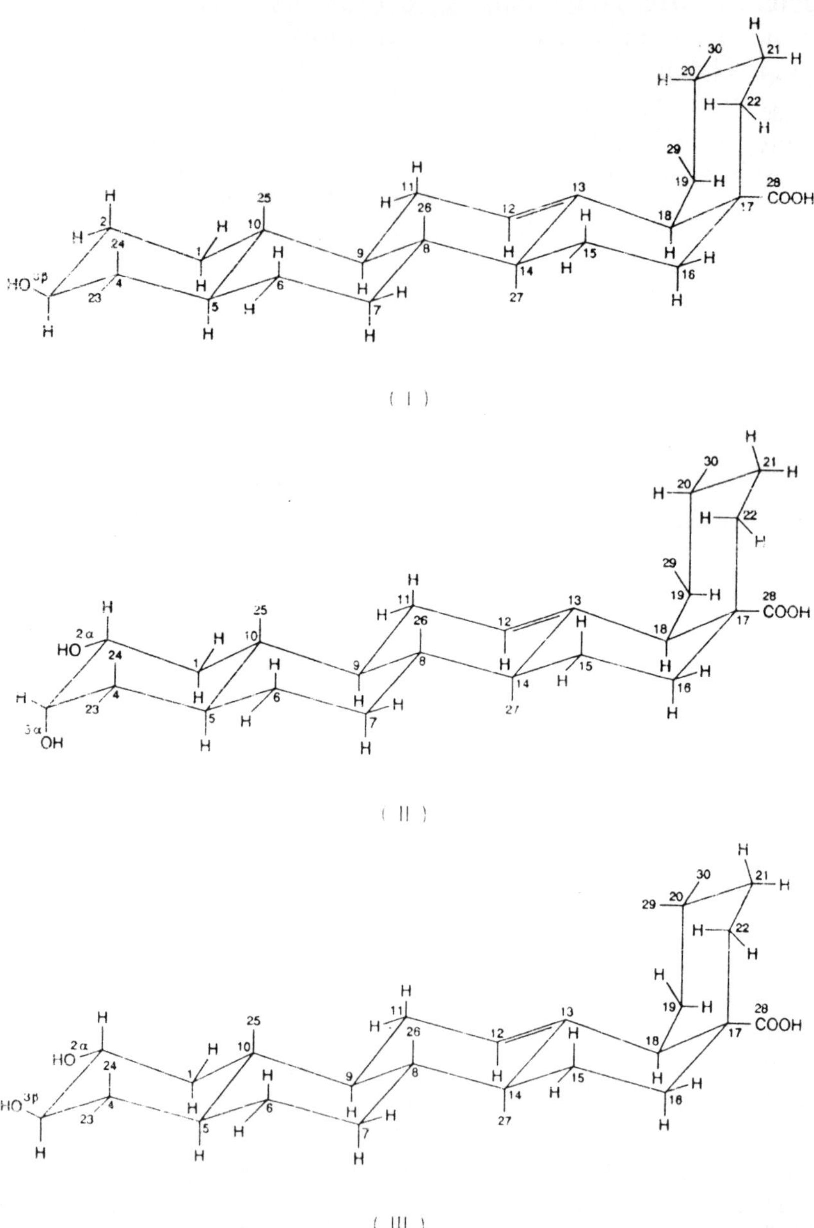

Figure 2. Structures of the three insect growth inhibitors isolated from *P. vulgaris*. I, Prunellin A (3β–hydroxyurs–12–en–28–oic acid); II, Prunellin B (2α,3α–dihydroxyurs–12–en–28–oic acid); III, Prunellin C (2α,3β–dihydroxyolean–12–en–28–oic acid).

Structure–activity relationships

With this study for the first time insect growth inhibitors have been isolated from *P. vulgaris*. It is proposed to name them as prunellins, because these pentacyclic triterpenoids represent a new type of bioactive compound. The three compounds significantly differ in their inhibitory activity against *E. varivestis* and *C. pipiens*.

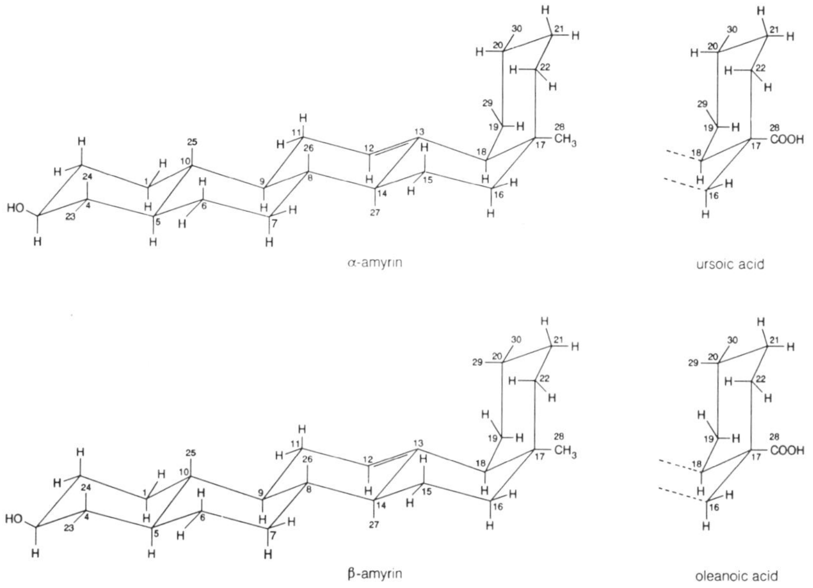

Figure 3.   Structures of α– and β–amyrin in comparison with ursoic and oleanoic acid.

Prunellin C has with MC_{50} = 433 ppm the highest activity in the *E. varivestis* assay, followed by Prunellin B (MC_{50}) = 1328 ppm) and, with less than 20% of C, Prunellin A (MC_{50} = 3139 ppm). One may speculate about the difference in ring A substitution to understand this reduced biological activity: there is indeed a stepwise decrease in growth inhibitory activity from Prunellins C (2α,3β–OH), B (2α,3α-OH), to A (3β–OH). An additional effect through the *geminal* methyl groups (29, 30) can also be discussed.

Interestingly, the growth inhibitory effect on *C. pipiens* larvae does not reflect the results from *E. varivestis*. In absolute terms, the much lower LC_{50} values can be understood from the continuous exposition of the larvae to the growth inhibitor. However, ring A substitution does not affect the biological activity any more: Prunellin C has the significantly lowest growth inhibitory effect on *C. pipiens* larvae, and Prunellins A and B have about the same activity. The most significant difference in chemical structure now seems to be the oleanenoic acid structure of Prunellin C (2 *geminal* methyl groups) in comparison with the ursenoic acid structure (2 *ortho* methyl groups) of Prunellins A and B.

The different biological activity of the three insect growth inhibitors against the two insect species can originate from very different reasons, like differences in permeability, target organs, receptor molecules, or metabolism. Although the activity of the prunellins can not compete with that of the azadirachtins, which is higher by about two orders of magnitude (Rembold, 1989), they add quite an interesting piece to the field of chemical insecticide design.

References

AL-SHAROOK, Z., BALAN, K., JIANG, Y. and REMBOLD, H. (1991). Insect growth inhibitors from two tropical Meliaceae. Effect of crude seed extracts on mosquito larvae. *Journal of Applied Entomology*, **111**, 425–430.

BRUNO, M., SAVONA, G., HUESO-RODRIGUEZ, J.A., PASCUAL, C. and RODRIGUEZ, B. (1987). Ursane and oleanane triterpenoids from *Salvea argentea*. *Phytochemistry*, **26**, 497–501.

CHANG, R.S. and YEUNG, H.W. (1988). Inhibition of growth of human immuno-deficiency virus in vitro by crude extracts of Chinese medicinal herbs. *Antivirus Research*, **9**, 163–176.

CHUNG Yao Ta Tzu Tien (1979). *Dictionary of Chinese Medicine, Science and Technology*. Publishing Co., Peking **2**, pp. 1731–1827.

HIGHNAM, K.C. (1979). Insect hormones. In *Topics in Hormone Chemistry* (W.R. Butt, ed), Vol. 1, pp. 216–250. John Wiley & Sons, Chichester.

KOJIMA, H. and OGURA, H. (1986). Triterpenoids from *Prunella vulgaris*. *Phytochemistry*, **25**, 729–733.

KOJIMA, H., TOMINAGA, H., SATO, S. and OGURA, H. (1987). Pentacyclic triterpenoids from *Prunella vulgaris*. *Phytochemistry*, **26**, 1107–1111.

LEE, K.H., LIN, Y.-M., WU, T.-S., ZHANG, D.-C., YAMA-GISHI, T., TOSHIMITSU, H., HALL, I.H., CHANG, J.-J., WU, R.-Y. and YANG, T.H. (1988). The cytotoxic principles of *Prunella vulgaris*, *Psychotria serpens* and *Hyptis capitata*: ursolic acid and related derivatives. *Planta Medica*, 308–311.

NAKATANI, M., MIYAZAKI, Y., IWASHITA, T., NAOKI, H. and HASE, T. (1989). Triterpenes from *Ilex rotunda* fruits. *Phytochemistry*, **28**, 1479–1482.

NOACK, S. and REICHMUTH, CH. (1978). Ein rechnerisches Verfahren zur Bestimmung von beliebigen Dosis-Werten eines Wirkstoffs aus empirisch ermittelten Dosis-Wirkungs-Daten. *Mitteilungen der Biologischen Bundesanstalt für Land) und Forstwirtschaft*, Heft 185.

REMBOLD, H. (1989). Azadirachtins. Their structure and mode of action. In *Pesticides of Plant Origin* (J.T. Arnason, B.J.R. Philogène and P. Morand, eds), pp. 150–163. ACS Symposium 387, American Chemical Society, Washington DC.

REMBOLD, H., SHARMA, G.K., CZOPPELT, CH. and SCHMUTTERER, H. (1980). Evidence of growth-disruption in insects without feeding inhibition by neem seed fractions. *Journal of Plant Diseases and Protection*, **87**, 290–297.

SHARMA, S.K. and WATTAL, B.L. (1982). Further studies on mosquito larvicidal potential of mucilaginous seeds. *Journal of Entomological Research*, **6**, 159–165.

SHIMANO, T., MIZUNO, M., OKAMOTO, H. and ADACHI, I. (1956). Studies on triterpenoids. IX. On the component of *Prunella*, ursolic acid. *Yakugaku Zasshi*, **76**, 974–975.

STORK, N. and GASTON, K. (1990). Counting species one by one. *New Scientist*, 11 August, 43–47.

TABBA, H.D., CHANG, R.S. and SMITH, K.M. (1989). Isolation, purification, and partial characterization of prunellin, an anti-HIV component from aqueous extracts of *Prunella vulgaris*. *Antivirus Research*, **11**, 263–274.

3
Effects of Brassinosteroids on Moulting Regulation in *Periplaneta americana* and Other Insects

K. RICHTER

Saxon Academy of Sciences Leipzig, Research Group Jena, P.O. Box 322, O-6900 Jena, Germany

Since their first detection by Grove *et al.* (1979), brassinosteroids, a group of phytosteroids, have been of high interest to the research on phytoeffectors as they act as growth hormones. Brassinosteroids show striking structural similarities with ecdysteroids (Figure 1) to which the moulting hormones of arthropods belong. Considering the intensive interaction of many insects with certain plant species as well as the first employment of brassinosteroids as plant growth-promoting regulators in agriculture, it is highly interesting to know whether brassinosteroids act in insects in a similar manner as insect ecdysteroids.

First results from investigations into this direction were published by Hetru *et al.* (1986). They investigated the interfering of brassinosteroids with 20-OH-ecdyson, the moulting hormone of insects, on the first differentiation steps (evagination) of the imaginal disks of the fly *Phormia terraenovae*. While 20-OH-ecdyson stimulates the evagination of the imaginal disks 100%, the activity of the moulting hormones is delayed when incubated with brassinosteroids (castasteron, 22S,23S-homobrassinolid) Figure 1, i.e. that these brassinosteroids show an antagonistic effect on ecdysteroids in bioassays.

Brassinosteroids are resorbed and show biological activity in insects when applied *per os* through insect diets. Activity depends on

Insecticides: Mechanism of Action and Resistance
© 1992   Intercept Ltd, P.O. Box 716, Andover, Hants SP10 1YG, UK

the type of brassinosteroids, on the doses incorporated, and on the metabolic capability of the insects used in the experiment. The feeding experiments on *Periplaneta americana* with 22S,23S-homobrassinolide resulted in a delayed moulting interval during the last larval instar by 11 days, i.e. by more than 30% of the normal time (Richter *et al.*, 1987).

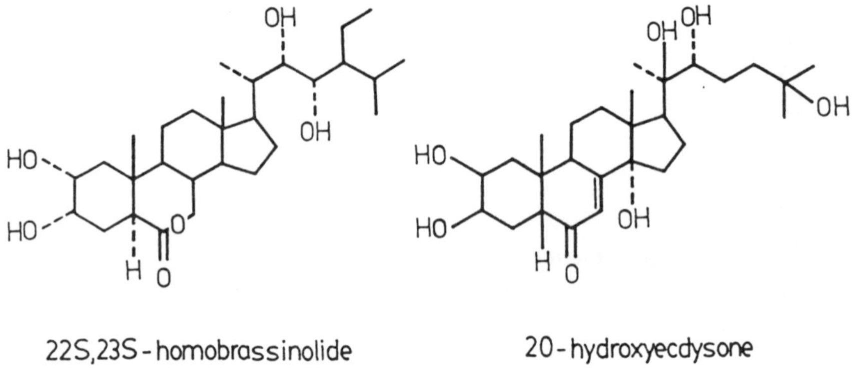

22S,23S-homobrassinolide   20-hydroxyecdysone

Figure 1.   Comparison of the structures of 22S,23S–homobrassinolide and 20–hydroxyecdyson.

Furthermore, brassinosteroids influence the spike activity of insect neurons. 22S,23S-homocastasterone and 22S,23S-homobrassinolide have neurodepressing effects as agonistic compounds in the *nervus corporis cardiaci II* (Ncc II) like the native 20-OH-ecdyson which suppresses the activity of Ncc II in *P. americana* (Richter and Adam, 1991). The spike activity of the Ncc II is a measurable expression of the transport processes of the prothoracotropic hormone from the neurosecretoric neurons in the brain to the effector sites in the corpora allata.

Receptor binding studies on nucleus fractions and on evagination stages on imaginal disks induced by 20-OH-ecdyson found in larvae of *Calliphora vicina* a competition of ecdysteroids with brassinosteroids on the binding sites of the hormone receptors (Lehmann and Koolman, 1988; Lehmann *et al.*, 1988). This reveals that brassinosteroids can interfere with the hormonal system of ecdysteroids interacting directly with receptor binding sites. This spectrum of activity comprises both agonistic and antagonistic effects.

Insect physiology takes a strong interest in phytosteroids as, on the one hand, they offer the possibility of a natural inhibition and differentiation, which is important for insecticide research, and, on the other hand, are valuable model substances for investigations into steroid receptors of insects.

A detailed summary stating the current knowledge of anti-ecdysteroid effects is given in Richter and Koolman (1991).

References

GROVE, M.D., SPENCER, G.F., ROHWEDDER, W.K., MANDAVA, N.B., WORLEY, J.F., WORTHEN, J.D., STEFFENS, G.L., FLIPPEN-ANDERSON, J.L. and COOK, J.C. (1979). Brassinolide, a plant growth-promoting steroid isolated from *Brassica napus* pollen. *Nature*, **281**, 216–217.

HETRU, C., ROUSELL, J.P., MORI, K. and NAKATANI, Y. (1986). Activité antiecdystéroids de brassinostéroides. *C. R. Acad. Sci. Saar. 2*, **302**, 417–420.

LEHMANN, M. and KOOLMAN, J. (1988). Ecdysteroid receptors of the blowfly *Calliphora vicina*: Partial purification and characterization of ecdysteroid binding. *Molecular Cell Endocrinology*, **57**, 239–249.

LEHMANN, M., VORBRODT, H.M., ADAM, G. and KOOLMAN, J. (1988). Antiecdysteroid activity of brassinosteroids. *Experientia*, **44**, 355–356.

RICHTER, K. and ADAM, G. (1991). Neurodepressing effect of brassinosteroids in the cockroach *Periplaneta americana*. *Naturwissenschaften*, **78**, 138–139.

RICHTER, K. and KOOLMAN, J. (1991). Antiecdysteroid effects of brassinosteroids in insects. In *Brassinosteroids* (H.C. Cutler, T. Yokota and G. Adam, eds). American Chemical Society Symposium Series 974, 265–278.

RICHTER, K., ADAM, G. and VORBRODT, H.M. (1987). Inhibiting effect of 22S,23S-homobrassinolide on the moult of the cockroach *Periplaneta americana* (L.) (Orthopt., Blattidae). *Journal of Applied Entomology*, **103**, 523–534.

4
The Biochemical Basis for the Insecticidal Properties of L-Canavanine, a Higher Plant Protective Allelochemical

GERALD A. ROSENTHAL

T.H. Morgan School of Biological Sciences and the Graduate Center for Toxicology, University of Kentucky, Lexington, Kentucky 40506, USA

Higher plants proliferate a myriad of potentially toxic metabolites which function in their defense against herbivores, pests, predators and pathogens (Rosenthal, 1979; Blum, 1981). These protective allelochemicals play a critical role in organismic interactions involving plants; yet, only in limited instances do we understand their mode of action at the biochemical level. This deficiency has motivated my strong interest in understanding how the arginine analogue, L-canavanine L-2-amino-4-guanidinooxy-butyric acid] (Figure 1) functions in the chemical defense of leguminous plants (Rosenthal, 1988; Rosenthal, 1991).

Canavanine is a structural analogue of arginine and its structural similarity accounts for the ability of arginyl-tRNA synthetase to activate and attach canavanine to the tRNA that normally carries arginine (Allende and Allende, 1964). Examination of the guanidinooxy group of canavanine and arginine's guanidino group reveals that pK_a of the former is 7.04 (Boyar and Marsh, 1982) while the pK_a for arginine is 12.48 (Greenstein and Winitz, 1961). Thus, canavanine is more acidic than arginine. Replacement of arginine in a protein by less basic canavanine can affect amino acid R group interactions and disrupt the tertiary and/or quaternary structure of the protein and ultimately alters its biochemical properties, e.g. solubility, stability, or enzymic activity.

Insecticides: Mechanism of Action and Resistance
© 1992   Intercept Ltd, P.O. Box 716, Andover, Hants SP10 1YG, UK

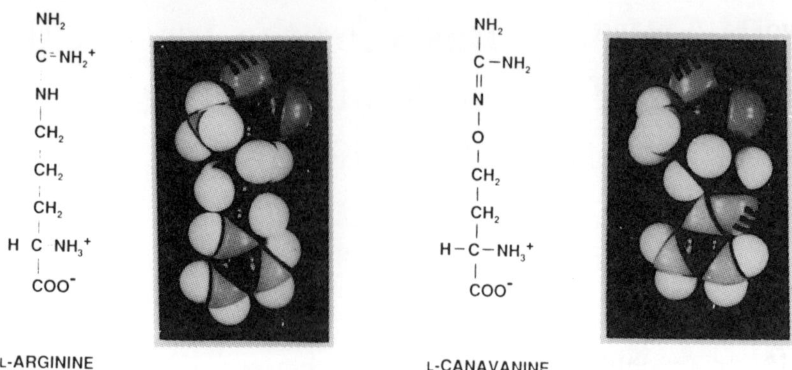

Figure 1   L-Arginine (left) and L-canavanine (right) are represented as they exist in solution at physiological conditions.

Larvae of the tobacco hornworm, *Manduca sexta* [Sphingidae], are sensitive to canavanine's antimetabolic effects. This antimetabolite attenuates growth and produces pupae and adults that exhibit massive developmental aberrations (Dahlman and Rosenthal, 1975). Often, canavanine-treated larvae perish in a futile attempt at metamorphosis. This publication provides a basic understanding of how canavanine manifests its antimetabolic properties in a canavanine-sensitive insect such as *M. sexta*.

Canavanine Incorporation into Protein

Experiments in which L-[*guanidinooxy*-^{14}C]canavanine was administered to *M. sexta* larvae have established that canavanine is assimilated readily into proteins during their synthesis. One can measure how readily canavanine substitutes for arginine in protein; the measurement can be expressed as the substitution error frequency (SEF). Under optimal analogue incorporation conditions, an average of at least one in three arginine residues of the newly synthesized proteins is replaced by canavanine (Rosenthal *et al.*, 1987). On the other hand, the bruchid beetle, *Caryedes brasiliensis* [Bruchidae], and the weevil, *Sternechus tuberculatus* [Curculionoidea], consume seeds that are richly laden with canavanine; these insects are adapted to a canavanine diet (Bleiler *et al.*, 1988). The SEF for *C. brasiliensis* and *S. tuberculatus* is 1 in 365 and 1 in 500–1000, respectively (Rosenthal *et al.*, 1987).

Other insects such as the tobacco budworm, *Heliothis virescens* [Noctuidae], do not normally eat canavanine-containing plants, but they nevertheless possess a high natural resistance to this arginine antagonist. This insect has a SEF of 1 in 65 (Rosenthal *et al.*, 1987). Thus, canavanine-resistant and canavanine-adapted insects avoid extensive canavanyl protein formation while canavanine-sensitive insects incorporate this arginine analogue into protein. These canavanine incorporation studies are important for two reasons. First, they reveal that a canavanine-sensitive insect can incorporate substantive amounts of canavanine into proteins. Second, comparisons of the SEF for canavanine-using or -resistant, as compared to canavanine-sensitive insects supports the contention that formation of structurally aberrant, canavanine-containing proteins is responsible for canavanine's toxicity.

Canavanyl Protein Conformation

The female migratory locust, *Locusta migratoria migratorioides* [Orthoptera], synthesizes vitellogenin for use by the developing oöcyte. Injection of the gravid female with canavanine yields a canavanyl vitellogenin in which 18 of the 200 arginine or about 1 residue in 225 amino acids is replaced by canavanine (Rosenthal *et al.*, 1989b). This level of canavanine incorporation elicits a pronounced structural change in vitellogenin that is effectively observed by electrophoretic analyses. Detergent treatment of canavanyl vitellogenin under reducing conditions yields an electrophoretic pattern in which the six bands seen for the native protein are shifted significantly relative to that of the native vitellogenin (Figure 2).

The altered electrophoretic pattern of canavanyl vitellogenin may result from changes in the post-translational assembly of vitellogenin. Vitellogenin, a massive protein in excess of 500,000 daltons, is thought to be assembled from two polypeptides encoded by separate structural genes (Chen *et al.*, 1978). The polypeptide units constituting vitellogenin may be constructed differently in canavanyl vitellogenin and changes in the assembly process may produce a macromolecule which detergent treatment under reducing conditions cleaves in a unique manner.

These two forms of vitellogenin were also treated with chemicals that are capable of reacting with surface-exposed amino acids to form new amino acids. For example, reaction with cyanate carbamylates surface-exposed lysine residues to homocitrulline. Such treatment reveals that about 25% of the lysyl residues normally buried in native vitellogenin are exposed in canavanyl vitellogenin and subject to carbamylation (Rosenthal *et al.*, 1989b). Similarly, the number of surface-exposed tyrosine residues is revealed to be nearly twice as great in canavanyl vitellogenin as in the native protein.

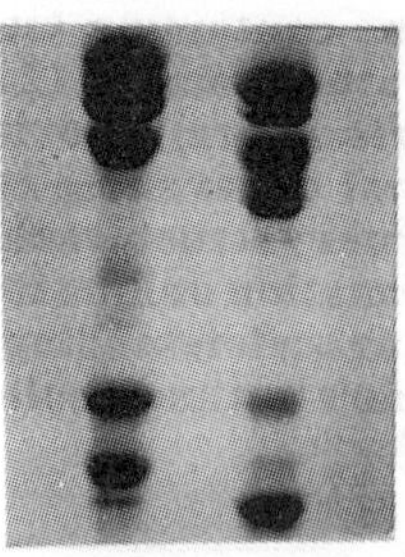

Figure 2 Electrophoretic analysis of native and canavanine-containing vitellogenin. Vitellogenin was analyzed by gradient polyacrylamide electrophoresis using a 7.5–15% acrylamide gradient with 4% stacking gel. The running buffer was 100 mM Tris/ glycine (pH 8.3). The protein was treated with sodium dodecyl sulfate in the presence of 2-mercaptoethanol for 3 minutes at 100°C prior to electrophoresis and stained with 0.5% Coomassie Blue. *Left lane*, native vitellogenin; *right lane*, canavanyl vitellogenin. Copyright *Journal of Biological Chemistry*.

Finally, a monoclonal antibody against canavanyl vitellogenin was found to react weakly with native vitellogenin epitopes. Analyses revealed that the intrinsic association constant for the canavanyl vitellogenin monoclonal antibody with canavanyl vitellogenin was 12.8 versus 4.3 litre $m^{-1} \times 10^{-6}$ for native vitellogenin; the antigen–antibody binding ratio decreased from 0.40 for canavanyl vitellogenin to 0.23 for native vitellogenin (Rosenthal *et al.*, 1989b). These three independent experimental approaches demonstrate that canavanine incorporation alters the three-dimensional conformation of a protein.

Loss of Function in Canavanine-containing Proteins

In response to microbial infection or mechanical injury, larvae of the meat-eating fly, *Phormia terranovae* [Diptera], produce antibacterial proteins, known trivially as the diptericins, that include: diptericin A, diptericin B, diptericin C, and peak V proteins (Keppi *et al.*, 1968). If canavanine is provided at the time of mechanical injury, it is assimilated into the newly produced diptericins and causes a nearly total loss of detectable biological activity for diptericin B, diptericin C, and the peak V protein; only diptericin A retains significant antibacterial potency (Rosenthal *et al.*, 1989a). Analysis of the biological efficacy of diptericin A, the only canavanine-containing diptericin whose antibacterial activity can be detected, reveals canavanine's ability to reduce antibacterial potency (Figure 3).

This investigation established that canavanine incorporation into a protein can impair its function.

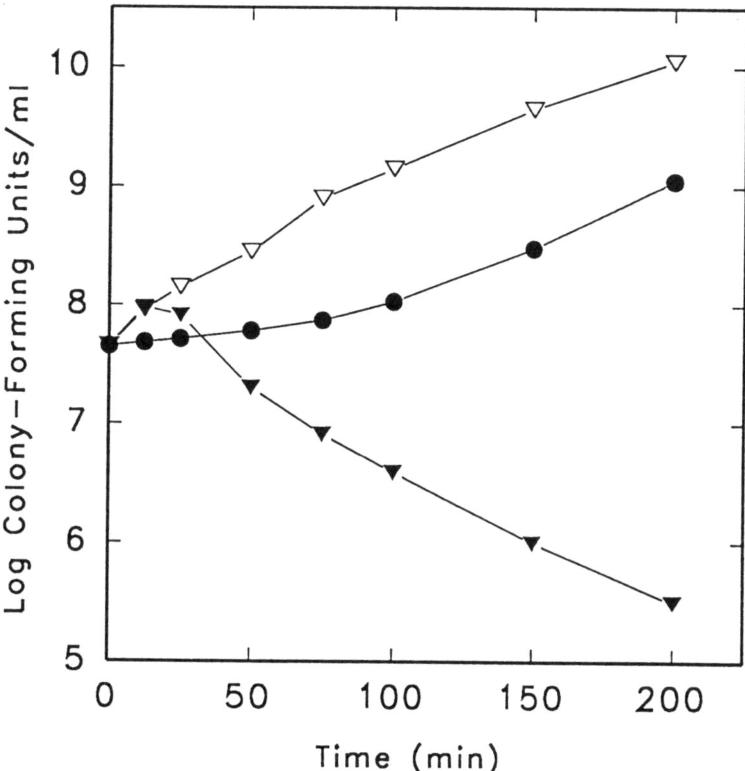

Figure 3   The biological activity of native and canavanine-containing diptericin A. Biological activity was determined by measuring the effect of a 3 μg ml^{-1} solution of native (▼) or canavanine-containing (●) diptericin A on the growth of *E. coli* D31 as described elsewhere (Rosenthal *et al.*, 1989a). Control cultures (▽) were provided 3 μg ml^{-1} of bovine serum albumin. The log of the colony forming units ml^{-1} reflects the growth capacity of a 5×10^7 cell ml^{-1} culture in exponential growth phase. Copyright *Journal of Biological Chemistry*.

Antimetabolic Effects of Other Arginine Analogues

The possible relationship between canavanyl protein formation and the severe developmental aberrations noted in pupae and adults developing from canavanine-treated *M. sexta* larvae was evaluated in a study employing several arginine analogues including: L-canavanine,   L-indospicine [L-2-amino-6-(amidino) hexanoic acid], L-homoarginine [L-amino-6-guanidinohexanoic acid],   L-2-amino-4-guanidinobutyric acid, and   L-2-amino-3-guanidinopropionic acid (Rosenthal and Dahlman, 1991a).

H_2N-C(NH_2) = N-O-CH_2-CH_2-CH(NH_2)-COOH
L-canavanine

H_2N-C(NH_2) = N-CH_2-CH_2-CH(NH_2)-C-OOH
L-indospicine

H_2N-C = (NH)N-$(CH_2)_4$-CH(NH_2)-COOH
L-homoarginine

H_2N-C = (NH)N-CH_2-CH_2-CH_2-CH(NH_2)-COOH
L-2-amino-4-guanidinobutyric acid

H_2N-C = (NH)N-CH_2-CH(NH_2)-COOH
L-2-amino-3-guanidinopropionic acid

Administration of 10.0 μmol canavanine per gram fresh larval weight by parenteral injection terminated larval growth. All of the treated larvae expired within 48 hours. In sharp contrast to canavanine's deleterious effects, a two- to three-fold greater dose of any of the other tested arginine analogues failed to affect growth adversely. All of the treated larvae ecdysed to pupae and then adults without exhibiting any adverse developmental effects.

In a separate experiment, employing radiolabelled analogues, the newly produced proteins of the body wall musculature and haemolymph taken from analogue-treated larvae were analysed for the presence of these radiolabelled compounds. As revealed in Table 1, only canavanine was present in a significant amount in the isolated proteins. A prior study of comparable *M. sexta* larvae, injected with [14C]canavanine) and [3H]arginine-containing haemolymph proteins, established that canavanine containing proteins were degraded preferentially relative to their native counterparts (Rosenthal and Dahlman, 1986). Thus, canavanyl protein formation was greater than the amount indicated by the data of Table 1. There was no significant stable incorporation of homoarginine, indospicine, 2-amino-3-guanidinopropionic acid nor 2-amino-4-guanidinobutyric acid into newly synthesized larval proteins (Table 1).

Table 1 Incorporation of ^{14}carbon-labelled arginine and certain of its analogues into protein by *Manduca sexta* larvae

Amino acid	Amino acid incorporation (% administered dose)	
	Haemolymph	Body wall
L-Arginine	8.4 ± .3	0.87 ± .04
L-Canavanine	3.2 ± .11	0.45 ± .02
L-Homoarginine	0.4 ± .12	0.07 ± .01
L-2-Amino-3-guanidino-propionic acid	nd	nd
L-2-Amino-4-guanidino-butyric acid	trace	nd
L-Indospicine	0.1 ± .06	0.03 ± .02

nd = not detected. See Rosenthal and Dahlman (1991a) for further experimental details.

Of the test arginine analogues, only canavanine was incorporated into the proteins of developing larvae and only canavanine elicited discernible developmental aberrations. These experimental findings strength the hypothesis that aberrant, canavanyl protein formation is responsible for the growth-inhibiting and adverse developmental effects of canavanine in this insect (Rosenthal and Dahlman, 1991a).

Formation of Canavanine-containing Lysozyme

A recent study of insect lysozyme (EC 3.2.1.17) provides additional insight into the relationship of canavanine incorporation and protein function (Rosenthal and Dahlman, 1991b). When *M. sexta* larvae are injected with fragments of the cell wall of *Micrococcus lutea*, the insects are stimulated to produce lysozyme. If these challenged larvae are also given 1.0 mg^{-1} fresh body weight canavanine, it is incorporated into their lysozyme. The ratio of canavanine to arginine in the lysozyme purified from such treated larvae is 1:3.8 ± 0.2. Assay of lysozyme obtained from such canavanine-treated insects reveals a 49.5% loss in catalytic activity as compared to comparably processed but arginine-treated larvae (Table 2). Other experiments establish that the reduced lysozyme activity is not caused by canavanine-mediated attenuation in lysozyme production. Comparable analysis of homogeneous lysozyme isolated from canavanine-treated pupae of the giant silkworm, *Hyalophora cecropia*, reveals a lysozyme canavanine to arginine ratio of 1:4.8 $\pm$ 0.3. Analysis of *H. cecropia* lysozyme fails to disclose a discernible loss in bacteriolytic potency; yet, experiments employing L-[*guanidino*-^{14}C]canavanine established unequivocally that canavanine is incorporated into the lysozyme of both insects (Rosenthal and Dahlman, 1991b). Another group of *M. sexta* larvae, administered 0.5 mg^{-1} canavanine, provided a purified lysozyme in which the ratio aqa S.E.M. of canavanine to arginine was 1:5.05 $\pm$ 0.3.

Table 2   The effect of L-canavanine on the lysozyme activity of *Manduca sexta* and *Hyalophora cecropia*
Haemolymph was collected from *M. sexta* larvae, provided 1.0 mg canavanine g^{-1} fresh body weight, or *H. cecropia* pupae as described in the text. Lysozyme was purified from the haemolymph and assayed as described in the text. Each value is the mean $\pm$ S.E.M. for three determinations.

Treatment	Lysozyme Activity	
	Manduca sexta	*Hyalophora cecropia*
	(units/mg^{-1} protein)	
Arginine	126.5 $\pm$ 7.9	183.4 $\pm$ 3.2
Canavanine	63.9 $\pm$ 5.8	185.6 $\pm$ 12.1

The canavanyl lysozyme of *H. cecropia* demonstrates that canavanine can be incorporated into a protein without affecting adversely its biological activity. Thus, even in a canavanine-sensitive insect such as *M. sexta*, where canavanine is pervasive in newly synthesized proteins (Rosenthal *et al.*, 1987), canavanine replacement for arginine does not necessarily cause the loss of function for a particular protein.

This difference in the effect of canavanine on the activity of lysozyme from these two insects may result from differences in the primary structure of their lysozymes (Table 3). Six of the nine arginyl residues of *M. sexta* lysozyme are shared by *H. cecropia* lysozyme. Arginyl residues at position 23, 42 and 107 in the *M. sexta* enzyme are replaced by serine, lysine and lysine, respectively in *H. cecropia* lysozyme.

Table 3   The primary structure of lysozyme obtained from *Manduca sexta* or *Hyalophora cecropia*

Manduca sexta	1	Lys-His-Phe-Ser-*Arg*-Cys-Glu-Leu-Val-His-Glu-Leu-*Arg*-*Arg*-Gln-Gly-Phe-Pro-Glu-Asn-
Hyalophora cecropia[1]		Lys-*Arg*-Phe-Thr-*Arg*-Cys-Gly-Leu-Val-Gln-Glu-Leu-*Arg*-*Arg*-Leu-Gly- Phe-Asp-Glu-Thr-
Manduca sexta	21	Leu-Met-*Arg*-Asp-Trp-Val-Cys-Leu-Val-Glu-Asn-Glu-Ser-Ser-*Arg*-Tyr-Thr-Asp-Lys-Val-
Hyalophora cecropia		Leu-Met-*Ser*-Asp-Trp-Val-Cys-Leu-Val-Glu-Asn-Glu-Ser-Gly-*Arg*-Phe-Thr-Asp-Lys-Ile-
Manduca sexta	41	Gly-*Arg*-Val-Asn-Lys-Asn-Gly-Ser-*Arg*-Asp-Tyr-Gly-Leu-Phe-Gln-Ile-Asn-Asp-Lys-Tyr-
Hyalophora cecropia		Gly-Lys-Val-Asn-Lys-Asn-Gly-Ser-*Arg*-Asp-Tyr-Gly-Leu-Phe-Gln-Ile-Asn-Asp-Lys-Tyr-
Manduca sexta	61	Trp-Cys-Ser-Asn-Gly-Ser-Thr-Pro-Gly-Lys-Asp-Cys-Asn-Val-Lys-Cys-Ser-Asp-Leu-Leu-
Hyalophora cecropia		Trp-Cys-Ser-Lys-Gly-Ser-Thr-Pro-Gly-Lys-Asp-Cys-Asn-Val-Thr-Cys-Asn-Gln-Leu-Leu-
Manduca sexta	81	Ile-Asp-Asp-Ile-Thr-Lys-Ala-Ser-Thr-Cys-Ala-Lys-Lys-Ile-Tyr-Lys-*Arg*-His-Lys-Phe-
Hyalophora cecropia		Thr-Asp-Asp-Ile-Ser-Val-Ala-Ala-Thr-Cys-Ala-Lys-Lys-Ile-Tyr-Lyr-*Arg*-His-Lys-Phe
Manduca sexta	101	Gln-Ala-Trp-Tyr-Gly-Trp-*Arg*-Asn-His-Cys-Gln-Gly-Ser-Leu-Pro-Asp-Ile-Ser-Ser-Cys
Hyalophora cecropia		Asp-Ala-Trp-Tyr-Gly-Trp-Lys-Asn-His-Cys-Gln-His-Gly-Leu-Pro-Asp-Ile-Ser-Asp-Cys

[1]The primary structure of *H.cecropia* lysozyme was taken from Bergstrom *et al.* (1985).

If canavanine replacement of arginine is uniform throughout *M. sexta* lysozyme, then 21% of every arginine residue is replaced by canavanine. The observed loss of 49.5% of the catalytic activity observed for this lysozyme is essentially the same as the prediction of a 50% loss in catalytic activity, if we assume that substitution of residues 23, 42, or 107 by canavanine results in the loss of all catalytic activity. This is, one half of the induced lysozyme molecules would lack canavanine in position 23, 42, or 107 and would be active; the remaining induced lysozyme molecules would have a canavanine in at least one of these three position and would be inactive. The observed loss of 39% of the catalytic activity noted with lysozyme obtained from larvae provided 0.5 mg canavanine is very close to the value of 42% predicted by the above hypothesis. Although this suggestion seems the simplest at this time, a number of other possible explanations cannot be excluded.

Similarly, examination of the primary structure of the diptericins may also provide a clue to canavanine's ability to cause loss of biological activity for diptericins B and C of *P. terranovae*, but not diptericin A. Diptericin B and C have an arginyl residue at position 38 which is replaced by histidine in Diptericin A (Rosenthal *et al.*, 1989a). Thus, it is not merely the replacement of arginine by canavanine that is of critical importance, but also the contribution of a given arginine residue to the native conformation of the macromolecule.

These insectan studies support the view that the biological effects of canavanine result from its incorporation into a protein, resulting in an alteration in protein conformation that leads ultimately to impairment of protein function. Of all the tested arginine analogues, only canavanine is incorporated into protein and only this analogue elicits antimetabolic effects. The studies considered in this review explain why canavanine-utilizing and -resistant insects evolved mechanisms avoiding scrupulously canavanine assimilation into their proteins. Canavanine-sensitive organisms lack such mechanisms and sustain the adverse effects of canavanine incorporation into protein. Our canavanine investigations also enhance our understanding of canavanine's protective efficacy against insects. Canavanyl protein formation and the associated disruption in protein function can reduce production of catalytically competent proteins and affect growth and developmental processes that reduce significantly an herbivore's overall ability to survive and reproduce.

Acknowledgements

The author acknowledges gratefully the support of the National Science Foundation Grant (DCB-89011749) and the Graduate School of the University of Kentucky for some of the studies presented in this communication.

References

ALLENDE, C.C. and ALLENDE, J.E. (1964). Purification and substrate specificity of arginyl-ribonucleic acid synthetase from rat liver. *Journal of Biological Chemistry*, **239**, 1102–1106.

BLEILER, J., ROSENTHAL, G.A. and JANZEN, D.H. (1988). Biochemical ecology of canavanine-eating seed predators. *Ecology*, **69**, 427–433.

BLUM, M.S. (1981). *Chemical Defenses of Arthropods*. Academic Press, New York.

BOYAR, A. and MARSH, R.E. (1982). L-Canavanine, a paradigm for the structures of substituted guanidines. *Journal of the American Chemical Society*, **104**, 1995–1997.

CHEN, T.T., STRAHLENDORF, P.W. and WYATT, G.R. (1978). Vitellin and vitellogenin from locusts (*Locusta migratoria*). Properties and post-translational modification in the fat body. *Journal of Biological Chemistry*, **253**, 5325–5331.

DAHLMAN, D.L. and ROSENTHAL, G.A. (1975). Non-protein amino acid-insect interactions. I. Growth effects and symptomology of L-canavanine consumption by the tobacco hornworm, *Manduca sexta* (L). *Comparative Biochemistry and Physiology*, **51A**, 33–36.

ENGSTROM, A., XANTHOPOULOUS, K.G., BOMAN, H.G. and BENNICH, H. (1985). Amino acid and cDNA sequences of lysozyme from *Hyalophora cecropia*. *EMBO Journal*, **4**, 2119–2122.

GREENSTEIN, J.P. and WINITZ, M. (1961). *Chemistry of the Amino Acids*. Wiley & Sons, New York.

KEPPI, E., ZACHARY, D., ROBERTSON, M., HOFFMAN, D. and HOFFMAN, J.A. (1986). Induced antibacterial proteins in the hemolymph of *Phormia terranovae* (Diptera). *Insect Biochemistry*, **16**, 395–402.

ROSENTHAL, G.A. (1979). Naturally occurring, toxic nonprotein amino acids. In *Herbivores: Their Interaction with Secondary Plant Metabolites* (G.A. Rosenthal and D.H. Janzen, eds), Academic Press, New York.

ROSENTHAL, G.A. (1988). Biochemical insight into the protective efficacy of L-canavanine, a toxic higher plant metabolite. *Bioscience*, **38**, 104–109.

ROSENTHAL, G.A. (1991). Nonprotein amino acids as protective allelochemicals. In *Herbivores: Their Interaction with Secondary Plant Metabolites* (G.A. Rosenthal and M. Berenbaum, eds), 2nd edition, Academic Press, New York.

ROSENTHAL, G.A. and DAHLMAN, D.L. (1986). L-Canavanine, insect protein synthesis, and higher plant chemical defense. *Proceedings of the National Academy of Sciences USA*, **83**, 14–18.

ROSENTHAL, G.A. and DAHLMAN, D.L. (1991a). Incorporation of L-canavanine into proteins and the expression of its antimetabolic effects. *Journal of Food and Agricultural Chemistry*, **39**, 987–990.

ROSENTHAL, G.A. and DAHLMAN, D.L. (1991b). Studies of L-canavanine incorporation into insectan lysozyme. *Journal of Biological Chemistry*, **266**, 15684–15687.

ROSENTHAL, G.A., BERGE, M.A., BLEILER, J.A. and RUDD, T. (1987). Avoidance of aberrant protein production and an organism's ability to utilize or tolerate L-canavanine. *Experientia*, **43**, 558–561.

ROSENTHAL, G.A., LAMBERT, J. and HOFFMAN, D. (1989a). Canavanine incorporation into the antibacterial proteins of the fly, *Phormia terranovae* (Diptera), and its effect on biological activity, *Journal of Biological Chemistry*, **264**, 9768–9771.

ROSENTHAL, G.A., REICHHART, J.-M. and HOFFMAN, J.A. (1989b). L-Canavanine incorporation into vitellogenin and macromolecular conformation. *Journal of Biological Chemistry*, **264**, 13693–13696.

5
Detoxification of L-Canavanine, a Higher Plant Protective Allelochemical, by the Tobacco Budworm, *Heliothis virescens* [Noctuidae]

GERALD A. ROSENTHAL

T.H. Morgan School of Biological Sciences and the Graduate Center for Toxicology, University of Kentucky, Lexington, Kentucky 40506, USA

Studies of the higher plant nonprotein amino acid, L-canavanine [L-2-amino-4-(guanidinooxy-butyric acid] have revealed its potent insecticidal properties in a wide variety of insects sensitive to this deleterious arginine analogue (see Rosenthal, this volume, p. 35). On the other hand, larvae of the bruchid beetle, *Caryedes brasiliensis* [Bruchidae] and the weevil, *Sternechus tuberculatus* [Curculionoidea] are examples of insects that feed upon seeds heavily laden with this normally poisonous antimetabolite (Rosenthal, 1983; Bleiler *et al.*, 1988). *Caryedes brasiliensis* is not only resistant to this toxicant but also it can utilize canavanine to support its dietary nitrogen needs (Rosenthal *et al.*, 1982); this insect is an herbivore that has evolved resistance to canavanine by a number of biochemical adaptations to canavanine's antimetabolic effects (Rosenthal, 1990a). I characterize such insects as canavanine-adapted.

Heliothis virescens Resistance to Canavanine

The marked sensitivity of most insects to canavanine stands in strong contrast to the innate ability of the tobacco budworm, *Heliothis*

Insecticides: Mechanism of Action and Resistance
© 1992   Intercept Ltd, P.O. Box 716, Andover, Hants SP10 1YG, UK

virescens [Noctuidae] to tolerate a high concentration of dietary canavanine (Berge *et al.*, 1986). While *H. virescens* is a major agricultural pest that consumes a diverse array of plants, these plants do not store significant canavanine (Lincoln, 1972). Thus, *H.virescens* has a *natural* resistance to canavanine which cannot be explained, as it can with *C. brasiliensis* and *S. tuberculatus*, in terms of an adaptive advantage derived from long-term consumption of canavanine-containing plants. *Heliothis virescens* represents a novel kind of canavanine consumer – one that has been termed *canavanine resistant*.

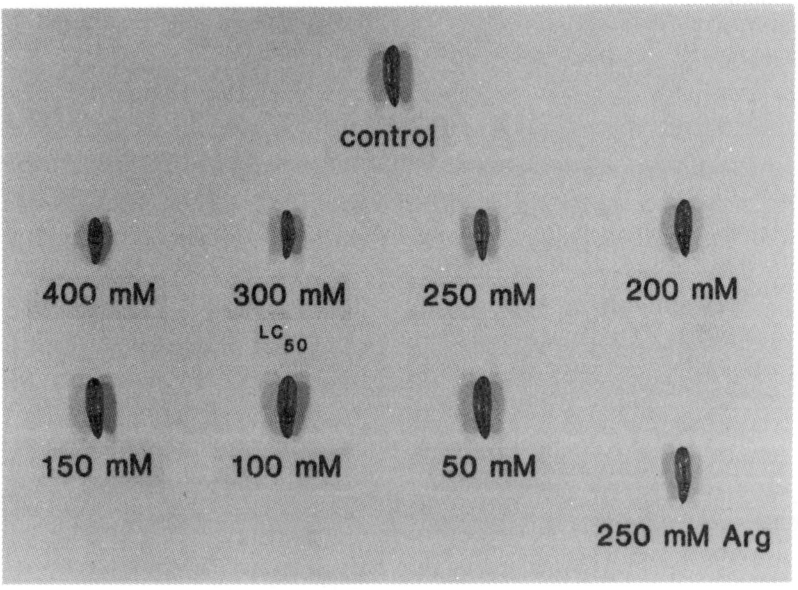

Figure 1   Dietary amino acids and larval development. L-Canavanine or L-arginine, at the indicated concentration, was incorporated into an agar-based diet used to rear the larvae. See Berge, Rosenthal and Dahlman (1986) for further experimental details. Copyright, The American Chemical Society

The LC_{50} for canavanine in the ultimate instar of *H. virescens* is 300 Mm (Berge *et al.*, 1986). This dietary value corresponds to 53,000 ppm wet weight or nearly 40% on a dry weight basis. In a canavanine-sensitive organism such as the tobacco hornworm, *Manduca sexta* [Sphingidae], larvae that consume as little as 2.5 mM canavanine-containing diet exhibited massive developmental aberrations (Dahlman and Rosenthal, 1975; Rosenthal and Dahlman, 1975). In contrast, *H. virescens* larvae, reared on an artificial diet supplemented with 400 mM canavanine failed to develop discernible developmental aberrations (Figure 1).

To determine if *H. virescens* excreted consumed canavanine, its faecal matter was analyzed. When this insect was reared on 150 mM canavanine-containing diet, it consumed 56 µmol of canavanine during the final day of a 3-day exposure period. Only 0.34 µmol or 0.6% of the consumed canavanine was in the frass. Examination of the haemolymph of such treated insects indicated a canavanine haemolymph concentration of 4.7 mM. Assuming a haemolymph volume of 90 µl, only 0.4% of the ingested canavanine was in the haemolymph. Administration of [*guanidinooxy*-^{14}C] canavanine to appropriate larvae revealed that they did not store significant canavanine in newly synthesized proteins (Rosenthal *et al.*, 1987). In addition, significant free canavanine was not found within the non-haemolymph portions of the larvae. Thus, faecal excretion and haemolymph, protein or other sequestration could not account for this generalist herbivore's ability to consume high levels of canavanine. Thus, it seemed most likely that *H. virescens* was actively metabolizing canavanine.

Canavanine Detoxification is Achieved by a Constitutive Enzyme

The ability of *H. virescens* to metabolize canavanine was assessed by examining larval capacity to clear the haemolymph of this amino acid. Canavanine-treated larvae responded to administration of 5 mg g^{-1} parenterally injected canavanine by clearing the haemolymph with a half life of 135 minutes. Several lines of experimental evidence indicate the enzyme(s) responsible for this biochemical ability to metabolize canavanine is part of the constitutive metabolic capacity of the larvae. First, the rate of canavanine clearance from the haemolymph of 135 minutes is the same when determined in an insect that has just been exposed to canavanine, and therefore requires time to induce the necessary degradative enzymes, with an animal maintained on a 150 mM canavanine-containing diet for 3 days prior to determining its canavanine clearance rate (Berge and Rosenthal, 1990). Second, the canavanine clearance rate was compared for larvae treated with 250 µg cycloheximide (per gram fresh body weight) and control animals. The cycloheximide-treated larvae have a t$_{1/2}$ for canavanine clearance that was not different from that of the control (Table 1). In another set of cycloheximide-treated animals, incorporation of L-[^{3}H]leucine into newly synthesized proteins was inhibited 80% as compared to the controls. Third, gel-electrophoretic analyses of extracts of the gut, fat body, malpighian tubules, and haemolymph of *H. virescens* larvae, exposed to 150 mM canavanine-containing diet for 3 days, did not disclose a novel protein band resulting from dietary canavanine consumption (Figure 2). Administration of ^{35}S-methionine to comparable canavanine-treated larvae also did not reveal any *de novo* synthesized proteins not also produced by control larvae. Thus, the detoxification system for canavanine is not induced in response to canavanine, but rather is constitutive (Berge and Rosenthal, 1990).

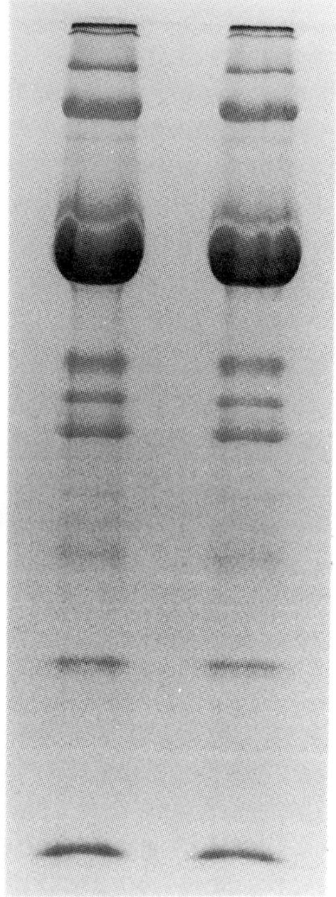

Figure 2 Electrophoretic profiles of *Heliothis virescens* haemolymph proteins from larvae exposed to control diet (left lane) or 150 mM canavanine-containing diet (right lane). See Berge and Rosenthal (1990) for additional experimental details. Copyright, The American Chemical Society

Table 1 Effect of inhibiting *de Novo* protein synthesis on canavanine clearance by *Heliothis virescens*

Treatment	$t_{1/2}$ Canavanine Clearance[1]
Sterile water	119 ± 5
Cycloheximide	121 ± 4

[1]Each value is the mean $\pm$ SEM. See Berge and Rosenthal (1990) for additional experimental details.

Table 2 Metabolism of L-[guanidinooxy-^{14}C]Canavanine by *Heliothis virescens* (% ^{14}Carbon recovered[1])

Fraction or substance	1h	2h	4h	6h	12h
canavanine	81.8 ± 0.6	67.1 ± 3.4	48.2 ± 1.3	32.1 ± 1.7	8.6 ± 5.6
guanidine	14.0 ± 2.0	22.8 ± 3.8	43.5 ± 6.4	53.5 ± 6.5	72.3 ± 8.3
urea	1.62 ± 0.4	3.7 ± 0.3	7.2 ± 0.5	7.6 ± 1.2	8.7 ± 1.0
respiratory $^{14}CO_2$	t[2]	t	t	t	t
total ^{14}carbon recovered	98	94	99	94	90

[1]Each value is the mean $\pm$ SEM (n = 3). See Berge and Rosenthal (1990) for additional experimental details.
[2]t = trace

Larval analyses of the NADPH-dependent cytochrome P_{450} monooxygenase or mixed function oxidase, glutathione transferase, hydrolases such as esterases (EC 3.1.1), glycosidases (EC 3.2) and ether hydrolases reveal that none of these enzymes, which usually function in the detoxification of xenobiotics, accounts for canavanine catabolism. It also appears that a conjunction reaction does not function in canavanine degradation. This insect metabolizes more than 50 µmol of canavanine per day. It is difficult to imagine what metabolite is produced in an adequate amount to conjugate so much canavanine without depleting other critical metabolic needs of the insect. The preponderance of evidence indicates that *H. virescens* is drawing upon a fortuitous, *preexisting* pathway for canavanine metabolism that is not part of the first line of defense against xenobiotics (Berge and Rosenthal, 1990).

Canavanine Catabolism

It is relevant to consider how other organisms catabolize canavanine. The sole pathway for canavanine catabolism by all eukaryotic life studied to date is by catalytic hydrolysis mediated by arginase (EC 3.5.3.1) to yield canaline and urea. For example, the rat has a marked ability to catabolize canavanine. Provided a 2 mg g^{-1} dose of L-[*guanidinooxy*-^{14}C]canavanine, this mammal excreted 75% of the administered radiolabelled carbon as urea. The only other radiolabelled catabolite of importance was [^{14}C]guanidine which accounted for only 5% of the ^{14}carbon provided to the rat (Thomas and Rosenthal, 1987). Higher plants that store and use canavanine as well as canavanine-adapted insects commonly degrade canavanine via hydrolysis to canaline and urea (Rosenthal, 1990b).

To evaluate the possible role of arginase in *H. virescens* larvae, L-[*guanidinooxy*-^{14}C]canavanine, supplemented with 5 mg g^{-1} (fresh body weight) carrier canavanine, was administered by parenteral injection. Analysis of the radiolabelled degradation products established that the principal ^{14}carbon-bearing catabolite was not urea but rather guanidine (Table 2). *Heliothis virescens* has only very weak arginase activity and virtually no urease activity. Hydrolytic cleavage of canavanine to canaline and urea does not figure significantly in canavanine processing by the larva. Thus, *H. virescens* appears to possess a reductase able to convert L-canavanine to L-homoserine and guanidine. This is the first report of such a degradative pathway in a plant or animal.

$$H_2N\text{-}C\text{-}(NH_2)=N\text{-}O\text{-}CH_2\text{-}CH_2\text{-}CH(NH_2)COOH \rightarrow HO\text{-}CH_2\text{-}CH_2\text{-}CH(NH_2)COOH \quad H_2N\text{-}C(=NH)\text{-}NH_2$$

        L-canavanine   L-Homoserine   guanidine

In order to verify that homoserine was a reaction product and to evaluate the metabolic fate of the aliphatic chain of canavanine, L-[*1,2,3,4*-^{14}C]canavanine was synthesized. *Heliothis virescens* efficiently catabolized L-[*1,2,3,4*-^{14}C]canavanine; only 4.5% of this radiolabelled compound remained after 6 half lives for canavanine degradation [this is very close to the 1.5% of the original

Table 3 Metabolism of L-[*1,2,3,4*-^{14}C]Canavanine in *Heliothis virescens*

Fraction or substances	% [14]Carbon Recovered[1]			
	135 min	270 min	405 min	810 min
I. Respiratory [14]CO$_2$	0.7	3.1	5.1	14.5
II. Acetone-soluble fraction	6.5	7.2	7.8	5.4
III. Water-soluble fraction				
a. neutral	5.8	6.5	5.6	4.9
b. charged				
aspartate/asparagine	0.6	0.4	2.2	2.4
glutamate/glutamine	4.3	3.6	7.0	6.3
proline	0.2	0.4	2.2	2.4
2-aminobutyrate	3.7	4.6	5.2	3.1
isoleucine	1.6	1.3	2.9	2.1
homoserine	10.4	28.6	39.3	38.2
ornithine	4.2	4.3	4.3	9.5
canavanine	60.3	30.8	14.8	4.5
total	85.3	74.0	76.5	67.2
IV. total [14]carbon recovered	98.3	90.8	95.0	92.0

[1]Each value is the mean of 2 independent determinations with a pooled group of 5 insects. See Berge and Rosenthal (1991) for additional experimental details.

L-[*1,2,3,4*-^{14}C]canavanine that should remain after 6 half lives] (Berge and Rosenthal, 1991). As anticipated, [^{14}C]homoserine was the most abundant degradation product (Table 3).

Evaluation of the various tissues of *H. virescens* disclosed that the enzyme responsible for canavanine degradation was located in the gut tissue. These tissues were used to prepare a homogenate which was analyzed for its canavanine-degrading ability. These studies revealed that the gut homogenate fostered an ATP-dependent reduction of L-canavanine, in which NADH served as the reductant, to generate L-homoserine and guanidine.

Subsequent purification of the gut homogenate for the canavanine-degrading protein resolved a single enzyme that mediated an ATP-dependent hydrolysis of L-canavanine to yield L-homoserine and hydroxyguanidine, not guanidine. Another enzyme is present in the gut tissue that fosters a NADH-dependent reduction of hydroxyguanidine to guanidine. These two enzymes working in consort account for the observed *in vivo* formation of guanidine (Berge and Rosenthal, 1990).

The experimental evidence gathered to date supports the proposal that the canavanine-degrading enzyme is a hydrolase that cleaves canavanine by adding water at the O–N bond. This protein is therefore unique as it is the only enzyme known that can attack an O–N bond by hydrolytic cleavage. The gut enzyme functions as an ATP-dependent hydrolase. This novel protein and particularly is substrate specificity is being investigated further at this time.

References

BERGE, M.A. and ROSENTHAL, G.A. (1990). Detoxification of L-canavanine by the tobacco budworm, *Heliothis virescens* [Noctuidae]. *Journal of Food and Agricultural Chemistry*, **38**, 2061–2065.

BERGE, M.A. and ROSENTHAL, G.A. (1991). Metabolism of L-canavanine and L-canaline in the tobacco budworm, *Heliothis virescens* [Noctuidae]. *Chemical Research in Toxicology*, **4**, 237–240.

BERGE, M.A., ROSENTHAL, G.A. and DAHLMAN, D.L. (1986). Tobacco budworm, *Heliothis virescens* [Noctuidae] resistance to L-canavanine, a protective allelochemical.*Pesticide Biochemistry and Physiology*, **25**, 319–326.

BLEILER, J., ROSENTHAL, G.A. and JANZEN, D.H. (1988). Biochemical ecology of canavanine-eating seed predators. *Ecology*, **69**, 427–433.

DAHLMAN, D.L. and ROSENTHAL, G.A. (1975). Non-protein amino acid-insect interactions. I. Growth effects and symptomology of L-canavanine consumption by the tobacco hornworm, *Manduca sexta* (L.). *Comparative Biochemistry and Physiology*, **51A**, 33–36.

LINCOLN, C. (1972). Distribution, abundance and control of *Heliothis* spp. in cotton and other host plants. In *Southern Cooperative Series Bulletin 169*, pp. 2–7, Oklahoma Agricultural Experimental Station, Oklahoma State University, Stillwater, Oklahoma.

ROSENTHAL, G.A. (1983). The adaptation of a beetle to a poisonous plant. *Scientific American*, **249**, 164–171.

ROSENTHAL, G.A. (1990a). Biochemical adaptations by the bruchid beetle, *Caryedes brasiliensis*. In *Bruchids and Legumes:*

Economics, Ecology and Coevolution (K. Fujii *et al.*, eds), pp. 161–170. Kluwer Acad. Publ., The Netherlands.

ROSENTHAL, G.A. (1990b). Metabolism of L-canavanine and L-canaline in leguminous plants. *Plant Physiology*, **94**, 1–3.

ROSENTHAL, G.A. and DAHLMAN, D.L. (1975). Non-protein amino acid-insect interactions. II. Studies of the effects of the canaline-urea amino acids on the growth and development of the tobacco hornworm, *Manduca sexta* (L.) (Sphingidae). *Comparative Biochemistry and Physiology*, **52A**, 105–108.

ROSENTHAL, G.A., HUGHES, C.G. and JANZEN, D.H. (1982). L-Canavanine, a dietary nitrogen source for the seed predator, *Caryedes brasiliensis* (Bruchidae). *Science*, **217**, 353–355.

ROSENTHAL, G.A., BERGE, M.A., BLEILER, J.A. and RUDD, T. (1987). Avoidance of aberrant protein production and an organism's ability to utilize or tolerate L-canavanine. *Experientia*, **43**, 558–561.

THOMAS, D.A. and ROSENTHAL, G.A. (1987). Metabolism of L-[*guanidinooxy*-^{14}C]canavanine in the rat. *Toxicology and Applied Pharmacology*, **91**, 406–414.

6
Effect of Amino Acid Diazoketone Derivatives on Insects

B. STYCZYNSKA[1], H. BARTOSZ-BECHOWSKI[2], J. NAWROT[3], W. SABOTKA[4] AND D. KONOPINSKA[2]

[1]*State Institute of Hygiene, ul. Chocimska 24, 00-791 Warsaw, Poland*
[2]*Institute of Chemistry Wroclaw University, ul. Joliot-Curie 14, 50-383 Wroclaw, Poland*
[3]*Institute for Plant Protection, ul. Miczurina 20, 60-318 Poznan, Poland*
[4]*Institute of Industrial Organic Chemistry, ul. Annopol 6, 03-236 Warsaw, Poland*

Introduction

In the search for new biorational insect control methods, one of the principal objectives is to find environmentally safe compounds which are harmless to vertebrates. During the past twenty years some interest has been shown in natural amino acid and peptide derivatives with insecticidal activity. Several substances of that type were isolated. These include microbial metabolites such as asprochacin (Myokei *et al.*, 1969), cyclodepsipeptide destruxin and analogues (Suzuki *et al.*, 1970), a toxic dipeptide from the defence glands of the Colorado beetle (Daloze *et al.*, 1986), and argiotoxins, a new class of toxins from the venom of the orb weaver spider (Adams *et al.*, 1987).

All these observations prompted us to undertake investigations into insecticidal (Konopinska and Sobotka, 1989) and antifeeding (Nawrot *et al.*, 1988; Nawrot *et al.*, 1991) effects of ester and amide peptide derivatives. Although, when externally applied against

Insecticides: Mechanism of Action and Resistance
© 1992   Intercept Ltd, P.O. Box 716, Andover, Hants SP10 1YG, UK

insects, they do not reveal any insecticidal properties, some of them, particularly containing an aromatic L-amino acid residue in the peptide chain, deter stored product pests for feeding.

It has been observed that the N-amino acid or peptide derivatives applied mainly as C-terminal esters or amides show no interference with insect cuticle. Therefore, attention was given to the more chemically reactive group of substances such as the diazoketone derivative of N-substitute amino acids with the general formula R-O-CO-NHCH(R')–COCHN$_2$ which comprise the following compounds: Boc-Tyr(p-OEt)–COCHN$_2$ (1), Boc-D-Tyr(p-OEt)-COCHN$_2$ (2), Boc-Phe(p-NO$_2$–COCHN$_2$ (3), Boc-Tyr(p-OBzl–COCHN$_2$ (4), Z-Phe-COCHN$_2$ (5), Z-γ–Abu-COCHN$_2$ (6), Z-β–Ala-COCHN$_2$ (7), Boc-Glu(O-t-But)–COCHN$_2$ (8), Z-Pro-COCHN$_2$ (9) and Boc-Ala-COCHN$_2$ (10) (*see Table 1 for references*).

Diazoketones, in comparison with esters of amide amino acid derivatives, show higher chemical reactivity due to the slow release in the photochemical decomposition process of the carbene intermediates, which may interfere with insect cuticle leading to metamorphosis disturbances or to instant insecticidal effects. The discovery of the inhibitory properties of diazoketone dipeptide derivatives, with respect to the cysteine proteinase catalytic centre (Grzonka *et al.*, 1990), was a factor influencing these studies.

Thus, the aim of this work was to determine the insecticidal effects of the amino acid diazoketones 1–10 against the housefly (*Musca domestica* L.) and to evaluate their influence (compounds 1, 3–10) on metamorphosis in the confused flour beetle (*Tribolium confusum* Duv.) larvae and pupae.

Materials and Methods

The diazoketone derivatives 1–10 were obtained via the reaction of N-substituted amino acid mixed anhydrides according to the previously described method (Bartosz-Bechowski and Konopinska, 1989).

The following abbreviations are used:

Ala	L-alanine	Boc	tert-butyloxy-carbonyl
β-Ala	β-alanine	Z	benzyloxy-carbonyl
γ-Abu	γ-aminobutyric acid	Et	ethyl
Glu	L-glutamic acid	Bzl	benzyl
Pro	L-proline	t-But	tert-butyl
Tyr	L-tyrosine	D-Tyr	D-tyrosine

Adult *Musca domestica* (3–4 days old), males and females, were raised under laboratory conditions (State Institute of Hygiene, Warsaw). *T. confusum* larvae (25–30 days old) were reared in the laboratory colony (Institute of Plant Protection, Poznan). The bioassays of *M. domestica* were performed by topical application of acetone solutions of the tested compounds **1–10** at doses of 0.0005 µg to 0.1 µg per insect. An acetone control was used. Biological activity was expressed as the percentage mortality (for 20 male and female houseflies in each experiment) after 24 hours exposure to the evaluated compound. All experiments consisted of three replications at 23°C and 70($\pm$5)% relative humidity. The results are presented in Table 1.

Table 1   Effect of N-protected amino acid diazoketone derivatives **1–10** on the mortality of the house fly (*Musca domestica* L.)

Compound number	Mortality % at a dose expressed in µg per insect[1]									
	0.1		0.01		0.001		0.0005		control	
	m	fm	m	fm	m	fm	m	fm	m	fm
1	96.6	0	80	6.6	96.6	6.6	96.6	90.0	10	0
2	60.0	16.6	63.3	0	60.3	13.3	96.6	6.6	6.6	0
3	40.0	6.6	30.0	3.3	43.3	3.3	36.6	0	0	0
4	56.6	13.3	56.6	23.3	33.3	13.3	70.0	16.6	0	0
5	3.3	0	6.6	0	0	3.3	0	0	0	0
6	23.3	80.0	16.6	16.6	16.6	6.6	33.3	6.6	0	0
7	6.6	6.6	3.3	3.3	0	3.3	0	3.3	0	0
8	3.3	0	3.3	0	3.3	0	3.3	0	0	0
9	10.0	0	13.3	3.3	23.3	20.0	6.6	0	0	0
10	0	0	6.6	0	6.6	0	6.6	0	0	0

[1]Mean value of three experiments.

Preliminary biological evaluation with *T. confusum* (10 larvae or 10 freshly formed pupae in each experiment) was performed by topical application of acetone solutions of the diazoketones **1** and **3–10** at a doses of 5 µg per larvae or pupa at ambient temperature and humidity. Larval development was observed until pupation (about 5 days). The treated pupae (1 day old) were next checked until adult eclosion. The results are shown in Table 2.

Table 2   Effect of N-protected amino acid diazoketone derivatives **1, 3–10** on the development of the stored product insect *Tribolium confusum* Duv. when applied at a dose of 5 µg per insect

Compound number	Developmental inhibition (%)	
	Larvae to pupae	Pupae to adults
1	0	0
3	0	12
4	10	10
5	10	0
6	0	0
7	10	0
8	75	10
9	0	48
10	25	65
Control	0	0

Results and Conclusions

Bioassays to determine the effect of ten diazoketone amino acid derivatives **1–10** (Table 1) revealed selective insecticidal activity against housefly males. A high mortality rate (70% to 96.9%), comparable to that for permethrine, was observed for diazoketones **1, 2** and **4** at the lowest dose applied (0.0005 µg per insect). It was also observed that compound **2** with a D-Tyr residue showed a high insecticidal effect at low concentrations. The pronounced insecticidal activity of Boc-Phe(p-NO$_2$–diazoketone (**3**) reached 40% mortality irrespective of the dose used. Considerably weaker insecticidal effect against housefly males was found for compounds **6** and **9** while the others were virtually inactive.

Mortality of the bioassayed housefly females was almost negligible although compound **1** revealed 90% insecticidal activity at the lowest dose, and compound **6** caused 80% mortality at the highest concentration.

The results show selective insecticidal activity of some diazoketone derivatives against housefly males, whereas females seem to be more resistant. Moreover, some structure–function relationship may be observed. Thus, compounds **1–4**, highly active against the males, have either configuration L (compounds **1, 3** and **4**) or D of the aromatic amino acid residue (compound **2**) substituted in the *para* position by the ethoxy (**1** and **2**), benzyloxy (compound **4**) or nitro group (compound **3**) while a lack of the substituent at the phenyl ring (compound **5**) leads to a loss in effectiveness. Among the aliphatic diazoketones **6–8** and **10**, Z-(gamma–Abu-diazoketone (**6**) revealed a noticeably selective action against housefly females at the dose of 0.1 µg per insect. The configuration of tyrosine derivatives (compounds **1** and **2**) does not seem to play a significant role in their activity against the males. The opposite is found for females where compound **1**, with L-tyrosine moiety, caused the highest mortality at the lowest dose applied. In conclusion, it might be assumed that the selective insecticidal effect observed against housefly males depends to a great extent on the presence of *para* substituted amino acid aromatic ring, particularly by an ethoxy or benzyloxy group.

In the course of studies on the diazoketones bioactivity we became interested in elucidating the influence of compounds **1** and **3–10** on *T. confusum* development (Table 2). Some noticeable effects were observed in the case of Boc-Glu(O-t-But–diazoketone (**8**), which substantially decreased the pupation rate of the larvae (by 70%), and Boc-Ala-diazoketone (**10**), which significantly inhibited (by 65%) the development from the pupal stage to the adult instar of the confused flour beetle.

The reported results indicate that diazoketone amino acid derivatives possess pronounced selective insecticidal activity. Their chemical decomposition into environmentally harmless components such as nitrogen and acyl amino acid derivatives offers encouraging new possibilities for further studies into this class of compounds. It has been noted that two of the bioassayed diazoketones **8** and **10**, structurally related to some peptide juvenoids, influence the metamorphosis of insects.

References

ADAMS, M.E., CARNEY, R.L., FU, E.T., JAMERA, M.A., LI, J.P., SCHOOLEY, D.A., SHAPIRO, M.J. and VENEMA, V.J. (1987). Structure and biological activities of three synaptic antagonists from orb weaver spider venom. *Biochemical, Biophysical Research Communications*, **148**, 678–683.

BARTOSZ-BECHOWSKI, H. and KONOPINSKA, D. (1989). Synthesis of new N-tert-butyloxycarbonyl-β–amino-(gamma–phenyl(p-substituted–L-butyric acid ('homo'-L-phenylalanyl) derivatives. *Journal für praktische Chemie*, **331**, 532–536.

DALOZE, D., BRAEKMANN, J.C. and PASTEELS, J.M. (1986). A toxic dipeptide from the defence glands of the Colorado beetle.

Science, **223**, 221–223.

GRZONKA, Z., KASPRZYKOWSKA, R., KASPRZYKOWSKI, F., GRUBB, A., ABRAHAMSSON, D., OLAFSSON, I. and TROJNAR, J. (1990). Synthesis of cysteine proteinase inhibitors structurally based on the proteinase-binding center of human cystatin C. In *Peptides: Chemistry, Structure and Biology, Proceedings of the 11th American Peptide Symposium* (J.E. River and G.R. Marshall, eds), pp. 375–377. ESCOM, Leiden.

KONOPINSKA, D. and SOBOTKA, W. (1989). Synthesis of permethroyl amides with potential deterrent and insecticidal activity. *Polish Journal of Chemistry*, **63**, 303–306.

MYOKEI, R., SAKURAI, A., CHANG, F.C., KODAORA, Y., TAKA-HASHI, N. and TAMURA, S. (1969). Structure of asprochacin, an insecticidal metabolite of *Aspergillus ochraceus*, *Tetrahedron Letters*, **10**, 695–698.

NAWROT, J., KONOPINSKA, D. and SOBOTKA, W. (1988). Deterrent properties of some amino acid derivatives for stored product insects. *Journal of Applied Entomology*, **106**, 413–416.

NAWROT, J., KONOPINSKA, D. and SOBOTKA, W. and BARTOSZ-BECHOWSKI, H. (1991). Further research of deterrent properties for stored product insects among some amino acid and peptide derivatives. *Journal of Applied Entomology*, **111**, 104–108.

SUZUKI, A., TABUCHI, H. and TAMURA, S. (1970). Isolation and structure elucidation of new cyclodepsipeptides, destruxin C and D and demethylodestruxin B, produced by *Metarrhizium anizoplie*. *Agricultural Biological Chemistry*, **34**, 813–816.

7
Novel Insecticidal Effects of Isothiocyanate Compounds

B. WEBER, D. MARTIN AND D. OTTO

Biological Research Centre Berlin, Stahnsdorfer Damm 81, 0-1532 Kleinmachnow, Germany

Introduction

Derivation of known biocidal natural compounds is a promising way to search for novel insecticides. The naturally occurring mustard oils represent a classic group of such compounds. Mustard oil derivatives of plant origin have been described both as semiochemicals, especially as host plant specific insect attractants (e.g. Pickett, 1988, 1989) and biocidal compounds. Matolcsy *et al.* (1986) and Ujvary *et al.* (1989) showed that several synthetic alkylene-bis-isothiocyanates have properties as insect growth regulators. Few references (Ujvary *et al.*, 1989; Qureshi *et al.*, 1982) describe the insecticidal properties of aromatic isothiocyanate compounds. Based on these considerations, several isothiocyanates, especially diverse aryl isothiocyanates, were synthesized and bioassayed for their possible biocidal effects on the larvae of *Agrotis segetum* Schiff. (Lepidoptera: Noctuidae) as well as for their influence on the development of *Sitophilus granarius* (resp. *S. oryzae*) populations (Coleoptera: Curculionidae).

Figure 1 shows the chemical structure of the aromatic isothiocyanate compounds (**1–10**) prepared and bioassayed for the present study. The isothiocyanates were synthesized according to known methods (Martin *et al.*, 1965; Hartmann, 1983; Habib and Rieker, 1984).

Insecticides: Mechanism of Action and Resistance
© 1992   Intercept Ltd, P.O. Box 716, Andover, Hants SP10 1YG, UK

ARYLISOTHIOCYANATES:

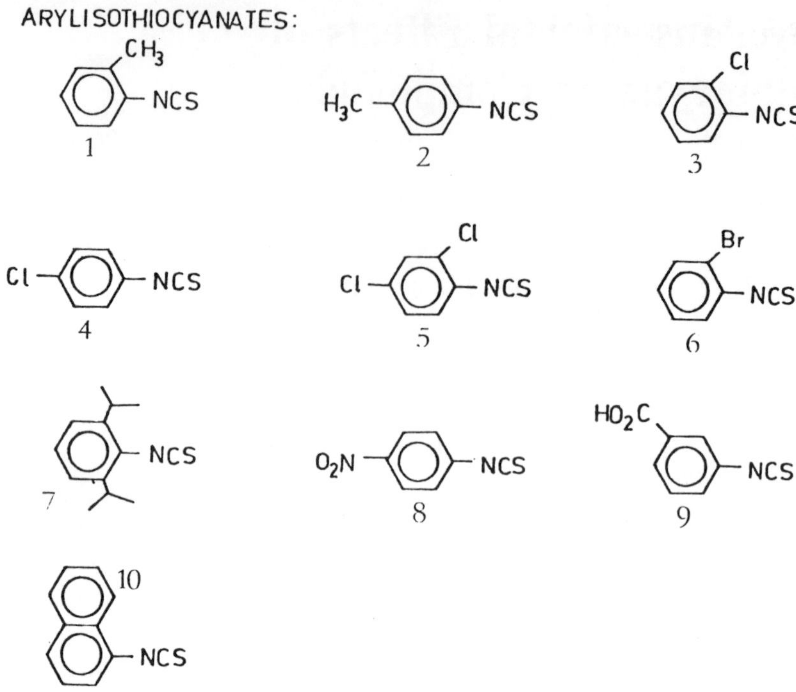

Figure 1 Chemical structure of the isothiocyanates investigated.

Material and Methods

Agrotis segetum (Development test)

Young larvae (first–second instar :L1/2) of the European cutworm (*A. segetum*) were reared in glass dishes (diam. 13cm) coated with filter paper. The larvae were fed until pupation with an artificial diet (Fischer and Otto, 1976). 200 ml units of diet were mixed with the compounds at various concentrations (Table 1). These diet preparations were stored in a refrigerator at 4°C until the experiment. The larvae were fed from this store by application of 5g portions of fresh diet every second day. The larval development was observed until the imaginal stage (maximum about 70–100 days after the first application of the treated diet) and compared to the development of an untreated control group (U.C.). Striking effects such as malformed larvae, pupae and imagoes, as well as the length of larval development until pupation, the behaviour of the larvae and the level of mortality were recorded.

Table 1 Biological effects caused by compounds **1–10**

Compound No.	concentration[1]	Affected stages of development				Effects	Mort. (%)	Days after after start of exp.
		L1/2	L3/4	Pupae	Iamgoes			
1	200	+	–	–	–	–	100	10
	100	+	+	(+)	(+)	m.P.	95	80
	50	+	+	(+)	(+)		100	100
2	200	+	–	–	–	L.ex.	100	10
	100	+	+	(+)	(+)		95	80
	50	+	+	(+)	(+)		100	100
3	200	+	–	–	–	–	100	5
	100	+	+	–	–	–	100	10
	50	+	+	+	(+)	m.P.	100	50
4	200	+	–	–	–		100	5
	100	+	–	–	–	L.ex.	100	10
	50	+	–	–	–		100	10
5	200	+	–	–	–	L.ex.	100	5
	100	+	–	–	–		100	10
	50	+	+	(+)	(+)	m.P./I.	75	10
6	200	+	–	–	–	L.ex.	100	5
	100	+	–	–	–		100	5
	50	+	+	+	+	m.P./I.	50	50
	30	+	+	+	+		46	70
7	200	+	+	(+)	–	L.ex.	35	100
8	200	+	+	–	–	L.ex.	50	14
9	200	–	–	–	–	–	0	–
10	200	+	+	–	–	L.ex.	100	10
U.C.	0	–	–	–	–	–	10	50

[1]Given in mg substance per 200ml diet; m.P. = malformed pupae/iamgoes; L.ex. = larvae died before pupation; + = predominantly affected stage; (+) = occasionally affected stage.

Sitophilus spp. (Development test)

Volume units of 300 ml untreated wheat were populated with 6 ml of fertile imagoes of *Sitophilus* spp., i.e. about 700 individuals of *S. granarius* or about 1000 individuals of *S. oryzae* respectively, and kept at 18–20°C under laboratory conditions. After 3 days the imagoes were separated from the wheat by sieving. This brood substrate (i.e. eggs in wheat grains) was then treated and mixed with the compounds in acetone at different concentrations (Table 2). After evaporation of the acetone, the different treatments were placed in 600 ml glasses and were kept under laboratory conditions. The influence on the further development and vitality of the emergent F1 generation was measured in comparison to an untreated control population (U.C.) by counting the emerging imagoes at the end of the experiment, approximately 100 days after application of the compounds.

Table 2   Influence of 2–bromophenylisothiocyanate on the development of cucurlionid populations living in grain

Concentration (mg comp. per 100 ml grain)	Number of imagoes (F1 generation)
Sitophilus granarius	
25	0
15	0
5	10
3	140
U.C.	450
Sitophilus oryzae	
25	0
15	20
5	70
3	430
U.C.	2000

Results

The aromatic isothiocyanates had several effects on the larval development as well as on the pupae and imaginal formation of *A. segetum*. Disruptions of moulting processes in older larvae were

Figure 2 Examples of malformed pupae (a, b, c) and imagoes with misshapen wings (d) of *A. segetum* Schiff. emerging after uptake of a diet containing 30mg 200ml^{-1} diet of compound **6** during their previous larval development. e = normally developed pupa of *A. segetum*.

observed as well as malformed pupae with striking vesicular formations and other deformations (Figures 2a–2d). Occasionally malformed imagoes with incompletely developed wings occurred (Figure 2d). Deformations in pupae were the predominant defects. The malformed pupae were all incapable of surviving and reaching the imaginal stage. Table 1 summarizes the observed biological effects caused by compounds **1–10** in *A. segetum*. Apart from compound **9**, which had no biological action even at the highest concentration, all other derivatives showed biocidal activity. They caused high mortality (95–100%) in larvae and/or pupae when applied at the higher doses (i.e. 100–200 mg a.i. in 200 ml diet = 500...1000 ppm), whereas malformations in pupae or imagoes were caused only by the compounds **1, 3, 5** and **6**, which have a halogene or methyl group in the ortho-position. These effects were observed at the lower concentrations (i.e. 30–50 mg a.i. per 200 ml diet = 150...250 ppm) and were connected with decreased larval mortality. Low concentrations of compound **6** (2-bromophenyl-isothiocyanate) in particular yielded numerous deformed pupae and imagoes of *A. segetum*.

Bioassay II simulated a practically used technique for the control of stored grain pest populations. The most effective derivative (**6**) was investigated by this technique for its possible influence on the development of *S. granarius* and *S. oryzae* populations. Treatment of grain with 2-bromophenylisothiocyanate at a concentration of 15 mg 100 ml^{-1} grain reduced the abundance of the emerging F1 generation to values of between 0% and 20% of an untreated control population. A concentration of 25 mg per 100 ml^{-1} disrupted the population development in both species (Table 2).

Discussion

Numerous synthetic thiol and isothiocyanate derivatives have been described as bioactive compounds (Zsolnai, 1975; Drobnica *et al.*, 1977). Nematicidal, helminthicidal, fungicidal and herbicidal effects have been reported in the past for the different alkyl and aryl isothiocyanates (Nakamura, 1961; Bakry *et al.*, 1968; US-Patent: 2.946.720, 1960; US-Patent: 3.366.538, 1968).

The alkylene-bis-isothiocyanates have, in particular, been described as a novel class of insect growth regulators (Matolcsy *et al.*, 1986) due to their influence on the spiracular epidermis and crochet formation on the prolegs of tobacco hornworm (*Manduca sexta*) caterpillars (Ujvary *et al.*, 1988). Ujvary *et al.* (1989) described the juvenile hormone activity (in *M. sexta*) of numerous synthesized isothiocyanate compounds. Other sulphur-containing and non-terpenoid carbamate (and isothiocyanate) derivatives having juvenile hormone activity are reported by Ujvary *et al.* (this volume, p. 147).

In the light of these recent findings we studied the influence of aromatic isothiocyanate compounds of related chemical structure on the development of noctuids and cucurlionids. As several of the weakly toxic and naturally occurring 2-phenylethylisothiocyanates from higher plant species are known to have antiinsect properties (Lichtenstein *et al.*, 1962), the search for related synthetic isothiocyanates with an aromatic structure seems to be a promising approach to new solutions in pest control.

Matolcsy *et al.* (1986) found that spiracle and crochet formation in

M. sexta were inhibited after topical application of acetonic solutions of isothiocyanates on the dorsal midline of the third instar larvae. They found a lethal effect caused by phenylisothiocyanate at 50 µg per larva, whereas the alkylene-bis-isothiocyanates generally caused the previously mentioned malformations in larvae at doses between 10 and 100 µg per individual.

In contrast, the noctuid species *A. segetum* suffered different morphological malformations after uptake of aryl isothiocyanates *per os* in our bioassay (I). The malformations were mainly recorded in pupae as well as during metamorphosis to the imaginal stages and seem to be restricted to the ortho-substituted aryl isothiocyanates at lower concentrations. All compounds investigated (except **9**) were toxic at higher concentrations and killed the larvae before pupation. In general, however, the compounds were less toxic than the aromatic bis-isothiocyanates found by Ujvary *et al.* (1989) in *M. sexta.*

The type of malformations observed in *A. segetum* differed from the effects described for *M. sexta* by Matolcsy *et al.* (1986). The laterally emerging bubble-formations, for example, occurring mainly during the late pupae stage (Figure 2a) immediately before the metamorphosis to the imaginal stages, seem to be typical of the action of the ortho halogene phenyl isothiocyanates. Similar malformations have been reported by Rosenthal (1988) for tobacco hornworms (*M. sexta*) fed during their larval development with an artificial diet containing L-canavanine, a non-protein amino acid occurring in several legume plants (see also Rosenthal, this volume, p. 35). Schmutterer (1989) reported similar malformations in pupae of *Pieris brassicae* after uptake of azadirachtin at low (sub-lethal) concentrations by the fifth instar larvae. These vesicular formations in pupae may be explained by the inhibition of the cuticular sclerotization processes during pupal development. Incomplete sclerotization in pupae may cause the striking bubble-like formations by increasing the haemolymph pressure during the final wing formation and stretching processes of the emerging imago within the pupae (S.E. Reynolds, personal communication). The occurrence of numerous imagoes having misshapen or incompletely developed wings (Figure 2d) also supports this assumption. Moreover dithiocarbamates (including isothiocyanates) are known to interfere with arthropod cuticular sclerotization by inhibiting the enzymes involved in catecholamine biosynthesis, e.g. tyrosinase (Brunet, 1980; Anderson, 1985; Kramer and Hopkins, 1987; Ujvary *et al.*, 1988, 1989). Ujvary *et al.* (1988) suggested that isothiocyanates may act as the active metabolites of N-mono-substituted dithiocarbamates. The -NCS compounds are known as alkylating agents capable of carbamoylating the nucleophilic groups of biopolymers. The bis-isothiocyanates can act as bifunctional alkylating agents which disrupt the replication of DNA and thus have cytotoxic properties (Lawley, 1984; Ujvary *et al.*, 1989). Several dithiocarbamates are also known as cytotoxicants. Histopathological effects in midgut epithelia as well as autolysis of the Malpighian tubes caused by the bis-dithiocarbamate fungicide Maneb® in the milkweed bug (*Oncopeltus fasciatus*) were described by McMullen (1965). We also found weak larvicidal side effects of Maneb® in larvae of the Colorado beetle (*Leptinotarsa decemlineata*) which were fed with Maneb® treated potato leaves (unpublished data). Ujvary *et al.* (1988) reported a high larval mortality caused by some long-chain and aromatic mono-isothiocyanates at higher doses, as well as strong cytotoxic local effects at lower doses in *M. sexta*, and proposed the inhibition of mitosis as the mechanisms of action. The acute toxicity of the substituted aromatic mono-isothiocyanates at higher concentrations in *A. segetum* larvae (Table 1) may reflect a similar mode of action.

A glutathione- and cystein-mediated cytotoxicity of allyl- and benzyl-isothiocyanates in mammals was reported, in which the allyl-structure was found to be a cancerogene, and the aromatic benzylisothiocyanate an anticancerogene agent (Bruggeman *et al.*, 1986). Finally, p-bromophenylisothiocyanate has been shown to inactivate several thiol enzymes, and other isothiocyanates are known to inhibit polypeptide synthesis in various types of bacteria. For a more detailed overview see Drobnica *et al.* (1977).

Two different types of effects in noctuids were demonstrated for the mono-isothiocyanates in the present study; the acute toxicity of most of the compounds at higher concentrations in *A. segetum* larvae and the disruption of pupal formation and metamorphosis caused, in particular, by the ortho halogene aryl isothiocyanates at lower doses. These results are generally consistent with the findings published by Ujvary *et al.* (1989) on the activity of several isothiocyanate derivatives in *M. sexta*. The available data do not, however, support a final and unambiguous conclusion regarding the mode of action of the investigated compounds. Detailed biochemical investigations could provide evidence clarifying the role of isothiocyanate xenobiotica in insect growth and development processes.

In order to obtain preliminary data on the range of biocidal activity of the aryl isothiocyanates (**1–10**), the influence of 2-bromophenylisothiocyanate (**6**) on the development of cucurlionid populations living in grain was investigated (bioassay (II): Table 2).

The laboratory bioassay (II) is suitable to evaluate the general activity of an insecticidal compound under simulated actual conditions. This bioassay technique provides no information regarding the mode of action or the prevailing affected stages of insect development (eggs, larvae, imagoes), but simply measures the populations abundance of the emerging F1 generation after treatment of the egg-containing grain in comparison to an untreated control population under standardized experimental conditions and after the same exposure time. Table 2 shows that 2-bromophenylisothiocyanate (**6**) is capable of interrupting the population development of these species even at relatively weak concentrations. The effective dose-range and the relatively weak mammal toxicity as well as the physico-chemical properties of such volatile ortho halogene phenylisothiocyanates makes them promising candidates as novel insecticides for the control of stored product pests (see also DD-Patent: WP A 01 N/333.633.7, 1989).

Further experiments should clarify whether 2–bromophenylisothiocyanate acts as an ovicidal, larvicidal and/or adulticidal agent against *S. granarius* and *S. oryzae*.

Conclusions

Naturally occurring mustard oils, synthetic isothiocyanates and several related compounds show a broad spectrum of different biological activities. The present study enlarges our knowledge of the biocidal effects caused by synthetic substituted aryl isothiocyanate derivatives in lepidopteran and coleopteran species. Several halogene aryl isothiocyanates interrupt the development of cucurlionid populations and cause morphological aberrations in pupae and imagoes of the noctuid species *A. segetum* after uptake of weak concentrations of the compounds during their larval development.

The results suggest a possible mode of action based on interference with the insect cuticular formation processes. These results are generally consistent with previous findings demonstrating juvenile hormone activities of related isothiocyanate compounds for practical pest control. Isothiocyanate and related compounds may be suitable key structures in this respect.

References

ANDERSON, S.O. (1985). Sclerotization and tanning of the cuticle. In *Comprehensive Insect Physiology, Biochemistry and Pharmacology*, (G.A. Kerkut and L.I. Gilbert, eds), Volume 3, pp. 59–74. Pergamon Press, Oxford.

BAKRY, N., METCALF, R.L. and FUKUTO, T.R. (1968). Organothiocyanates as insecticides and carbamates synergists. *Journal of Economical Entomology*, **61**, 1303–1309.

BRUGGEMAN, J.M., TEMMINK, J.H.M. and BLADEREN, P.J. VAN (1986). Glutathione- and cystein-mediated cytotoxicity of allyl- and benzyl-isothiocyanate. *Toxicity and Applied Pharmacology*, **83**, 349–359.

BRUNET, P.C.J. (1980). The metabolism of aromatic amino acids concerned in the cross-linking of insect cuticle. *Insect Biochemistry*, **10**, 467–500.

DD-PATENT. Insektenentwicklungsbeeinflussende Mittel. (B. Weber, D. Martin, D. Otto: Institut für Pflanzenschutzforschung der Akademie der Landwirtschaftswissenschaften der DDR), WP A 01 N/333.633.7 (17.10.1989).

DROBNICA, P., KRISTIAN, P. and AUGUSTIN, J. (1977). The chemistry of the -NCS group. In *The Chemistry of Cyanates and their Thio Derivatives* (S. Patai, ed), Part 2, pp. 1003–1221. Wiley & Sons, New York.

FISCHER, G. and OTTO, D. (1976). Die Aufzucht von *Agrotis (Scotia) segetum* Shiff. und *Mamestra (Barathra) brassicae* L. auf semisynthetischer Diät. *Archiv für Phytopathologie und Pflanzenschutz, Berlin*, **12**, 117–126.

HABIB, S.N. and RIEKER, A. (1984). Metathesis of aryl isothiocyanate: A novel method for the synthesis of sterically hindered aryl isothiocyanates. *Synthesis*, 825.

HARTMANN, A. (1983). Isothiocyanate (Senföle). In *Huben-Weyl: Methoden der Organischen Chemie* (H. Hagemann, ed), Volume E4 (Kohlensäurederivate), pp. 834–858, Verlag G. Thieme.

KRAMER, K.J. and HOPKINS, T.L. (1987). Tyrosine metabolism for insect cuticle training. *Arch. Insect Biochemistry and Physiology*, **6**, 279 pp.

LAWLEY, P.D. (1984). Carcinogenesis by alkylating agents. in *Chemical Carcinogenesis* (C.E. Searly, ed), *ACS Monograph* 182, 2nd edition, Volume 1, 331 pp.

LICHTENSTEIN, E.P., STRONGF, F.W. and MORGAN, D.C. (1962). Identification of 2-phenylethylisothiocyanate as an insecticide occurring naturally in the edible part of turnips. *Journal of Agriculture Food Chemistry*, **10**, 30–33.

MATOLCSY, G., UJVARY, I., RIDDIFORD, L.M. and HIRUMA, K. (1986). Alkylene-bis-isothiocyanates: Novel insect growth regulators. *Zeitschrift für Naturforschung*, **41c**, (11/12), 1069–1072.

MARTIN, D., BEYER, E. and GROSS, H. (1965). Notiz über eine

einfache Methode zur Darstellung von Senfölen. *Chemische Berichte*, **98**, 2425–2426.

McMULLEN, R.D. (1965). Symptoms and histopathology of poisoning by Maneb in *Oncopeltus fasciatus*. *Canadian Entomologist*, **97**, 1200 pp.

NAKAMURA, T. (1961). Ovicides against *Ascaris* eggs. *Osaka Shiritsu Daigaku Igaku Zasshi*, **10**, 381–385.

PICKETT, J.A. (1988). Chemical pest control – the new philosophy. *Chemistry in Britain*, February 1988, 137–142.

PICKETT, J.A. (1989). Towards zero pesticides residues: The biomanagement of pests and diseases of oilseed rape. In *IACR Report 1989*, 79–81.

QURESHI, S.A., MOHIUDDIN, S., PARVEEN, R. and KEMAL, R. (1982). Larvicidal activity of aryl-substituted isothiocyanates against the yellow-fever mosquito, Aedes aegypti (L.). *Pakistan Journal of Zoology*, **14**, 91 pp.

ROSENTHAL, G.A. (1988). The protective action of higher plant toxic product. *BioSciences*, **38**, 104–109.

SCHMUTTERER, H. (1989). Environmental sound pest control by application of neem (*Azadirachta indica*–based natural pesticides. *Tagungsberichte der Akademie der Landwirtschaftswissenschaften der DDR* (Berlin), **274**, 134–144.

UJVARY, I., MATOLCSY, G., HIRUMA, K. and RIDDIFORD, L.M. (1988). Novel type of insect growth regulators affecting spiracle and crochet formation: Structure-activity studies. In *Endocrinological Frontiers in Physiology Insect Ecology* (F. Sehnal, A. Zabza and D.L. Denlinger, eds), Vol. 1, pp. 593–600, Wroclaw Technical University Press, Poland.

UJVARY, I., MATOLCSY, G., RIDDIFORD, L.M., HIRUMA, K. and HORWATH, K.L. (1989). Inhibition of spiracle and crochet formation and juvenile hormone activity of isothiocyanate derivatives in the tobacco hornworm *Manduca sexta*. *Pesticides-Biochemistry and Physiology*, **35**, 259–274.

US-PATENT (1960). Heterocyclic methyl-isothiocyanates as nematicides. (N.J. Lewis. Monsanto Chemical Co.). US-2.946.720 (26.7.1960).

US-PATENT (1968). Biocidal isothiocyanates. (H. Werres, Schering AG). US-3.366.538 (30.1.1968).

ZSOLNAI, T. (1975). *Die Chemotherapeutischen und Pestiziden Wirkungen der Thioreagenzien*, pp. 244–259. Akademiai Kiado, Budapest.

8
The Neem-Extractor and the Biological Activity of the Product NeemAzal-S

HUBERTUS KLEEBERG

Trifolio-M GmbH, Sonnenstr. 22, D-6335 Lahnau 2, Germany

Introduction

For plant protection in non-industrialized countries the availability of cheap but efficient insecticides is of prime importance for an increase in the yield and quality of agricultural products.

The seeds of the neem tree (*Azadirachta indica* A. Juss), which grows abundantly in many tropical countries, contain considerable amounts of highly insecticidal substances like Azadirachtin (Broughton *et al.*, 1986; Jones *et al.*, 1989; Rembold *et al.*, 1987).

Aqueous extracts of neem kernel powder have proven to be a valuable tool for pest control (Dreyer, 1984; Siddig, 1987; Hellpap, 1989). In order to retain the advantages of this possibility but to eliminate certain drawbacks, a simple apparatus has been developed. This Neem-Extractor allows the extraction of active ingredients from neem kernels by a new extraction method on the level of village technology. Even under unfavourable conditions, Neem-Extractors with daily turnover of more than 15 kg of neem kernels may be run using solar energy and manual labour only.

For the extraction of the insecticidal ingredients of the neem kernels a specially developed extracting fluid, 'AzaSolv', is used, which serves as formulating agent at the same time. Depending on the quality of the neem kernels the liquid insecticide which is produced according to this method – called 'NeemAzal-S' – contains between 1000 to 10 000 ppm azadirachtin.

Insecticides: Mechanism of Action and Resistance
© 1992   Intercept Ltd, P.O. Box 716, Andover, Hants SP10 1YG, UK

The storage stability of NeemAzal-S depends considerably on the quality of the neem seed kernels used. Usually the Azadirachtin content has decreased by not more than 20% after 3 months. The addition of microbiocides usually prevents the decomposition of the active ingredient of NeemAzal-S for more than a year.

The Biological Activity of NeemAzal-S

We have investigated the activity of NeemAzal-S with larvae of *Pieris brassicae*, *Pieris rapae* and of *Tenthredinidae*.

First tests showed the characteristic phenomena after treatment with NeemAzal-S. Figure 1 shows 15-day-old larvae of *P. brassicae* which have been fed with cabbage leaves treated with NeemAzal-S (corresponding to 0.5 ppm Azadirachtin) for eight days (left) or with untreated leaves (blank: right). The treated larvae are obviously smaller and hardly consumed any leaf material.

Figure 1   Larvae of *P. brassicae* feeding upon treated (NeemAzal–S: 0.5 ppm Azadirachtin for 8 days: left) or untreated cabbage leaves (blank: right).

Larvae of *P. rapae* were fed for two days with treated cabbage leaves (dipped for 2 minutes into an aqueous solution of the corresponding content of NeemAzal-S and subsequently air dried) and after that with untreated leaves. From the size of the leaves before and after feeding the leaf area consumed was determined approximately every two days. From correlations between the leaf area and the leaf weight the consumption of leaf mass may be estimated.

The leaf consumption during the first two days (treated leaves) depended only slightly on the Azadirachtin content applied (see Figures 2, 3 and 5). After that time two factors contribute to the decreased leaf consumption of treated leaves in comparison to untreated ones or the blank (i.e. treatment with all compounds of the NeemAzal-S formulation except substances originating from neem) – (A) decrease in the number of living larvae and (B) decrease in leaf intake per living larvae.

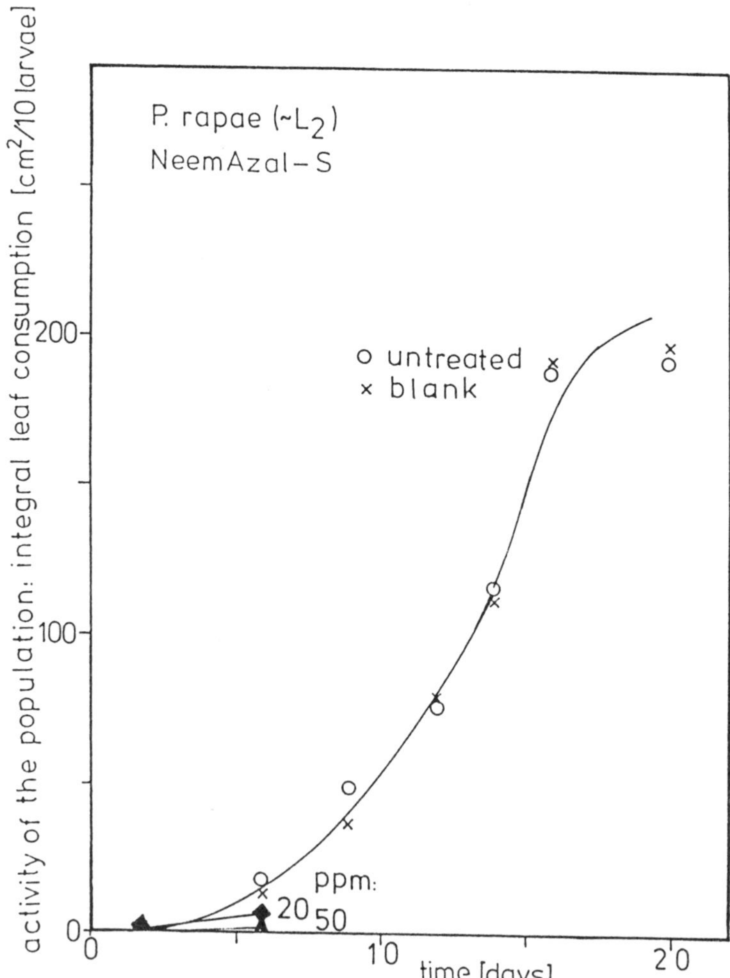

Figure 2 Leaf consumption of second instar larvae of *P. rapae* starting with 2 days feeding of treated leaves. The activity of the population corresponds to: 10 [(digested leaf area)/(number of larvae at the beginning of the test)]. Temperature: 25°C (night) to 32°C (day).

Figure 2 shows the integral leaf consumption corresponding to the total population of *P. rapae* second instar larvae after the beginning of the experiment. For the ease of comparison the leaf area is expressed in cm^2/10 larvae (initial population). After five days there is no feeding activity of the larvae which have digested leaves treated with NeemAzal-S corresponding to 5 to 50 ppm Azadirachtin.

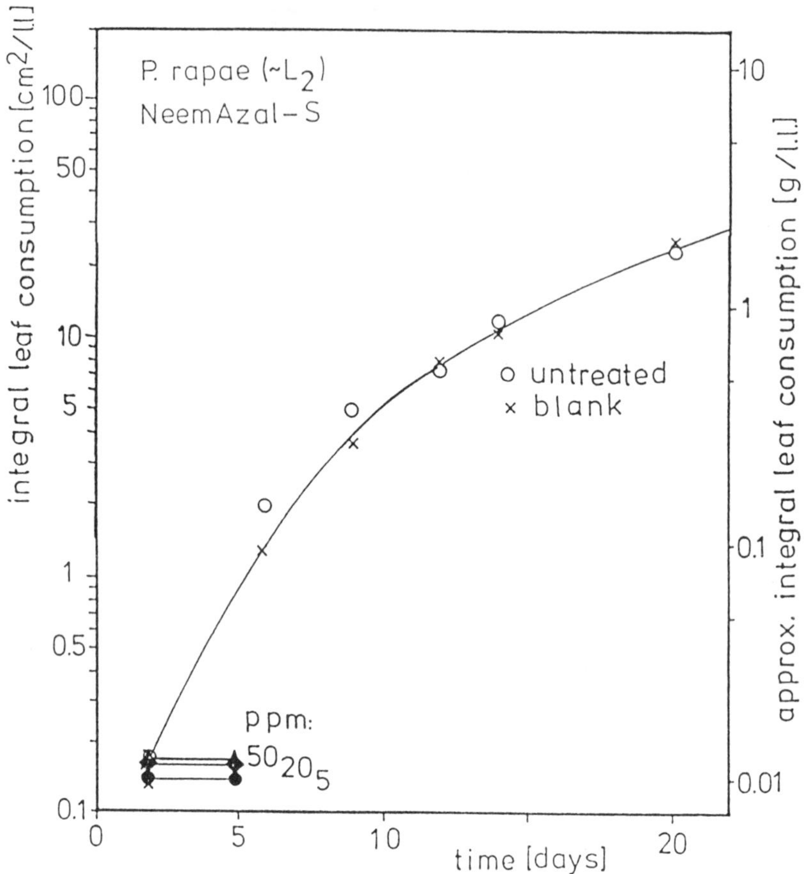

Figure 3   The integral leaf consumption per living larvae (l.l.) of *P. rapae* is given by: (digested leaf area)/(number of living larvae). Temperature: 25°C (night) to 32°C (day).

Figure 3 shows that larvae which survive the treatment with NeemAzal-S practically do not take up food after the second day. However, as shown in Figure 4, after six to eight days all the larvae which fed upon treated leaves have died.

For investigations with older larvae pupae formation occurred and the number of living or dead pupae was included in the evaluation. For larvae treated with NeemAzal the amount of malformed adults – if they emerged at all – is considerable. The number of malformed adults was added to the number of dead larvae and pupae, since these adults will not be in a position to survive and propagate.

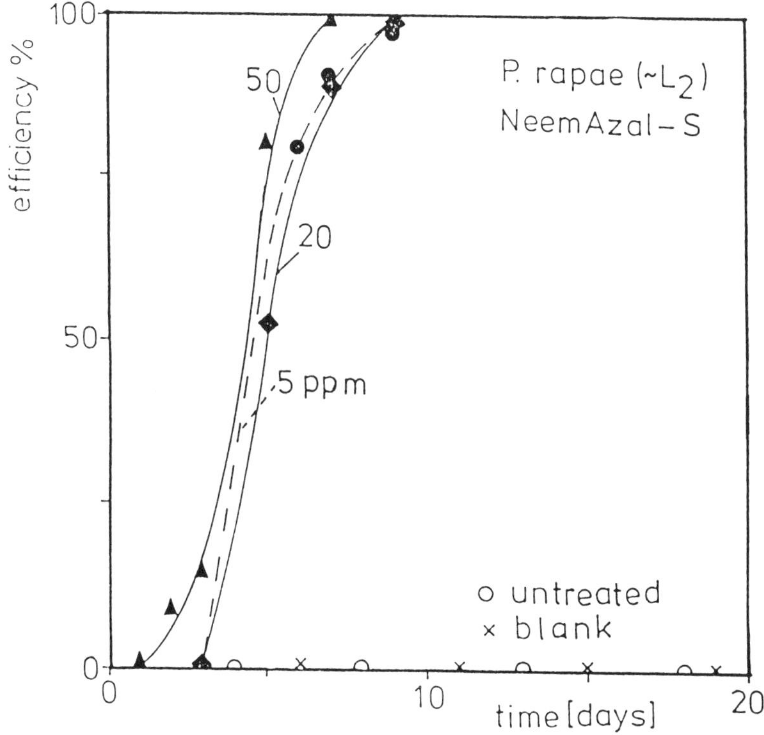

Figure 4 The efficiency indicates the percentage of dead larvae, pupae and malformed adults divided by the total number of larvae initially present (see also Figures 2 and 3).

These results show that the lethal dose of NeemAzal-S corresponds to an Azadirachtin content below 5 ppm for second instar (see Figure 4). Treatment corresponding to 5 ppm Azadirachtin reduces the leaf consumption of the initial population to a few percent of the control value. Results for first instar larvae of *P. rapae* are very similar to Figures 2 and 4.

For older larvae (approx. fifth instar) held between 20°C (night) to

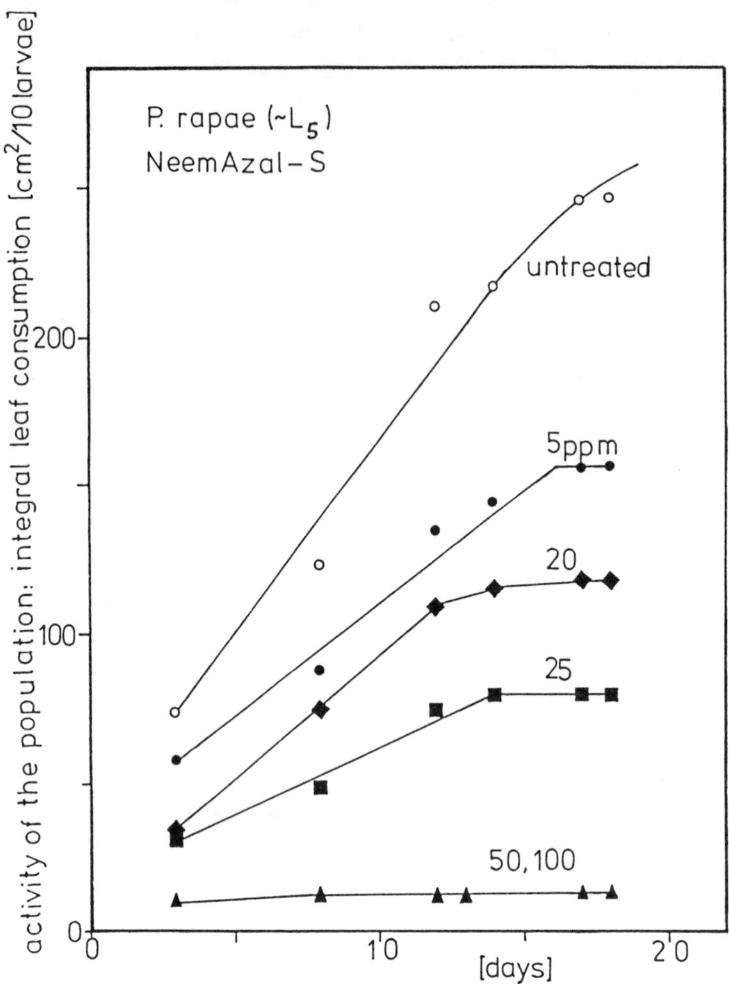

Figure 5   Leaf consumption of fifth instar larvae of *P. rapae* starting with 2 days feeding of treated leaves. The activity of the population corresponds to: 10 [(digested leaf area)/(number of larvae at the beginning of the test)]. Temperature: 20°C (night) to 28°C (day)

28°C (day), the decrease in the activity of the population with increasing NeemAzal-S treatment is obvious (see Figure 5). In this case only doses corresponding to approximately 50 ppm Azadirachtin reduce the integral leaf consumption to a few percent of the control.

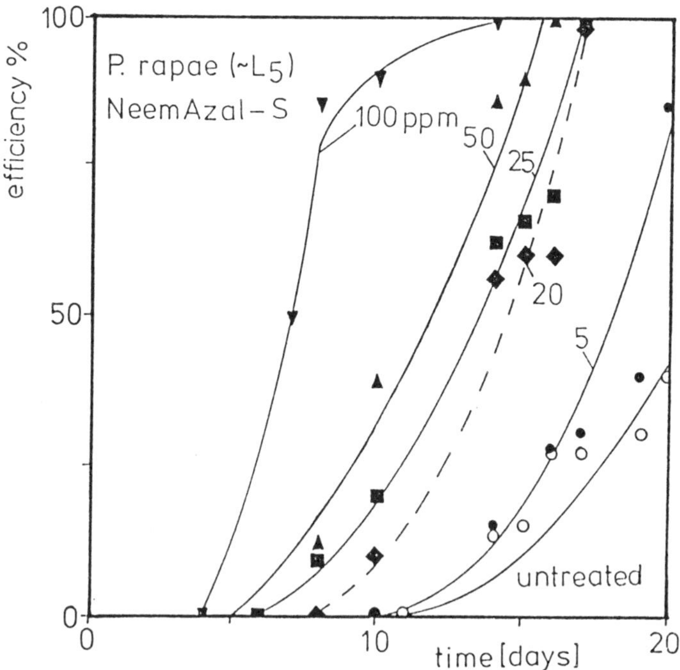

Figure 6 Efficiency of the influence of NeemAzal-S on fifth instar larvae of *P. rapae*. Experiments were performed with an initial treatment with NeemAzal-S for 2 days. Temperature: 20°C (night) to 28°C (day).

Figure 7 Differences in the amount of cabbage leaves digested between day 3 and day 8 (see Figure 5) by fifth instar *P. rapae* are obvious. treatment (from left to right): untreated, blank, and corresponding to 5, 20, 25 and 50 ppm Azadirachtin of NeemAzal-S.

The efficiency of the protection of crop by NeemAzal-S becomes obvious from Figure 7. While untreated or blank leaves are nearly totally digested, the protection with NeemAzal-S increases obviously from 5 to 50 ppm Azadirachtin. The latter concentration gives very satisfactory results, even with fifth instar larvae.

High efficiency of NeemAzal-S is clearly reached after a longer time in the case of older larvae (see Figure 6) than in the case of younger ones (see Figure 4). Only to some extent is the lower temperature of the experiment with older larvae (compare Figures 2 and 5) expected to postpone the influence of NeemAzal-S.

The experiments with *P. rapae* indicate that the results corresponding to 20 to 50 ppm Azadirachtin are very satisfactory. From the amount of Azadirachtin consumed with the leaves, and the average body weight of the larvae of the different experimental series, we may calculate an 'effective dose' of Azadirachtin. Figure 8 shows that the amount of Azadirachtin taken up does not vary considerably with the age of the larvae although their weight increases about 100-fold. Roughly speaking, we may say that less than 5 mg/kg larvae are sufficient to reduce the activity of *P. rapae* populations to a satisfactory degree.

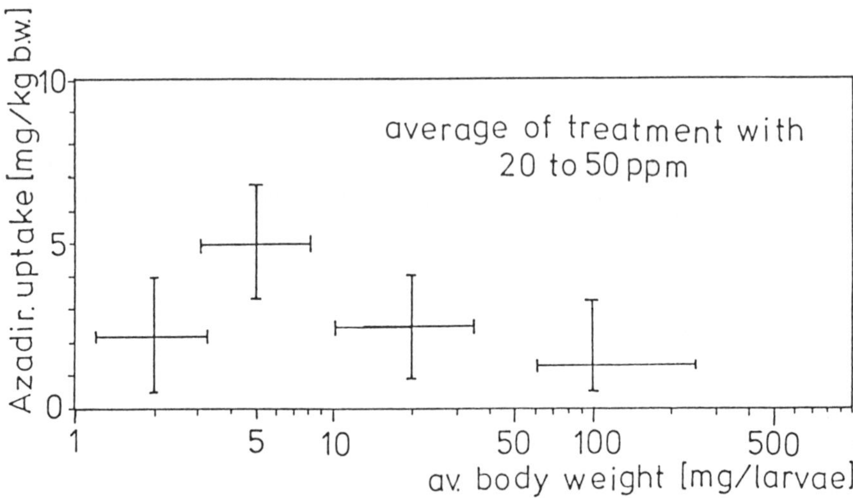

Figure 8   From the leaf area digested by larvae of *P. rapae* at different instar, the weight of the larvae under these conditions, and the amount of NeemAzal-S solution adhering to the leaf after treatment, the Azadirachtin intake per kg body weight (b.w.) of the larvae can be determined.

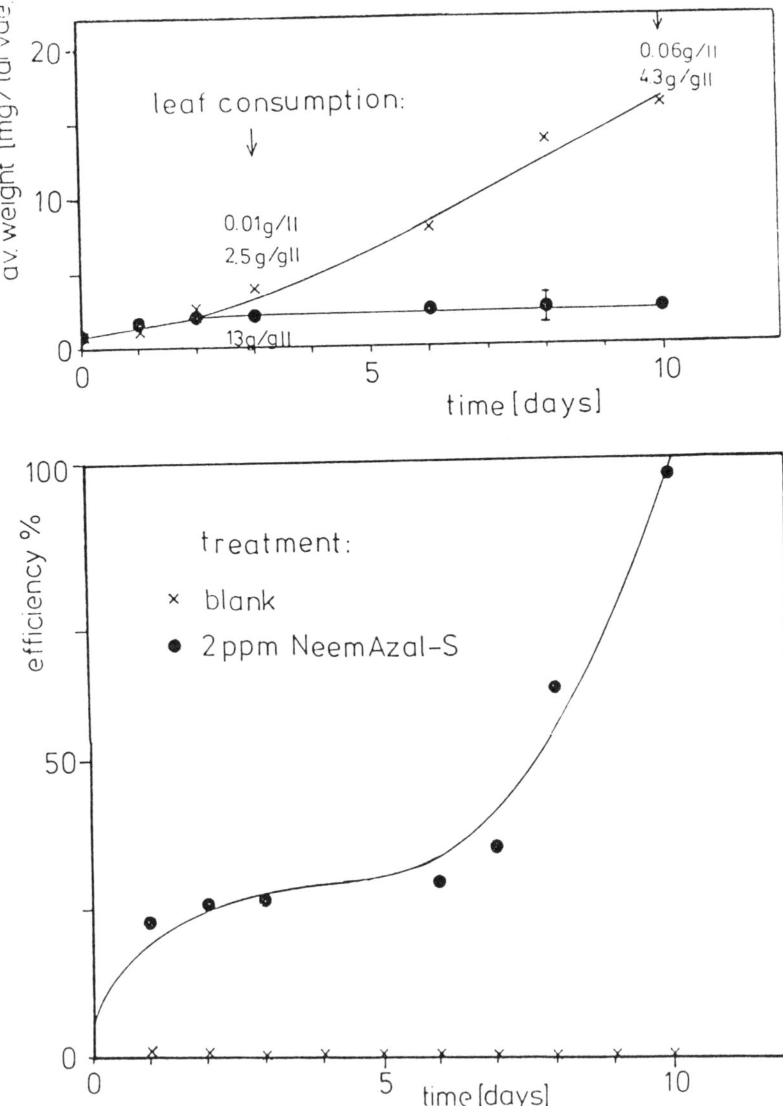

Figure 9   Change of the average body weight of the living larvae of *Pontania proxime* after treatment with NeemAzal-S (● corresponding to an Azadirachtin content of 2 ppm; × blank: the same treatment but without ingredients from neem seeds) (upper part). The efficiency of the treatment is indicated in the lower part of the figure. the larvae fed all the time of the experiment on the treated leaves. Temperature 8°C (night) to 18°C (day).

Our results indicate that the efficiency of NeemAzal-S depends on the temperature. However, even under comparatively cold (central European) conditions, very satisfactory effects are observed – even after two days of application.

We investigated this more carefully under field conditions with *Pontania proxima* on small *Salix caprea* tress at temperatures between approximately 8°C (night) and 18°C (day). NeemAzal-S corresponding to 2 ppm Azadirachtin was applied to the plant with two-day-old larvae. Application of a solution containing all other ingredients at the corresponding concentration of NeemAzal-S except compounds from neem seeds served as a control (blank). The average weight of the larvae was estimated from their volume, which was determined in approximately two-day intervals with a microscope. A preceding test showed that this is possible within comparatively small error limits.

It is obvious from Figure 9 that the larvae stop growing two days after treatment with NeemAzal-S. Their body weight remains constant while that of larvae feeding on leaves treated with blank solution increases steadily.

After approximately twenty-five days pupae were formed by the blank population. The body weight of the larvae had increased from 2mg/larvae at the beginning of the test to more than 100 mg/larvae and the leaf area consumed had been more than 20 times that of the population treated with NeemAzal-S.

As indicated in Figure 9, the leaf consumption increases naturally with increasing age of the larvae. After 10 days they have digested leaves corresponding to more than four times their body weight and after twenty-five days more than seven times their body weight. After treatment with NeemAzal-S the damage is only about 5% of the control population. These results speak in favour of an application of NeemAzal-S even under comparatively cold central European conditions.

It is typical for the mode of action of NeemAzal-S that a very satisfactory reduction of the leaf consumption is obtained, at most, two days after treatment, but that the infestation seems to persist (though inactive) for more than one week.

Conclusions

The results demonstrate that NeemAzal-S may contribute to an ecologically sound and economically beneficial plant protection in non-industrialized countries. The possibility to produce its own insecticide will save foreign currency and give work to several people in addition to improving their food supply.

References

BROUGHTON, H.B., LEY, S.V., SLAWIN, A.M.Z., WILLIAMS, D.J. and MORGAN, E.D. (1986). *Chem. Communs.*, 46–47.

DREYER, M. (1984). In *Natural Pesticides from the Neem Tree and Other Tropical Plants* (*Proceedings* of the Second International Neem Conference) (H. Schmutterer and K.R.S. Ascher, eds), pp. 435–443. GTZ, Eschborn, Germany.

HELLPAP, C. (1989). In *'89 Integrated Pest Management in Tropical and Subtropical Cropping Systems* (DLG, ed), pp. 711–716. CTA, Ede, The Netherlands.

JONES, P.S., LEY, S.V., MORGAN, E.D. and SANTAFIANOS, D. (1989). *Focus on Phytochemical Pesticides: Vol. 1, The Neem Tree* (M. Jacobson, ed), pp. 19–45. CRC Press Inc., Boca Raton, Florida.

REMBOLD, H., FORSTER, M., and CZOPPELT, CH., (1987). In *Natural Pesticides from the Neem Tree and Other Tropical Plants* (Proceedings of the Third International Neem Conference) (H. Schmutterer and K.R.S. Ascher, eds), pp. 149–160. GTZ, Eschborn, Germany.

SIDDIG, S.A. (1987). In *Natural Pesticides from the Neem Tree and Other Tropical Plants* (*Proceedings* of the Third International Neem Conference) (H. Schmutterer and K.R.S. Ascher, eds), pp. 449–459. GTZ, Eschborn, Germany.

9
Properties of NeemAzal-F – a New Botanical Insecticide

HUBERTUS KLEEBERG

Trifolio-M GmbH, Sonnenstr. 22, D-6335 Lahnau 2, Germany

Introduction

For ecologically sound plant protection, the search for naturally occurring active principles and modes of action is an important aim in order to increase the yield and quality of agricultural products.

Different conceptions are applied for the protection of plants against insect pests. They range from ideologically innocent positions – 'do not disturb the course of nature and hope that no serious pest will set in' – to radical positions of 100% control with chemical insecticides. In addition to a limited use of synthetic insecticide according to the principle of integrated pest management, the potential of mechanical and bio-technological methods seems to increase. Obviously, the application of biological principles – such as the use of pheromones, pests' own natural enemies and diseases, or the cultivation of pest-resistant breeds – would find a desirable enlargement if the defensive substances of plants could be made available in larger scale. According to climatic and economic conditions, for example, these strategies or a combination of them may be applicable.

Considerable research is done in biological strategies, and new practically relevant results may be expected in the future. In order to become a valuable tool for future pest management a conception should fulfil the following requirements

1.Efficiency.

Insecticides: Mechanism of Action and Resistance
© 1992   Intercept Ltd, P.O. Box 716, Andover, Hants SP10 1YG, UK

2. Reliance on the success of its application.
3. Considerate influence on beneficial insects.
4. Minimization of the appearance of resistances.
5. (Toxicologically) safe for the user.
6. (Toxicologically) safe for the consumer.
7. Environmentally sound (bio-degradable).
8. Toxicologically and environmentally safe means of production.
9. Inexpensive production.
10. World-wide availability of standard quality.

According to the large body of information on properties of pesticides from the neem tree (*Azadirachta indica* A. Juss) (Schmutterer *et al.*, 1980; Schmutterer and Ascher, 1984, 1987; Jacobson, 1989; Menn, 1983) – especially with respect to the first seven requirements – it seems promising to develop technical methods for the extraction of the biologically active principle from the seeds of this tree.

The major insecticidal ingredient is Azadirachtin (Broughton *et al.*, 1986; Jones, *et al.*, 1989). According to the present knowledge, this substance and similar ones usually extracted together with azadirachtin appear to be non-toxic to mammals. This is, of course, a necessary prerequisite for the safe application of an insecticide. Due to high efficiency of these substances (i.e. low amounts applied to plants for their protection) and its fast biodegradability (see below) no risk for consumers of products treated with this material is to be expected.

Due to the occurrence of neem in a large number of non-industrialized countries, the production of a plant protectant from this renewable source seems advantageous in these countries. This would also support the world-wide availability of standard quality.

Thus we have started to develop simple extraction procedures which permit a toxicologically and environmentally safe, inexpensive production of a standard quality of the active material from the seeds of the neem tree.

We developed two different isolation procedures:

(A)   NeemAzal: a powder with an Azadirachtin content of about 40%; the formulation of NeemAzal yields NeemAzal-F, a storage-stable, ready-to-use botanical insecticide with an Azadirachtin content of 5%. Some results on the degradation of NeemAzal-F are discussed below.

(B)   NeemAzal-S: a ready-to-use solution for the protection of plants against various insect pests. For the production of

NeemAzal-S under village technology conditions the 'Neem-Extractor' has been developed for which neither electric power nor special chemical solvents are necessary (see this volume, p. 75). The extraction medium used serves as a formulation additive at the same time (Kleeberg 1991a, b; Karelina *et al.*, 1991).

As a by-product of both processes neem oil and neem cake may be obtained. For example, neem oil is traditionally used in India for soap making and sales of neem cake (in pure form as well as in the form of neem-coated urea) as a manure have considerably increased during recent years (see Ketkar and Ketkar, 1989).

Results and Discussion

Detailed studies of the efficacy of NeemAzal show that its biological activity is comparable to that of pure Azadirachtin (see Figure 1). At a concentration of 0.5 or 1 ppm the efficiency of pure Azadirachtin

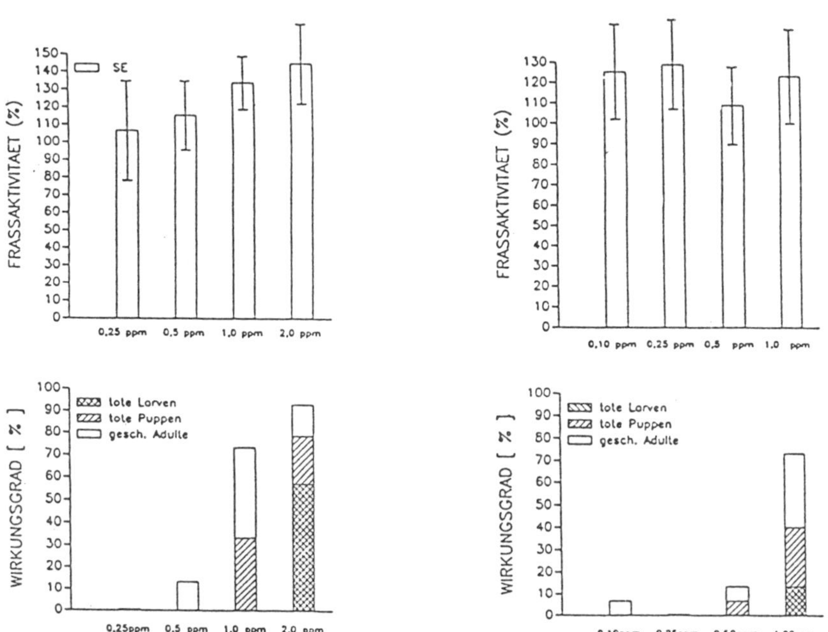

Figure 1 Laboratory tests of NeemAzal (left) and Azadirachtin (right) with *Epilachna varivestis* show that the feeding activity (upper part) is similar and independent from concentration. The efficacy of both substances is similar at 0.5 and 1 ppm of substance applied.

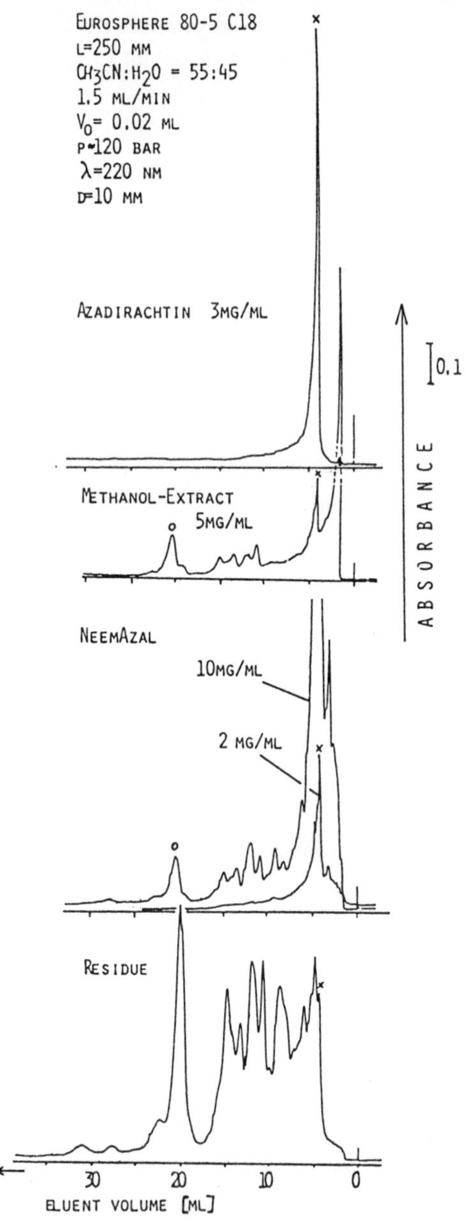

Figure 2   HPLC-analysis shows that the Azadirachtin content of neem-extracts increases in the series: Residue < Methanol-extract < < NeemAzal. (Peaks of Azadirachtin and Salannin are indicated by x and o respectively.)

(Figure 1, right) is very similar to that of the 'crude' Azadirachtin NeemAzal (Figure 1, left). In both cases the feeding activity (Figure 1, upper part) is the same and does not decrease with increasing concentration of Azadirachtin. This shows that repellent effects can hardly contribute to the efficacy of the compounds in the case of *Epilachna varivestis* at the given concentrations.

Apart from the promising results of laboratory tests, field studies demonstrate that high efficacy of NeemAzal-F applied at concentrations corresponding to about 10 to 50 ppm Azadirachtin (Karelina *et al.*, 1991).

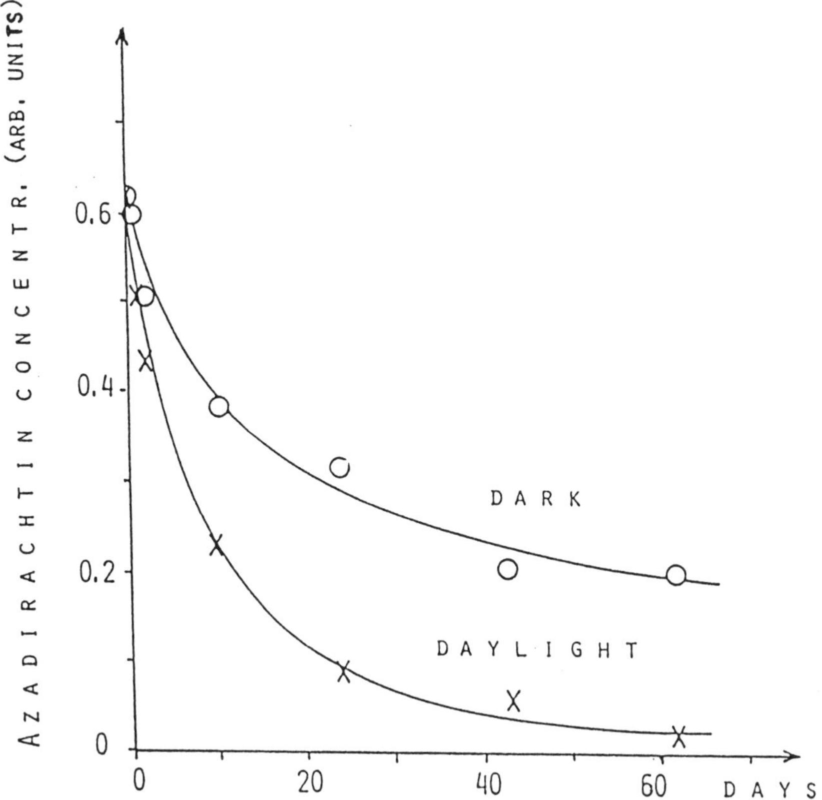

Figure 3 Degradation of Azadirachtin as indicated by HPLC-analysis of aqueous solutions of NeemAzal-F is more than three times as rapid in daylight than in the dark.

The presence of additional insecticidal substances present in NeemAzal (see Figure 2) will minimize the occurrence of resistancies.

Analytical HPLC results (Figure 2) show that alcoholic neem seed kernel extracts contain much more neem ingredients than NeemAzal. Separation of the less biologically active from a Methanol extract of neem kernels (see Figure 2, bottom: 'residue') results in a fraction with low Azadirachtin and high Salannin content. This material may be regarded more or less as ballast for plant protection purposes.

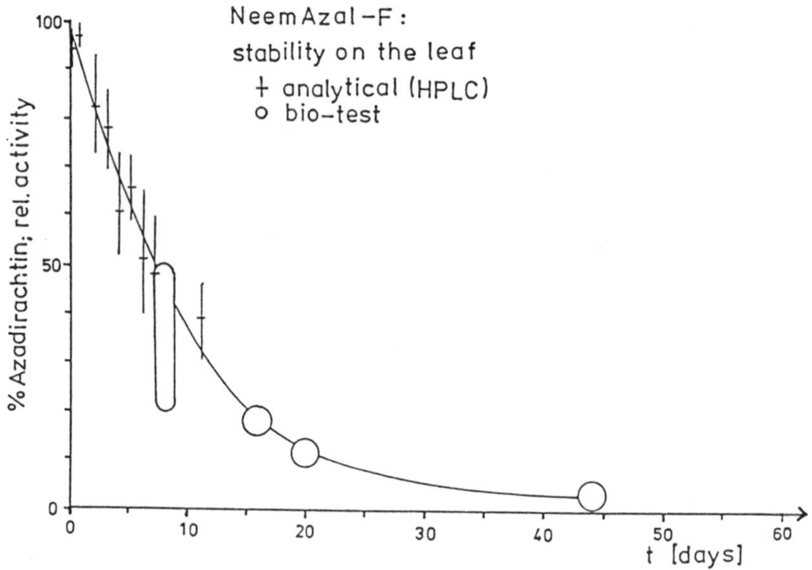

Figure 4   Degradation of the active substance of NeemAzal-F after treatment of leaves was determined by HPLC-analysis as well as in bio-tests with *Pieris rapae*. The results of both tests fit favourably on one common curve.

Investigations of the degradation of the active ingredient of NeemAzal indicate that Azadirachtin decomposes more than three times as rapidly in the light than in the dark (Figure 3). The half life of Azadirachtin in aqueous solution (Figure 3), after application to the leaf (Figure 4) and the in the earth (Figure 5), is similiar: seven-to ten days (compare Figures 3, 4 and 5). As may be expected, the elution of Azadirachtin from the earth is very slow; thus the comparatively rapid degradation will prevent the transportation of the active material in the soil.

The ecologically sound properties of NeemAzal-F for plant protection purposes are completed by an efficient, environmentally safe production.

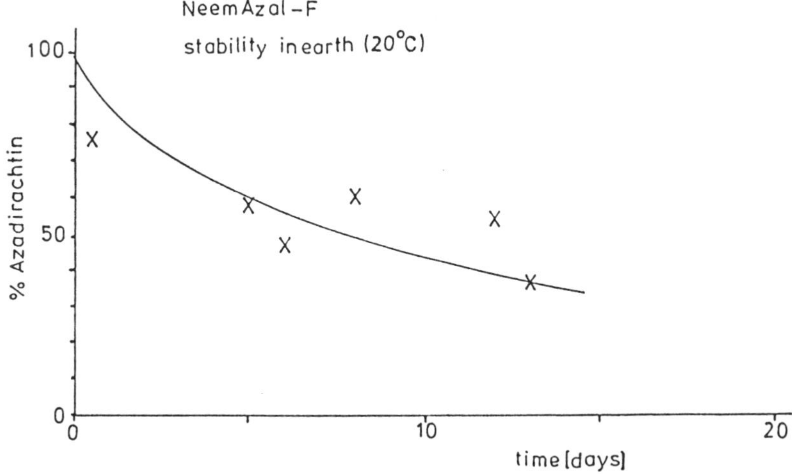

Figure 5 HPLC-analysis show that the Azadirachtin content in earth treated with NeemAzal-F decreases rapidly.

All these results demonstrate that the biologically active ingredients of neem seed kernels as extracted by the methods developed may contribute to an ecologically sound plant protection in the framework of concepts like 'Integrated Pest Management' or 'Organic Farming'. Additionally, the production and application of NeemAzal will be economically very beneficial especially for non-industrialized countries, where neem trees grow abundantly.

Acknowledgements

The author is grateful to Dr Ermel for bio-tests of NeemAzal with *Epilachna varivestis* and for the skilful technical assistance of A. Schlicht and I. Schaafer.

References

BROUGHTON, H.B., LEY, S.V., SLAWIN, A.M.Z., WILLIAMS, D.J. and MORGAN, E.D. (1986). *Chem. Communs.*, 46–47.
JACOBSON, M. (ed) (1989) *Focus on Phytochemical Pesticides: Vol. 1, The Neem Tree*. CRC Press Inc., Boca Raton, Florida.
JONES, P.S., LEY, S.V., MORGAN, E.D. and SANTAFIANOS, D. (1989). *Focus on Phytochemical Pesticides: Vol. 1, The Neem Tree* (M. Jacobson, ed), pp. 19–45. CRC Press Inc., Boca Raton, Florida.
KARELINA, T.N., FILIPPOV, N.A., KLEEBERG, H.,

KOVALEV, B.G. and PUHALSKYA, N.A. (1991). See this volume, p. 95.

KETKAR, C.M. AND KETKAR, M.S. (1989) In *'89 Integrated Pest Management in Tropical and Subtropical Cropping Systems* (DLG, ed), pp. 689–709. CTA, Ede, The Netherlands.

KLEEBERG, H. (1991a). In press.

KLEEBERG. H. (1991b). See this volume, p. 75.

MENN, J.J. (1983). In *Natural Products for Innovative Pest Management* (D.L. Whitehead and W.S. Bowers, eds), p. 5. Pergamon Press, Oxford.

SCHMUTTERER, H., ASCHER, K.R.S. and REMBOLD, H. (1980). *Natural Pesticides from the Neem Tree* (*Proceedings* of the First International Neem Conference). GTZ, Eschborn, Germany.

SCHMUTTERER, H., and ASCHER, K.R.S. (1984). *Natural Pesticides from the Neem Tree and Other Tropical Plants* (Proceedings of the Second International Neem Conference). GTZ, Eschborn, Germany.

SCHMUTTERER, H., and ASCHER, K.R.S. (1987). *Natural Pesticides from the Neem Tree and Other Tropical Plants* (Proceedings of the Third International Neem Conference). GTZ, Eschborn, Germany.

10
Evaluation of the Biological Activity of NeemAzal and NeemAzal-S Against *Mamestra brassicae*, *Pieris rapae* and *Heliothis armigera*

T.N. KARELINA[1], N.A. FILIPPOV[1], H. KLEEBERG[2], B.G. KOVALEV[1] and N.A. PUHALSKYA[1]

[1]*All Union Institute of Biological Methods for Plant Protection, 277072 Kishinev, USSR*
[2]*Trifolio-M GmbH, Sonnenstr. 22, D-6335 Lahnau 2, Germany*

Introduction

In the last ten years intensive studies have been carried out in search for ecologically safe insecticides, in particular among substances of plant origin. This has been determined first, by disadvantages objectively characterizing standard insecticides such as their high prices, as well as by the wish to find new structures among natural substances with high biological activity easily synthesized and transformed. Great attention is being paid to seed extracts of the neem tree *Azadirachta indica* (Schmutterer and Ascher, 1983, 1986), containing Azadirachtin as the main active ingredient.

We studied the biological activity of NeemAzal (powder with 40% content of Azadirachtin) and NeemAzal-S (liquid formulation with 0.3% content of Azadirachtin) under laboratory and field conditions. Both preparations were made by Trifolio-M.

The powder was applied in 0.001, 0.01, and 0.1 and 1% concentrations in ethanol. Before use NeemAzal-S was diluted with water in the ratios 1:50, 1:100, 1:150, and 1:200. Artificial

Insecticides: Mechanism of Action and Resistance
© 1992   Intercept Ltd, P.O. Box 716, Andover, Hants SP10 1YG, UK

nutritional diet served as food for larvae of *Mamestra brassicae* and *Heliothis armigera*, while cabbage leaves served as food for *Pieris rapae*. Insects were maintained in biochambers with a temperature of 23–24°C, relative humidity of 75–80% and fourteen hours photoperiod.

In greenhouses larvae of *Mamestra brassicae* and *Pieris rapae* were released in the second and third instar onto treated cabbage plants in the phase of 6–8 leaves (10–30 larvae/plant).

Mamestra brassicae

Introduction of different instar larvae onto the food treated with NeemAzal in 0.001–1% concentrations revealed one type of response from the insects, which increases with concentration. In all cases feeding activity was reduced and this parameter continued to decrease with concentration increase (Figure 1). Larvae became less mobile with badly expressed colouring and drawing pattern. Their growth and development was considerably retarded (Figure 2).

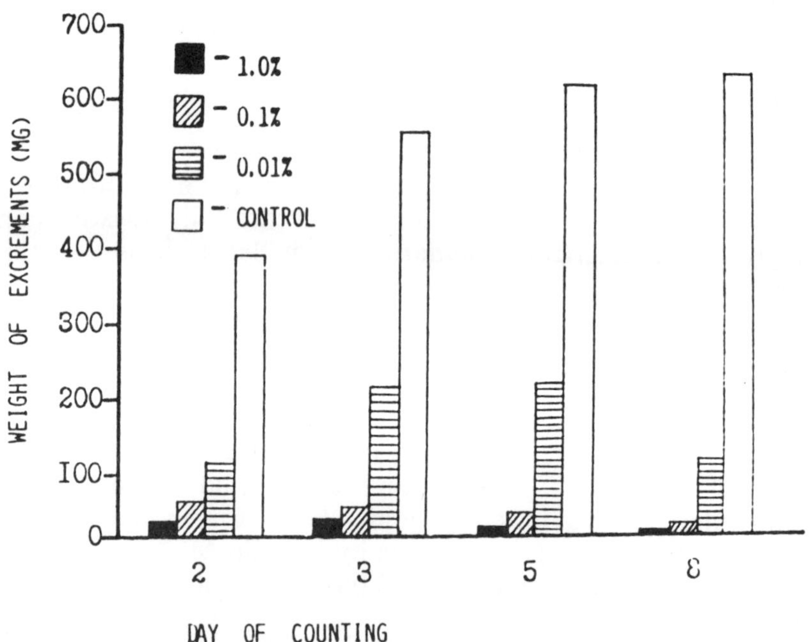

Figure 1   The influence of NeemAzal (introduced onto treated food) on the feeding activity of third instar larvae of *Mamestra brassicae*.

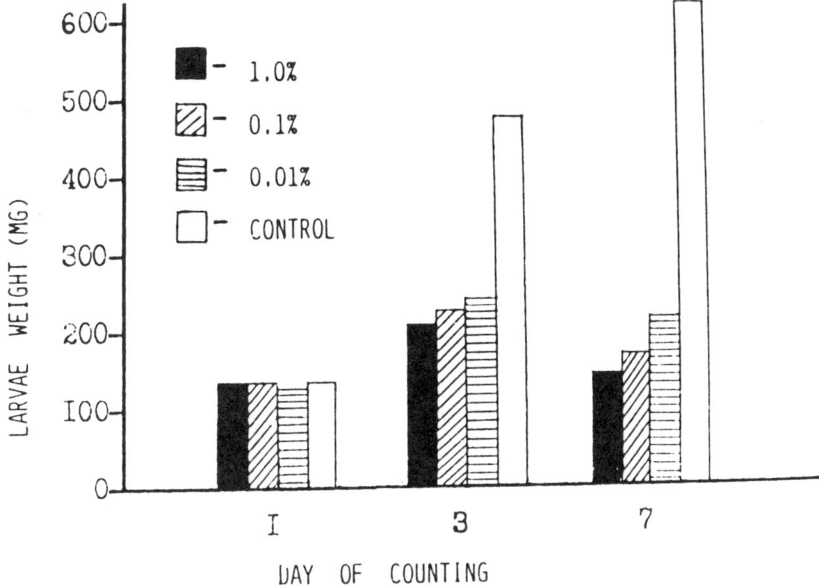

Figure 2 The influence of NeemAzal (introduced onto treated food) on the growth intensity of *Mamestra brassicae* larvae (third instar).

Observed loss in weight was followed by insect mortality which (in the experiment with third instar larvae) constituted 23.5–76.6% on the day 5 and 36.6–100% on day 8 (Table 1). However, larvae of early instars were most sensitive. Thus, food treatment (in experiments with larvae of the first and second instars) with 0.01% concentrations resulted in the mortality of half of the insects (43.3–53.3%) on the fifth day of counting and 93.3–100% and the eighth day. Larvae of the third instar were more resistant to the preparation in the mentioned concentration. Their response to the preparation is retarded and the mortality constitutes only 26.6–46.6% on the fifth day (Table 1). 100% mortality is reached in this variant only on the twelfth day.

Larvae surviving this period reach the fourth instar though the quantity of food consumed is 3.5 times reduced in comparison to that in the control (Figure 3). However, their vital activity is disadvantageous, since the larvae penetrate into the cabbage heads and contaminate the product with excretions reducing its quality. Hence, to prevent harmfulness of the third instar larvae, it is probably necessary to use higher concentrations of NeemAzal. In our experiments a concentration increase of up to 0.1–1% can reduce the number of the first and third instar larvae to 66.6–93.3% on the

Table 1 The influence of NeemAzal on the *Mamestra brassicae*. First to third instar larvae were fed with treated diet. each test started with 30 larvae.

Concen- tration (%)	Mortality (%) of larvae on day of counting:				
	3	**5**	**8**	**10**	**12**
			— First instar —		
1	60.0	93.3	100	–	–
0.1	46.6	86.6	100	–	–
0.01	20.0	53.3	100	–	–
0.001	13.3	26.6	86.6	93.3	93.3
control	0	0	6.6	6.6	13.3
			— Second instar —		
1	76.6	93.3	100	–	–
0.1	53.3	76.6	100	–	–
0.01	26.6	43.3	93.3	100	–
0.001	13.3	23.3	60.0	63.3	73.3
control	0	0	6.6	6.6	13.3
			— Third instar —		
1	53.3	76.6	100	–	–
0.1	23.3	66.6	93.3	100	–
0.01	0	26.6	46.6	70	100
0.001	6.6	23.3	36.3	46.6	80.0
control	0	0	6.6	6.6	13.3

fifth day and 93.3% on the eighth day of counting (Table 1).

The response of senior instar larvae to the effect of the preparation was similar. Nutritional deterrancy limiting insect feeding prolonged the period of larval development for 5–12 days. It resulted in 36.6–100% mortality of the larvae (Table 2). Pupated insects (13.4–63.6%) formed non-viable pupae depending on the concentration applied. Autopsy of blackened pupae has shown that all tissues and internal organs of exposed insects were lysed turning into brown mass. Some pupae (5–17%) had morphogenetic defects which manifested themselves either with features of the larval phase in the region of the head and thorax or with the deformed thorax segments resulting in deep cavity. No adults emerged in all of these experiments with larvae of the different instars.

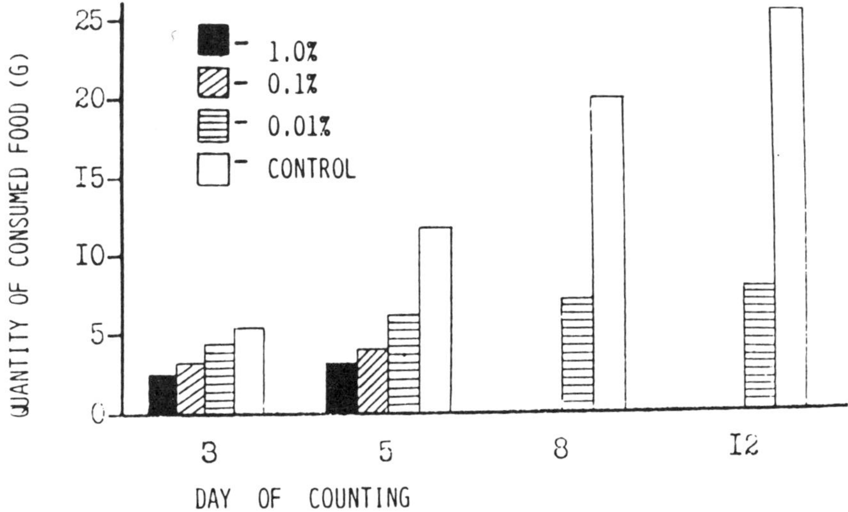

Figure 3 The influence of NeemAzal on feeding (nutritional diet) of third instar larvae of *Mamestra brassicae*.

Table 2 The influence of NeemAzal on the development of *Memestra brassicae* when larvae of the sixth instar were fed with treated diet. Each test started with 30 larvae.

concen-tration (%)	— mortality (%) — on day of counting:						pupae (%)	imago (%)
	3	5	8	10	12	20		
1	0	0	23.3	86.6	96.6	100	0	0
0.1	0	0	0	43.3	63.3	86.6	13.4	0
0.01	0	0	0	26.6	43.3	60.0	40.0	0
0.001	0	0	0	23.3	30.0	36.6	63.4	0
control	0	0	6.6	p u p a t i o n			93.4	82.7

Releases of larvae onto greenhouse cabbage plants treated with NeemAzal confirmed the results of laboratory studies. It is already on the third day that the response of the most sensitive early instar larvae to the preparation (in 0.001–0.01% concentrations) was noticeable: many individuals left the plants.

By the fifth day mortality constituted 23.3–46.6% and by the eighth day 73.2–100% respectively (Table 3). Application of higher concentrations (0.1–1.0%) resulted in evidently toxic effects, and mortality of the second instar larvae constituted 66.6–88.3% on the third day. By the fifth day of counting there were no individuals left. Plant damage by larvae constituted (depending on the concentration; see Table 3) 11.3–84.6% on the third day of the experiment when 95.3% of the leaves in the control had been skeletized. Thus, NeemAzal can be rather effective at concentrations of 0.01–0.1% against early instar larvae of *Mamestra brassicae*.

Table 3 The influence of NeemAzal on the viability and harmfulness of *Mamestra brassicae* when second instar larvae were introduced onto treated cabbage leaves. Each test started with 30 larvae. (Least significant difference mortality – 0.05%; damage – 6.2%).

concen- tration (%)	mortality (%) on day of counting:			damage of leaves on third day (%)
	3	**5**	**8**	
1	83.3	100	–	11.3
0.1	66.6	100	–	11.7
0.01	26.6	46.6	100	39.3
0.001	6.6	23.3	73.2	84.6
control	3.3	6.6	6.6	95.3

The effect NeemAzal-S on *Mamestra brassicae* larvae was similar to the effect of NeemAzal in the expression. They consumed 9.5–16.7g of the nutritional diet instead of 68.8g in the control (Figure 4). This reduced consumption hampered the transition to the subsequent larval instar of early developmental stages (Figure 5). Exposure of the last instar larvae to the treated food prolonged the larval stage from fourteen to fifteen days (Figure 6). Exposed insects died (Figure 7). A 4.1–7.2 times reduction in the quantity of the consumed food as compared to that in the control (Figure 4) was accompanied by minimum release of dehydrated excretions. Reduction of feeding intensity from the third to the fifteenth day of insect feeding (see Figure 4) is evidence of an increasing effect of the preparation and its ability of bringing the populations vitality to zero.

Treatment of greenhouse cabbage plants with NeemAzal-S followed by the release of the second instar larvae has demonstrated the effectiveness of the preparation. Leaf surface damage in the experiment (depending on the concentration applied) constituted

17.8–25.5% by the third day and 18.1–48.4% on the seventh day, while in the control half the leaf surface has already been damaged by the third day and almost all tested leaves have been skeletized by the seventh day (Figure 8).

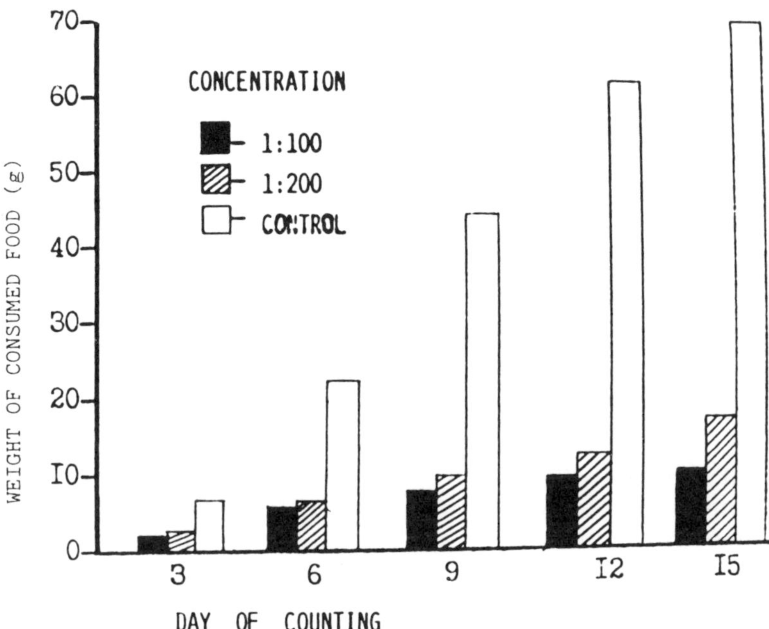

Figure 4 Feeding activity of second instar larvae of *Mamestra brassicae* after treatment of food with NeemAzal-S at different dilutions.

Pieris rapae

Pieris rapae were released into greenhouses together with *Mamestra brassicae*. The results obtained were similar to those described above. Just like *Mamestra brassicae*, some *Pieris rapae* larvae left the plants. They could be found on the soil, on the cage frame and walls. Larvae which remained developed more slowly in the experiments as compared with those in the control, became weak and finally died.

The degree of plant damage (depending on the concentration applied) varied within 11.2–34.5% by the third day and 18.4–55.2% by the seventh day of counting (in the control 58.8–81.6% respectively; see Table 4).

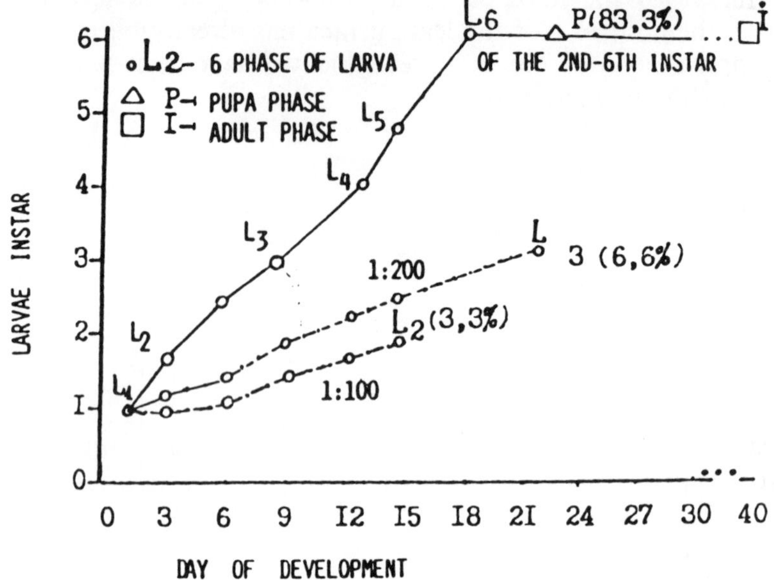

Figure 5   The influence of NeemAzal-S treatment of the diet on the development of *Mamestra brassicae*. Larvae were fed with the diet starting with first instar.

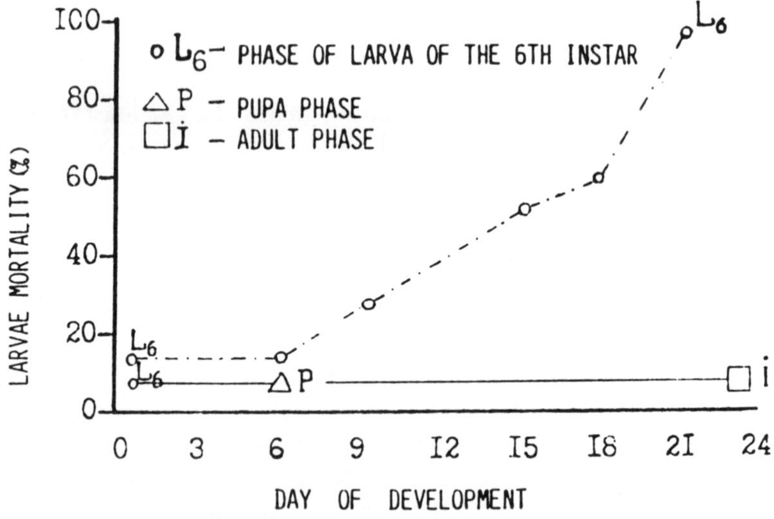

Figure 6   The influence of NeemAzal-S on the development and survival of *Mamestra brassicae* when larvae of the last instar were fed with treated food. ————=control (untreated food); —·—· =treated food (dilution of NeemAzal-S: 1:50).

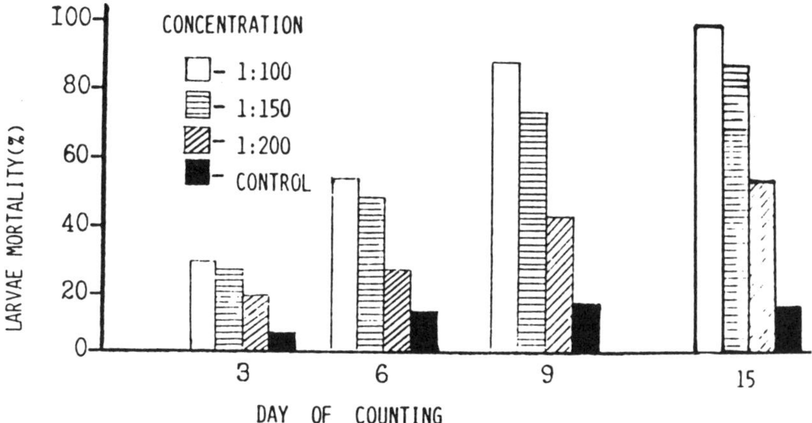

Figure 7 Mortality of *Mamestra brassicae* larvae fed from the first instar with diet treated with NeemAzal-S.

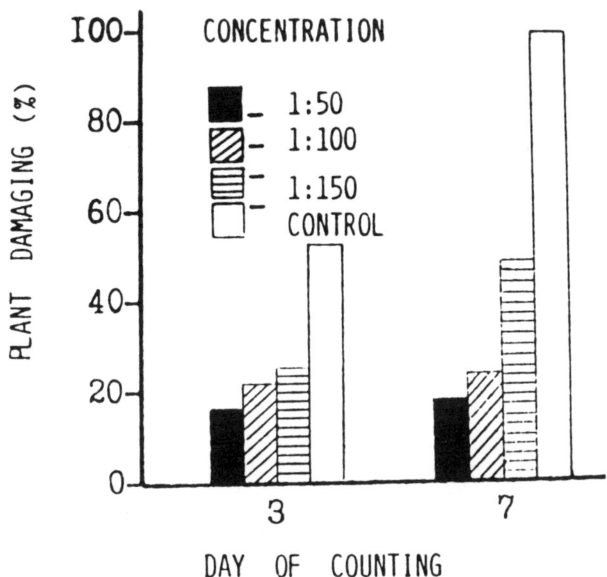

Figure 8 The degree of damage of NeemAzal-S treated cabbage plants by second instar larvae of *Mamestra brassicae*.

Table 4   The influence of NeemAzal-S on the harmfulness of *Pieris rapae* when second instar larvae were introduced onto treated cabbage leaves. Each test started with 30 larvae. (Least significant difference in the degree of plant damage (%): day – 12.8%; day 7 – 11.3%.)

Dilution with water:	Degree of plant damage (%) on:	
	day 3	**day 7**
1:50	11.2	18.4
1:100	23.9	27.7
1:150	34.5	55.2
control	58.8	81.6

The best results with NeemAzal-S were obtained when cabbage plants were treated with the preparation diluted 1:50. However, the high effectiveness is accompanied by unfavourable effects on the treated plants at high concentrations: top young leaves were burnt which retarded plant growth. Thus it is necessary to take into account possible phytotoxicity of high concentrations of NeemAzal-S when the preparation would be applied commercially.

Figure 9   Cabbage plants damaged by *Mamestra brassicae* on the fifth day after release of second instar larvae. Top: plant treated with NeemAzal-S (dilution 1:150). *Bottom*: control (untreated).

Heliothis armigera

Giving food to larvae as nutritional diet treated with NeemAzal revealed their response to the preparation already on the second to third days after feeding. Although food was present in excess, larvae were aggressive: they demonstrated cannibalism while attacking each other. Falling behind in growth and development prolonged the period of early instar larvae resulting in mortality. The most effective were 1.0 and 0.1% concentrations of NeemAzal as in the experiments with *Mamestra brassicae* (53.3–55.6% were dead by the third day; 86.6–93.3% were dead by the twelfth day; Table 5). No pupation took place in these experiments. In rare cases, at low concentrations non-viable pupae were obtained. Results of the comparison of larvae weight parameters before pupation (Table 5) demonstrates that their energy potential was not sufficient for subsequent transformation in pupal and adult metamorphosis.

Table 5 The influence of NeemAzal on the viability of *Heliothis armigera* when second instar larvae were introduced onto treated food (nutritional diet). Each test started with larvae.

concen-tration (%)	mortality (%) of larvae on day of counting:			weight (mg) of excretions (fifth day)		weight of larvae on day (mg):	
	3	5	12		1	3	12
1	56.6	73.3	93.3	18	20	16.2	–
0.1	53.3	66.6	86.6	99	20	26.8	20
0.01	40.0	53.3	73.3	110	20	28.3	158
0.001	36.6	46.6	66.6	213	20	41.5	250
control	16.6	23.3	36.6	507	20	63.8	635

11
Behavioural Responses of German Cockroaches (*Blattella germanica* L.) Induced by Plant Repellents

I. SCHEFFLER and M. DOMBROWSKI

Universität Potsdam, Park Sanssouci, Villa Liegnitz, O-1570 Potsdam, Germany

Introduction

The long and widespread utilization of insecticides has resulted in changes in insect populations. One of the best known changes is the development of physiological resistance. Another response to insecticides is the learned modification of behaviour. Many insecticides possess high repellency against the German cockroach, as shown by avoidance responses. Both resistance and repellency are fundamental reasons for the decreasing efficacy of blatticides. We tested the repellent effects of 350 plant extracts against the German cockroach to gain knowledge of the behaviour of this species and to evaluate the potential use of repellent plant material in pest control.

Method

We evaluated every plant extract to test the repellent effect in 45 petri-dishes in a preference test. The cockroaches, separated into small (LS), medium (LM), large (LL) larvae, females and males (450 animals, 10 per dish), can choose between a filter paper with extract (E) and the neutral control filter paper (K). The animals are of a breed from our institute. Extract production: 3g fresh leaf material was extracted in 50ml ethanol of 96%, 70% and 50%. 0.5ml of the

Insecticides: Mechanism of Action and Resistance
© 1992   Intercept Ltd, P.O. Box 716, Andover, Hants SP10 1YG, UK

extract was given onto a filter paper. A control filter paper was impregnated with 0.5ml ethanol (70%).

Quantative analysis: The Repellency Value ($RV = K/K + E$) is a measure of the repellent effect of the plant extract.

Results and Discussion

The Repellency of Selected Species of some Plant Families

In figure 1 to 5 the repellency Values from some species are documented which were found within the first 2 hours after an adaption phase of 1 hour (short-time-test). There are no attractive effects of plant extracts. We could establish neutral (n), weekly (sr), middle (mr) and strongly (str) repellent effects in all families. Considerable fluctuations of the values also occur in genera (Euphorbia, RV = 46–0.79; Fiscus, RV = 0.40–0.70). To complete the figures, we wish to describe the very striking effects of two other species: *Citrus sinensis*, RV = 0.92 and *Allium sativum*, RV = 0.98 (after two days; 10 g of bulb material) were shown to be very strongly repellent.

The fact that there are no attractive effects in these families is remarkable. Dietz and Busenberry (1989) have tested plant extracts against nematodes and obtained similar results. The repellency values or the repellency indices ($RI = E-K/E + K$) of the strongly repellent extracts are comparable with the values of synthetic repellents or insecticides (Burden, 1975; Bret and Ross, 1985; Randell and Bower, 1986; Apple and Mack, 1989).

Frequency of the Distribution of Repellency Values of Larvae and Sexes

All instars and sexes tend towards neutral (n) and repellent (r) responses to the plant extracts (RV > 0.60) (Figures 6 and 7). The attractive effects show a lower frequency. The curves of small (LS) and medium (LM) larvae largely correspond. But, the bell-shaped frequency breaks in increasing order from the large larvae (LL), to the females and finally the males. The curve of the males shows different peaks within the whole spectrum. This suggests that males have a high olfactory sensitivity combined with an increased exploratory behaviour to vegetable repellents. A sensitive reaction to odorants is necessary for the reception and the processing of information gained in exploration (Denzer *et al.*, 1988).

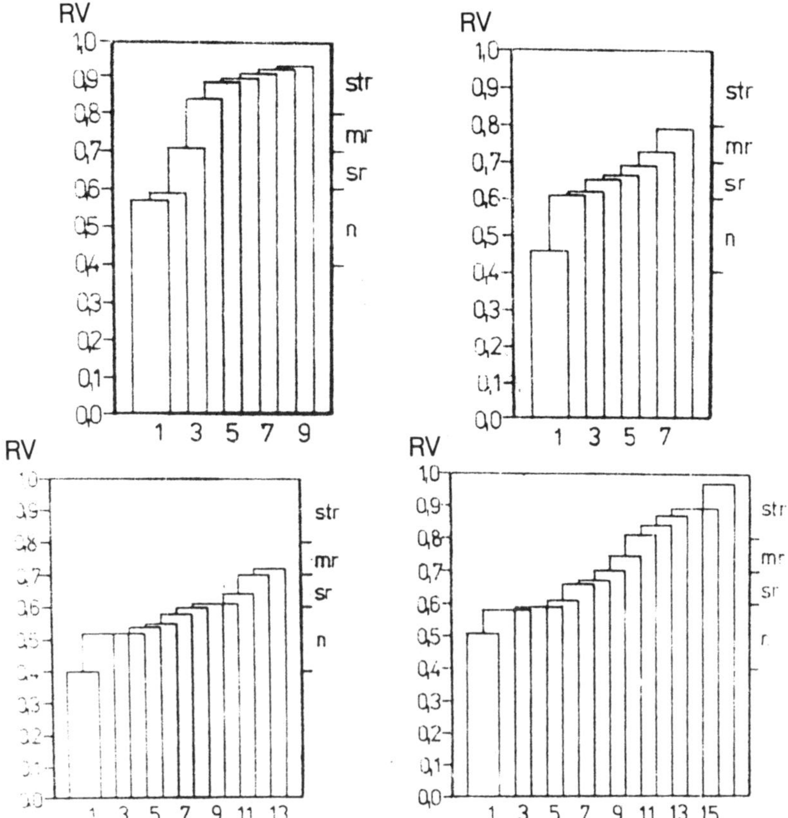

Figure 1   Repellency values of species of Araceae in short-term test. 1, *Anthurium crystallinum*; 2, *Monstera deliciosa*; 3, *Anchomanes difformis*; 4, *Dieffenbachia maculata*; 5, *D. maculata 'picturata'*; 6, *Xanthosoma sagittifolium*; 7, *Zantedeschia aethiopica*; 8, *Philodendron erubescens*; 9, *P. bipinnatifidum*. **Figure 2**   Repellency values of species of Euphorbiaceae in short-term test. 1, *E. myrsinites*; 2, *E. tirucalli*; 3, *E. plychroma*; 4, *Securinega suffroticosa*; 5, *E. milli*; 6, *E. lathyris*; 7, *Adrachne cholchica*; 8, *E. cyparissias*. **Figure 3**   Repellency value of species of Moraceae in short-term test. 1, *Ficus buxifolia*; 2, *F. retusa*; 3, *F. lyrata*; 4, *rubiginosa*; 5, *F. religiosa*; 6, *F. sagittata*; 7, *F. nekbudu*; 8, *F. cyathistipula*; 9, *F. elastica*; 10, *F. pumila*; 11, *F. carica*; 12, *F. benjamina*; 13, *Cannabis sativa*. **Figure 4**   Repellency values of species of Rosaceae in short-term test. 1, *Rubus idaeus*; 2, *Cotoneaster bullatus*; 3, *Sorbaria sorbifolia*; 4, *Rosa rugosa*; 5, *Geum urbanum*; 6, *Rosa multiflora*; 7, *Fragaria ananassa*; 8, *Rubus bifrons*; 9, *Sorbus intermedia*; 10, *C. adpressus*; 11, *Spiraea media*; 12, *Crataegus monogyna*; 13, *Padus serobnia*; 14, *Spiraea x. vanhouttei*; 15, *Malus baccata*; 16, *Sorbus aucuparia*.

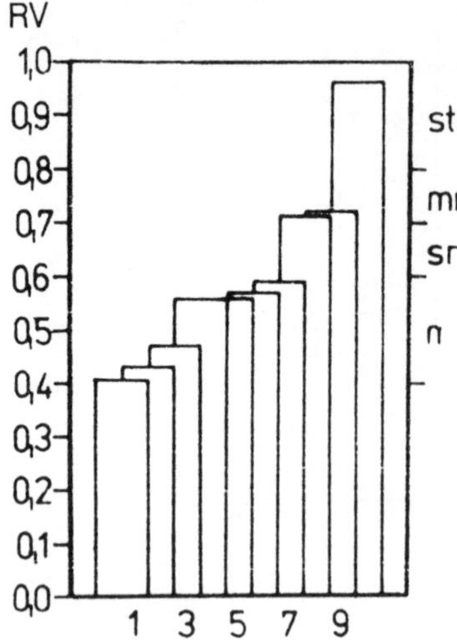

Figure 5   Repellency values of species of Solanaceae in short-term test. 1, *Physalis alkekengi*; 2, *Nicotina rustica*; 3, *Solanum alatum*; 4, *Petunia axillaris*; 5, *Atropa bella-donna*; 6, *S.luteum*; 7, *S. cornutum*; 8, *Lycopersicon lycopersicum*; 9, *Hysocymus niger*; 10, *Cestrum aurantiacum*.

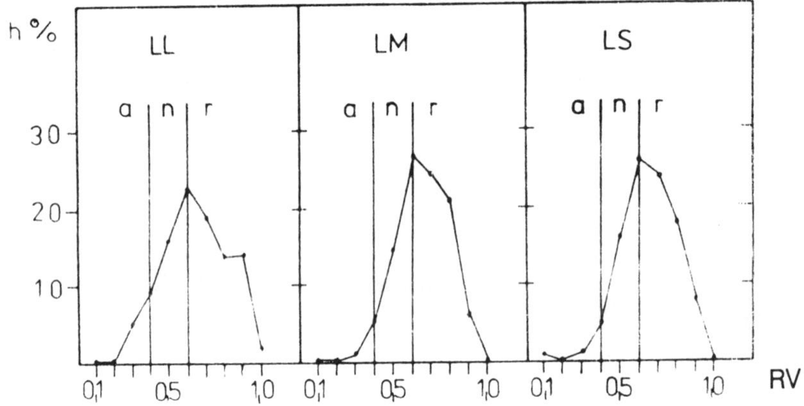

Figure 6   Frequency of the distribution of repellency values of plant extracts against larvae established in 350 short-term tests. LS, small larvae; LM, medium larvae; LL, large larvae; RV, repellency value; h%, frequency of distribution in per cent; a; attractive; n, neutral; r, repellent.

To gain knowledge about the possibilities of the utilization in pest control it is necessary to check the long-term effects of the plant extracts.

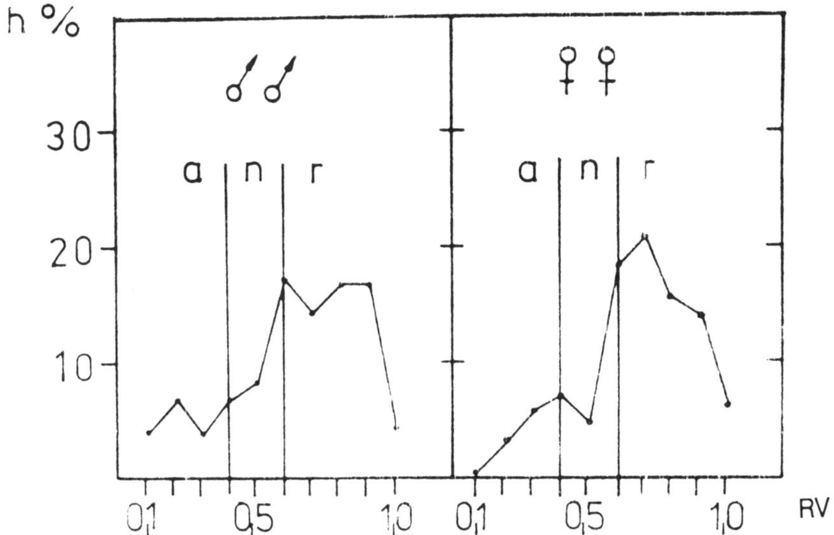

Figure 7 Frequency of distribution of repellency values of plant extracts against males and females established in 350 short-term tests. RV, repellency value; h%, frequency of distribution in per cent; a; attractive; n; neutral; r, repellent.

Long-term Test with Extracts of *Citrus sinensis* and *Allium sativum*

Figure 8 shows the development of repellent effects of *Citrus sinensis* and *Allium sativum* against small, medium and large larvae, females and males. The strongly repellent effects (str) of these components decreases after the sixth day (*Citrus sinensis*) and the sixteenth day (*Allium sativum*) of the experiment. In Figures 9 and 10, the single curves of larvae (the mean of small, medium and large larvae), females and males in the same experiment show the different effects on the three groups.

With both extracts the rapid decrease of the repellent effect on the males are striking. This corresponds to findings on the olfactory sensitivity and the exploratory behaviour of the males (Bret and Ross, 1985; Denzer *et al.*, 1988).

The curve in Figure 11 shows the development of the repellent effect of extract of *Citrus sinensis* against small, medium and large larvae, females and males (the mean) with application of filter paper with fresh extract after the fifth, twelfth and eighteenth day. The fresh extract filter paper cannot reach the same repellent effects as the fresh extracts at the start of the experiment but are short breaks

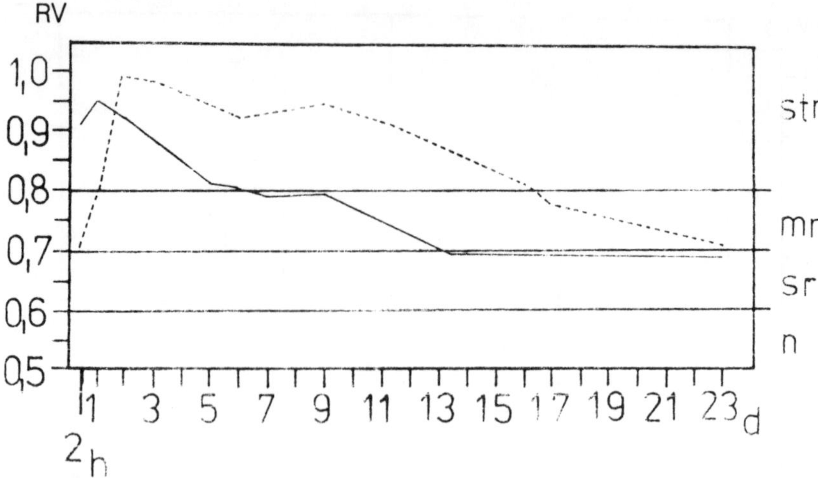

Figure 8 The development of repellency values (mean values) of *Citrus sinensis* (continual line) and *Allium sativum* (interupted line) against small, medium and large larvae, and females and males. RV, repellency value; h, hour; d, day; n, neutral; sr, weak repellency; mr, medium repellency; str, strong repellency

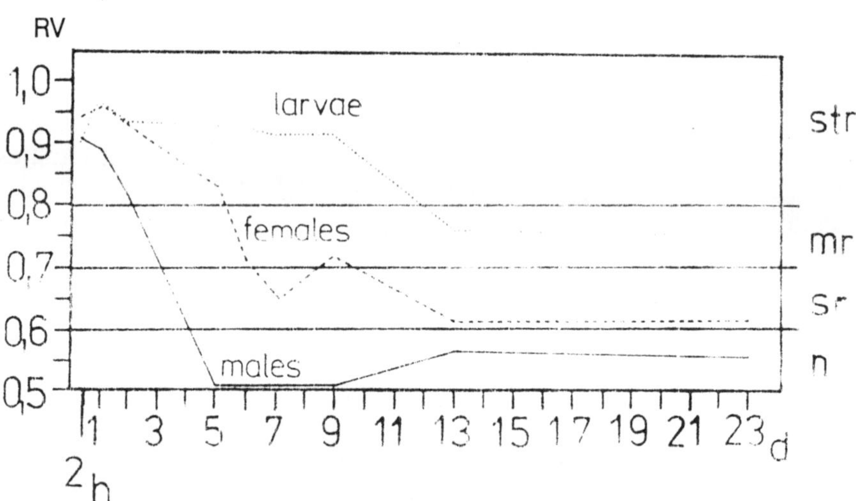

Figure 9 The development of repellency values (mean values) of *Citrus sinensis* against small, medium and large larvae, and females and males. RV, repellency value; h, hour; d, day; n, neutral; sr, weak repellency; mr, medium repellency; str, strong repellency.

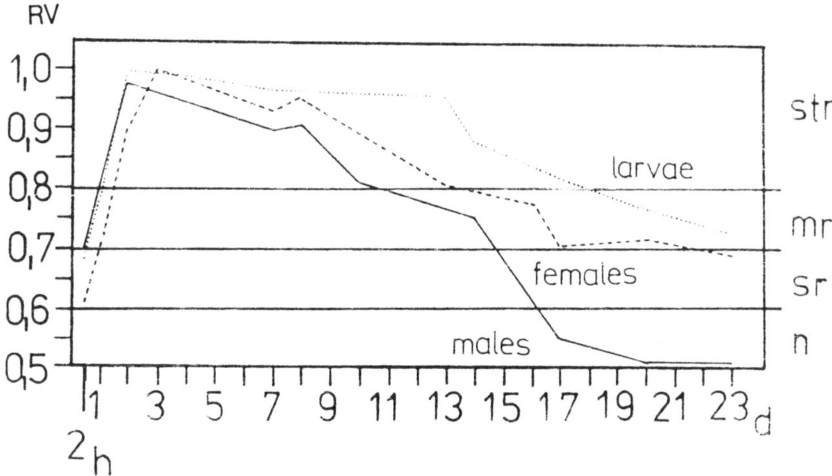

Figure 10   The development of repellency values (mean values) of *Allium sativum* against small, medium and large larvae, and females and males. RV. repellency value; h, hour; d, day; n, neutral; sr, weak repellency; mr, medium repellency; str, strong repellency.

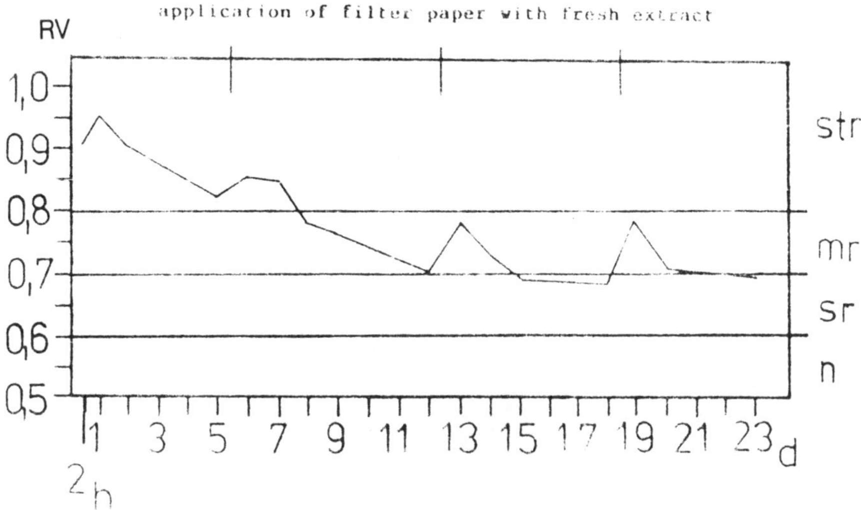

Figure 11   The development of repellency values (mean values) of *Citrus sinensis* against small, medium and large larvae, and females and males with application of filter paper with fresh extract. RV, repellency value; h, hour; d, day; n, neutral; sr, weak repellency; mr, medium repellency; str, strong repellency.

in the normal development after application. It is interesting to see the increase in the repellent effects of used filter papers from the fifth up to the eighteenth day. The increasing role of the aggregation pheromone in the faeces on the control filter paper after eighteen days and the role of avoidance of new extract filter papers without faeces after application have to be noted. The rapid reduction of the repellent effect against males and also females is not only a result of the decreasing quality of the extract but of the behavioural adaptation. In fact, extract filter paper of *Citrus sinensis* that was taken for five days shows the same repellent effect as fresh extract paper on untested animals. 16-day-old extract filter paper of *Allium sativum* shows similar effects (Figure 12).

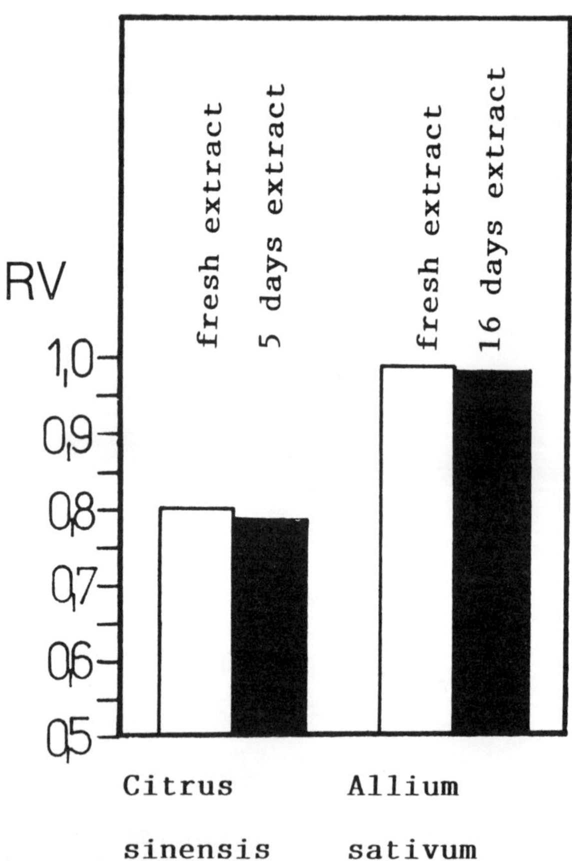

Figure 12   The repellency values of fresh extract filter paper and of extract filter papers which were laid up for 5 days (*Citrus sinensis*) and 16 days (*Allium sativum*). RV, repellency value.

Important factors that influence the repellency test are the quantity and quality of the agents of the extracts, the behavioural modifications of the instars and sexes (avoidance and exploration) and the quantity of the faeces and aggregation pheromone in long-term tests.

Resistance and repellency are the two sides of the same coin in the use of insecticides which can influence the effect, considerably. The appearance of repellency is more behavioural-biologically determined and can be described as a special case of complex dynamic decision processes. The general understanding of those optional processes shows still big deficits. One of the main problems which is also shown in our inquiries is the unforesaying of behavioural reactions of individuals. Even in strongly repellent substances repellency is a statistic figure.

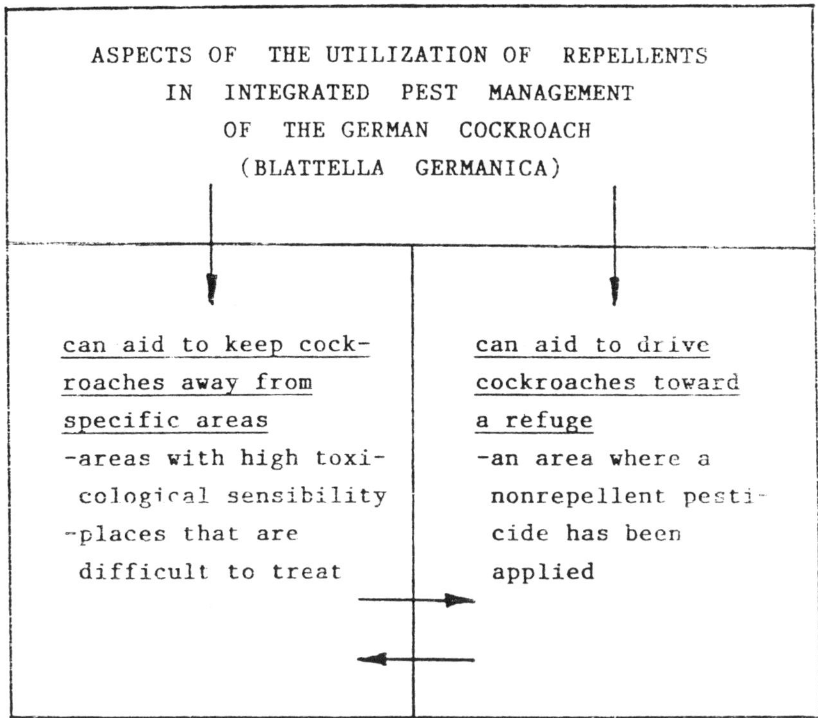

Figure 13   Repellents in integrated pest management.

At present models of the unlinear dynamics and the chaos give the biggest explaining potential for processes like these. After that the processes within the border line of two attractors of a dynamic system are comparable with behavioural decisions. This means that it is, for instance, difficult or even impossible to find untoxical substances which are durable and 100 per cent repellent under all conditions. Nevertheless, plant extracts with a high repellency could be an effective part of integrated control programmes against the animals, doubtlessly (Figure 13).

References

APPEL, A.J. and MACK, T.P. (1989). Repellency of milled aromatic eastern red cedar to domiciliary cockroaches (Dictyoptera: Blattellidae). *Journal of Economic Entomology*, **82**, 152–155.

BRET, B.L. and ROSS, M.H. (1985). Insecticide-induced dispersal in the German cockroach, *Blattella germanica* (L.) (Orthoptera: Blattellidae). *Journal of Economic Entomology*, **78**, 1293–1298.

BURDEN, G.S. (1975). Repellency of selected insecticides. *Pest Control*, **43**, 16–18.

DENZER D.J., FUCHS, M.E.A. and STEIN, G. (1988). Zum Verhalten von *Blattella germanica* L.: Aktionsradius und Refugientreue. *Journal of Applied Entomology*, **105**, 330–343.

DIETZ, J.A. and BUSENBERY, D.B. (1989). Repellent of roodknot nematodes from exudat of host roots. *Journal of Chemical Ecology*, **15**, 2445–2456.

RANDALL, G.B. and BROWER, D.O. (1986). A new method to determine repellent, neutral or aggregative properties of chemicals on *Blattella germanica* (Dictyoptera: Blattellidae). *Journal of Medical Entomology*, **23**, 251–255.

12
A New Class of Antifeedants Against Stored Product Insects

W. SOBÓTKA[1], D. KONOPIŃSKA[2] and J. NAWROT[3],

[1]Institute of Industrial Organic Chemistry, ul. Annopol 6, 03-236 Warsaw, Poland
[2]Institute of Chemistry Wroclaw University, ul. Joliot-Curie 14, 50-383 Wroclaw, Poland
[3]Institute for Plant Protection, ul. Miczurina 20, 60-318 Poznań, Poland

Introduction

In the search for biorational methods of modern insect pest control we became interested in amino acid and peptide derivatives as potential feeding deterrents against stored product insects. Since these compounds are generally nontoxic to vertebrates and highly species-specific, elucidation of their antifeeding character may afford new, environmentally acceptable means for safer management of stored product pests. Most of the antifeedants isolated so far in low quantities from natural sources have complicated chemical structures. Therefore, the search for novel, simpler and more easily available insect antifeedants seemed to be justified.

Antifeedants can be divided into synthetically prepared compounds and substances derived from natural sources. In this paper the gustatory repellents obtained by chemical modifications of predominantly naturally occurring amino acids are presented. In spite of much research activity in the area of feeding deterrents, no commercially acceptable product has yet emerged. Antifeedants inhibit the taste reception of the mouth region, so that without proper gustatory stimulus the insect fails to recognize the treated

Insecticides: Mechanism of Action and Resistance
© 1992   Intercept Ltd, P.O. Box 716, Andover, Hants SP10 1YG, UK

object. A remarkable feature of the deterrent receptors is their sensitivity to compounds which are not present in the insect's environment. Receptors for amino acids exist in many insect species.

The well-known insect deterrent activity of 5-hydroxytryptamine (serotonin) (Rosenthal and Bell, 1979), a compound with the indole heterocyclic ring system, attracted our attention towards L-tryptophan derivatives.

We attempted to discover whether amino acid derivatives acylated at the N-terminal amino group by permethrinic acid would reveal pyrethroid-like antifeeding properties.

This study also describes the gustatory perception of the insect neuropeptides: tetra- and pentapeptide derivatives of proctolin.

Table 1 Feeding deterrent activity of L-tryptophan derivatives

Compound	Total coefficient of antifeeding activity			
	Sitophilus granarius (beetles)	*Tribolium confusum* (beetles)	*Trogoderma granarium* (larvae)	*Tribolium confusum* (larvae)
HCl.Trp-OMe (1)	148	101	57	62
Ac-Trp-OH (2)	114	200	70	100
Boc-Trp-OH (3)	59	133	34	106
Z-Trp-OH (4)	77	40	65	65
Z-Trp-NHBzl (5)	88	101	14	12
Ac-Trp-AN (6)	136	132	139	148
Z-Trp-AN (7)	132	100	144	80
Bz-Trp-AN (8)	118	124	157	108
Perm-Trp-OH (9)	46	98	120	77
Perm-Trp-OMe (10)	–	118	–	–
Z-Val-Trp-OMe (11)	56	91	51	91
Z-D-Val-Trp-OMe (12)	113	123	52	83
Z-Trp-Leu-OMe (13)	77	167	65	53
Z-Trp-D-Leu-OMe (14)	131	74	85	130
c-Trp-Gly (15)	77	19	39	44
c-Trp-Ala (16)	67	103	118	47
c-Trp-D-Ala (17)	124	80	65	81
c-Trp-Val (18)	50	1	45	-2
c-Trp-D-Val (19)	82	121	181	53
c-Trp-Leu (20)	113	135	72	80
c-Trp-D-Leu (21)	123	124	135	112

This work describes three types of amino acid derivatives (Tables 1–3) bioassayed for their antifeeding action against *Sitophilus granarius* and *Tribolium confusum* beetles as well as *Trogoderma granarium* and *T. confusum* larvae, all pests of stored products.

The compounds comprise (Table 1) 21 L-tryptophan mono-(**1–10**) and dipeptides (**11–14**) substituted at the N- or C-terminal by

Table 2 Feeding deterrent activity of some amino acid and dipeptide derivatives

Compound	Total coefficient of antifeeding activity			
	Sitophilus granarius (beetles)	*Tribolium confusum* (beetles)	*Trogoderma granarium* (larvae)	*Tribolium confusum* (larvae)
Phe(p-NO$_2$)-OH (**22**)	123	67	20	51
NO$_2$-Guanidino-Phe(p-NO$_2$)-OH (**23**)	78	74	51	69
HCl.Phe-OMe (**24**)	95	75	11	47
Z-Phe-OH (**25**)	28	121	63	67
HCl.Tyr-OMe (**26**)	112	105	45	77
Perm-Tyr-OMe (**27**)	85	95	63	84
Boc-Tyr(3'-NO$_2$)-NHNH (**28**)	117	55	-17	53
Boc-Tyr(OBzl)-OH (**29**)	92	78	63	36
Boc-Tyr(OBzl)-NH(Me)OMe (**30**)	–	124	–	72
Boc-β-homo-tyr(OBzl)-OMe (**31**)	91	129	11	90
Z-Tyr-OEt (**32**)	145	113	166	93
HCl.L-Dopa-OMe (**33**)	99	38	13	84
HBr.Ala-AN (**34**)	125	42	45	86
Boc-Ala-AN (**35**)	135	119	125	84
Perm-Ala-AN (**36**)	89	104	69	110
Perm-Val-AN (**37**)	134	128	160	126
Perm-D-Val-AN (**38**)	66	87	46	84
Perm-AN (**39**)	72	65	105	122
Z-D-Val-AN (**40**)	32	102	96	73
Leu-4-NO$_2$-anilide (**41**)	44	121	159	77
NO$_2$-Gav(Z)-OH (**42**)	127	70	77	106
Boc-Thz-Thr-OBzl (**43**)	108	134	82	81
Unox (**44**)	138	98	121	78
N$^\alpha$,N$^\delta$-di-Perm-Orn-OMe (**45**)	111	122	119	122
(N$^\delta$-Perm-Orn) Cu$_2$^{II} (**46**)	127	51	138	59
H-Glu(α-Met-OH)-OH (**47**)	83	107	107	77
Boc-Leu-Thz-OH (**48**)	121	54	56	78

protecting groups such as acetyl, tert-butyloxycarbonyl, benzyloxy-carbonyl, benzoyl and methyl. Two compounds with L-tryptophan moiety were acylated at the amino group by (1R,S) (cis/trans)permethrinic acid (**9–10**). Aromatic systems of benzylamine and ethyl 4-aminobenzoate occur in compounds of **5** and **6–8** respectively. Cyclic 2,5-dioxapiperazines derived from L-tryptophan containing dipeptides are represented by the compounds **15–21**.

Among twenty-seven amino acid and dipeptide derivatives (**22–48**) (Table 2) twenty-three contain an aromatic ring system either in the amino acid residue or in the blocking group. Unox (**44**) is an unsaturated disubstituted azlactone derived from the N-benzoylgly-cine cyclization product followed by condensation with carbonic aldehyde ethyl ester.

Protected tetrapeptide (**57**) and eight pentapeptide derivatives (**49–56**) of the insect neuropeptide proctolin are summarized in Table 3.

Table 3 Feeding deterrent activity of proctolin and analogue derivatives

Compound	Total coefficient of antifeeding activity			
	Sitophilus granarius (beetles)	*Tribolium confusum* (beetles)	*Trogoderma granarium* (larvae)	*Tribolium confusum* (larvae)
Boc-Arg(NO$_2$)-Tyr(OBzl)- Leu-Pro-Thr-OBzl (**49**)	117	128	61	103
N$^\alpha$,N$^\delta$-di-Z-Orn-Tyr(Obzl)- Leu-Pro-Thr-OBzl (**50**)	128	129	112	54
N$^\alpha$,N$^\delta$-di-Z-Lys-Tyr(OBzl)- Leu-Pro-Thr-OBzl (**51**)	5	-4	51	15
NO$_2$-Gav(z)-Tyr(OBzl)- Leu-Pro-Thr-OBzl (**52**)	123	92	54	94
NO$_2$-Gac(Z)-Tyr(OBzl)- Leu-Pro-Thr-OBzl (**53**)	106	65	77	86
Boc-Phe(p-NO$_2$)-Tyr(OBzl)- Leu-Pro-Thr-OBzl (**54**)	66	43	28	13
NO$_2$-Guanidino-Phe(p-NO$_2$)- Tyr(OBzl)-Leu-Pro- Thr-OBzl (**55**)	101	28	-11	76
Z-Arg(NO$_2$)-Tyr(OBzl)-Leu- Hpr(4-OBzl)-Thr-(OBzl (**56**)	65	31	66	67
Boc-L-Dopa(di-OBzl)-Leu- Pro-Thr-OBzl (**57**)	108	61	52	55

Material and Methods

Amino acid and peptide derivatives were obtained by standard-methods of amino acid and peptide chemistry (Bodanszky and Bodanszky, 1985; Konopińska *et al.*, 1986, 1990; Konopińska and Sobótka, 1989). The following abbreviations are used.

Ala	L-alanine
Arg	L-arginine
L-Dopa	3,4-dihydroxy-L-phenylalanine
Gac	2-guanidino-6-amino-L-caproic acid
Gav	2-guanidino-5-amino-L-valeric acid
Gly	glycine
Glu	L-glutamic acid
Hpr	L-4-hydroxyproline
Leu	L-leucine
Lys	L-lysine
Met	L-methionine
Orn	L-ornithine
Phe	L-phenylalanine
Pro	L-proline
Thr	L-threonine
Trp	L-tryptophan
Thz	L-thiazolidine-4-carboxylic acid
Tyr	L-tyrosine
Val	L-valine
Ac	acetyl
AN	ethyl 4-aminobenzoate (anesthesin)
Boc	tert-butyloxycarbonyl
Bz	benzoyl
Bzl	benzyl
c	cyclo
Et	ethyl
Me	methyl
NO_2	nitro
Z	benzyloxycarbonyl
Perm	(1R,S)(cis/trans)-2,2-dimethyl-3(2'-dichlorovinyl) cyclopropane carboxylic acid residue (cis:trans ratio = 1:2)
Unox	2-phenyl-4-[1'-methylidene(2',2'-dimethyl-3'-R,S-ethoxycarbonyl)cyclopropyl]-oxazolone-5

Bioassays were carried out by the previously described wafer disc method (Nawrot *et al.*, 1986). The values of the total coefficient of antifeeding activity are expressed on a scale between 0 and 200 and are divided into four groups of intervals to measure the antifeeding activity of each tested compound: I = 0–50 weak; II = 51–100 medium; III = 101–150 good; IV = 151–200 very good. The value 0 denotes an inactive compound, whereas the score 200 denotes an ideal antifeedant.

Results

From the data presented in Table 1 four L-tryptophan derivatives with very good antifeeding properties can be distinguished. These are acetyl-L–tryptophan (**2**), substituted amino acid with anesthesin moiety (**8**), and two dipeptides with L-leucine (**13**) and D-valine residues (**19**). Both compounds **2** and **13** were active against *T. confusum* beetles while compounds **8** and **19** revealed their deterrent effects against *T. granarium* larvae. The following sixteen amino acid and dipeptide derivatives can be considered to be good antifeedants: **1–3, 5–10, 12, 14, 16–17** and **19–21**. Compounds **6, 8**, and **21** revealed a broad range of feeding deterrent properties against all the insects tested.

Table 2 gives the results of the antifeeding activity of twenty-seven various amino acids and peptides. Only Z-tyrosine ethyl ester (**32**), permethroyl-valyl-p-aminobenzoic acid ethyl ester (**37**) and leucyl-4-NO-anilide (**41**) demonstrates high gustatory repellent properties (very good) against *T. granarium* larvae. The following twenty-one other compounds, **22, 25–26, 28, 30–32, 34–37, 39–48** can be included in group III of antifeedants. The most sensitive insect species to this category (good) of feeding deterrents were adult *S. granarius* and *T. confusum*, as was also observed for the L-tryptophan derivatives presented in Table 1. Compound **45**, revealing a broad range of antifeeding activity, can be considered as a model system in which two permethroyl units in a linear arrangement are separated by four carbon atoms of the L-ornithine skeleton, perhaps enabling a better fit to the insect gustatory receptor.

Among the nine proctolin derivatives (**49–57**) shown in Table 3, five oligopeptides (**50, 52–53, 55** and **57**) and protected proctolin (**49**) have good feeding deterrent properties. Pentapeptide **50** with an L-ornithine residue revealed antifeeding activity against three bioassayed insects, with *S. granarius* beetles being the most sensitive to the tested compounds. Weak feeding attractant properties were noticed in the case of compounds **51** and **55**.

Conclusions

Amino acids and peptides (**1–57**) described in this paper belong to groups of very good (IV), good (III) and medium (II) feeding deterrents when tested against four stored product infesting insects. Of seven compounds with excellent but selective antifeeding activity (**2, 8, 13, 19, 32, 37** and **41**) against *T. confusum* beetles and *T. granarium* larvae, four compounds (**2, 8, 13** and **19**) contain the L-tryptophan residue. The total coefficient of feeding deterrent activity against adult *T. confusum* reached the highest value of 200 with acetyl-L-tryptophan (**2**).

Forty-three amino acid and dipeptide derivatives revealed good, but mostly selective, antifeeding activity except for the compounds **6, 8, 21** and **45** which possess feeding deterrent properties against all bioassayed insects. Among the group of good antifeedants, sixteen compounds contain the L-tryptophan residue. Insertion of permethroyl moiety markedly increases antifeeding effectivity (**10, 36, 37, 45** and **46**).

Four compounds show feeding-attractant properties (**18, 28, 51** and **55**).

So far no structure-activity relationships can be inferred from the biodata reported here. However, the apparent increase in the deterrent effect due to the presence of L-tryptophan or L-ornithine (**45, 46** and **50**) seems to warrant further research into gustatory repellents from amino acid and peptide derivatives.

References

BODANSZKY, M. and BODANSZKY, A. (1985). *The practice of peptide synthesis*, pp. 1–263. Akademie-Verlag, Berlin.

KONOPIŃSKA, D. and SOBÓTKA, W. (1989). Synthesis of N-permethroyl amides with potential deterrent and insecticidal activity. *Polish Journal of Chemistry*, **63**, 303–306.

KONOPIŃSKA, D., SOBÓTKA, W., LESICKI, A., ROSIŃSKI, G. and SUJAK, P. (1986). Synthesis of proctolin analogues and their cardioexcitatory effect on cockroach. *Periplaneta americana* L., and yellow mealworm, *Tenebrio molitor* L. **27**, 597–603.

KONOPIŃSKA, D., ROSIŃSKI, G., BARTOSZ-BECHOWSKI, H., LESICKI, A., SUJAK, P. and SOBÓTKA, W. (1990). Role of guanidine group at the N-terminal proctolin chain of cardioexcitatory effects in insects. *International Journal of Peptide and Protein Research*, **35**, 12–16.

NAWROT, J., BLOSZYK, E., HARMATHA, J., NOVOTNY, L. and DROZDZ, B. (1986). Action of antifeedants of plant origin on beetles infesting stored products. *Acta Entomologica Bohemoslovaca*, **83**, 327–335.

ROSENTHAL, G.A. and BELL, E.A. (1979). Naturally occurring, toxic nonprotein amino acids. In *Herbivores: their interaction with secondary plant metabolites* (G.A. Rosenthal and D.H. Janzen, eds), pp. 353–386. Academic Press, New York.

Part 2
Moulting and Developing Processes and Their Hormonal Regulation as Targets for Insecticide Actions

13
Selective Insect Control Agents – Mechanisms and Agricultural Importance

I. ISHAAYA

Department of Entomology, Agricultural Research Organization, The Volcani Center, Bet Dagan 50250, Israel

The discovery of the insecticidal activity of benzoylphenyl ureas pointed to new pathways in insect control strategies. The first commercial compound, diflubenzuron, disrupts the moulting process thereby serving as an insect growth regulator. It acts mainly by ingestion, but also through contact in certain species. A portion of its activity appears to be associated with inhibition of biochemical processes leading to chitin formation (Ishaaya and Casida, 1974; Post *et al.*, 1974; Degheele, 1990), but it is probably not a direct inhibitor of chitin synthetase (Cohen and Casida, 1980; Mayer *et al.*, 1981). An interesting suggestion related to the biochemical mode of action of benzoylphenyl ureas was put forward by Mitsui *et al.* (1984), who showed that DFB inhibited the transport of UDP-N-acetylglucosamine across the midgut epithelium in the cabbage armyworm. These authors proposed that the catalytic site of chitin synthetase is located in the outer surface of the membrane, and, as such, the substrate must cross the plasma membrane in order to interact with the enzyme.

The continuous search for more potent acylureas led to three new compounds with a much higher potency on agricultural pests than diflubenzuron: the Ishihara compound chlorfluazuron (known as Atabron, IKI-7899, CGA-112,913) (Haga *et al.*, 1982), the Celamerk compound teflubenzuron (known as Nomolt, CME-134) (Becher *et al.*, 1983) and the Dow compound hexaflumuron (known as Consult,

---

Insecticides: Mechanism of Action and Resistance
© 1992   Intercept Ltd, P.O. Box 716, Andover, Hants SP10 1YG, UK

XRD-473) (Sbragia *et al.*, 1983). Studies carried out in our laboratory revealed that these compounds exhibited similar toxicity on both malathion-susceptible and -resistant strains of *Tribolium castaneum*. On the other hand, diflubenzuron was considerably less toxic to the resistant strain (Ishaaya and Yablonski, 1987; Ishaaya *et al.*, 1987). According to mortality curves and LC_{50} values the toxicity of the test compounds on both *Tribolium* strains was hexaflumuron > teflubenzuron > chlorfluazuron > diflubenzuron. The high potency of the recent benzoylphenyl ureas on *Tribolium* and on other agricultural pests such as *Spodoptera* and *Heliothis* species, along with their low mammalian toxicity, render these compounds potential components in integrated pest management programmes.

Parallel to the development of benzoylphenyl ureas, a novel chitin synthesis inhibitor, buprofezin (Applaud, 2-tert-butylimino-3-iso-propyl-5-phenyl-3,4,5,6-tetrahydrothiadiazine-4-one) has been developed (Kanno *et al.*, 1981). It acts specifically on some homopteran pests such as the greenhouse whitefly (*Trialeurodes vaporariorum*) (Yasui *et al.*, 1985, 1987), the sweetpotato whitefly (*Bemisia tabaci*) (Ishaaya *et al.*, 1988; Ishaaya, 1990), the brown planthopper (*Nilaparvata lugens*) (Izawa *et al.*, 1985; Nagata, 1986), and the citrus scales (*Aonidiella aurantii* and *Saissetia oleae*) (Yarom *et al.*, 1988; Ishaaya *et al.*, 1989). Whiteflies are important pests of cotton and vegetables. In some cases they transmit viruses and are considered limiting factors for growing agricultural crops. The brown planthopper is an important pest of rice in the far East. Most of the pests have developed resistance to conventional insecticides and buprofezin is an important addition enabling continuing production of various agricultural commodities. Studies carried out in our laboratory indicated that buprofezin suppresses embryogenesis and progeny formation of *B. tabaci* (Ishaaya *et al.*, 1988). The estimated concentration for 50% inhibition of egg hatch was 0.0015% a.i. in the spray solution and 0.006% a.i. for 50% cumulative larval mortality. Hence the compound exerts its effect on the egg hatch and on the larval stage. It has no ovicidal activity, but suppresses embryogenesis through adults. The length of exposure of the whitefly females to buprofezin corresponds well with the suppression of egg hatch (Table 1). Adult females exposed for up to 5 hours to cotton seedlings treated with 62.5 ppm laid eggs with fertility similar to that of the control, but those exposed for a period of > 24 hours laid infertile eggs. A good correlation between the length of adult exposure to buprofezin and egg fertility was observed also with *T. vaporariorum* (Yasui *et al.*, 1987). Part of buprofezin's efficiency resulted from its vapour phase. Laboratory assays carried out with *B. tabaci* (De Cock *et al.*, 1990) indicated that 96%, 62% and 39% of first-instar larvae died after they were placed at a distance of 2cm, 4cm and 6cm respectively from leaves treated with 450 ppm buprofezin. Cumulative larval mortality of 90%, 83% and 65% was obtained when infested seedlings were introduced at 1, 9 and 16 days respectively in cotton field treated with 500g a.i./hectare (Table 2). These results indicate that high-vapour phase toxicity of buprofezin persists at least ten days after application, enabling good control of whitefly larvae which are present on the lower surface of the leaves and are difficult to control under standard spray conditions of other insecticides. Buprofezin is harmless to aphelinid parasites such as *Encarsia formosa* and *Cales noaki* (Garrido *et al.*, 1984; Wilson and Anema, 1988) and to predacious mites (Anon., 1987), and as such it is considered a selective insecticide.

Table 1   Effect of length of exposure of *B. tabaci* females to buprofezin on egg hatch

Oviposition period of adults exposed to treated and untreated plants, hours	Egg hatch, %
0–5, untreated	75 ± 4a
0–5, treated	67 ± 6a
6–10, untreated	67 ± 5a
6–10, treated	44 ± 5b
11–24, untreated	79 ± 7a
11–24, treated	16 ± 6c
25–48, untreated	76 ± 5a
25–48, treated	Od

Whitefly females were exposed for various periods to cotton seedlings sprayed until runoff with 62.5 ppm buprofezin. Data are averages of 5-10 replicates of 30-40 whitefly females. Figures followed by the same letter do not differ significantly at $P > 0.05$ (Ishaaya *et al.*, 1988).

Other insect growth regulators (IGRs) of agricultural importance are the triazine compound cyromazine, the juvenoid analogue fenoxycarb and pyriproxyfen and the thio urea compound diafenthiuron known as Pegasus. Cyromazine, extremely effective against dipteran species, acts at the apolytic stage, affecting thereby the ecdysis process, but has no effect on chitin synthesis (Friedel and McDonell, 1985). Fenoxycarb and pyriproxyfen act as juvenile hormones, affecting specifically scale insects and egg fertility (Masner *et al.*, 1987; Peleg, 1982, 1988; Ascher and Eliyahu, 1988). The diversity of selective insecticides available today are potential components to be used along with natural enemies in integrated pest management programmes in various field crops.

Table 2   Effect of buprofezin vapor phase on *B. tabaci* larvae under field conditions

Infested seedlings introduced into a treated cotton field, at various periods after application, days	Cumulative mortality until pupation, %
1	90 ± 3
9	83 ± 2
16	65 ± 63

Part of a cotton field was sprayed with 500 g a.i./hectare, the other part kept untreated as control. Cotton seedlings infested with first instar larvae were introduced for 7 days into a treated or untreated field, at various periods after application. Cumulative mortality until pupation (corrected according to Abbott's formula) was then determined. Each treatment was carried out with ten replicates of 50–200 larvae each and presented as means with their SE values.

Acknowledgements

The authors thank Sara Yablonski and Zmira Mendelson for expert technical assistance, and appreciate the financial support of the Israel Cotton Board, Nihon Nohyaku Co., Tokyo and Makhteshim Chemical Co., Beer Sheva. This article is contribution no. -E, 1991 series, from the Agricultural Research Organization, Bet Dagan, Israel.

References

ANONYMOUS (1987). Applaud, new pesticide (insect growth regulator), technical information. Nihon Nohyaku, Tokyo, Japan.
ASCHER, K.R.S. and ELIYAHU, M. (1988). The ovicidal

properties of the juvenile hormone mimic Sumitomo S-31183 (SK-591) to insects. *Phytoparasitica*, **16**, 15–21.

BECHER, H.M., BECKER, P., PROKIC-IMMEL, R. and WIRTZ, W. (1983). CME, a new chitin synthesis inhibiting insecticide. *Tenth International Congress of Plant Protection*, Brighton, pp. 408–415.

COHEN, E. and CASIDA, J.E. (1980). Inhibition of *Tribolium* gut chitin synthetase. *Pesticide Biochemistry and Physiology*, **13**, 129–136.

DE COCK, A., ISHAAYA, I., DEGHEELE, D. and VEIEROV, D. (1990). Vapor toxicity and concentration-dependent persistence of buprofezin applied to cotton foliage for controlling the sweetpotato whitefly (Homoptera: Aleyrodidae). *Journal of Economic Entomology*, **83**, 1254–1260.

DEGHEELE, D. (1990). Chitin synthesis inhibitors: effects on cuticle structure and components. In *Pesticides and Alternatives* (J.E. Casida, ed.), pp. 377–388. Elsevier Science Publisher B.V., Amsterdam.

FRIEDEL, T., and McDONELL, P.A. (1985). Cyromazine inhibits reproduction and larval development of the Australian sheep blow fly (Diptera: Calliphoridae). *Journal of Economic Entomology*, **78**, 868–873.

GARRIDO, A., BEITIA, F. and GRUENHOLZ, P. (1984). Effects of PP618 on immature stages of *Encarsia formosa* and *Cales noaki* (Hymenoptera: Aphelinidae). *British Crop Protection Conference – Pest and Diseases*, pp. 305–310.

HAGA, T., TOBI, T., KOYANAGI, T. and NISHIYAMA, R. (1982). Structure activity relationships of a series of benzoylpyridyloxyphenyl-urea derivatives. Abstract, *Fifth International Congress of Pesticide Chemistry* (IUPAC), p. IId–7.

ISHAAYA, I. (1990). Buprofezin and other insect growth regulators for controlling cotton pests. *Pesticide Outlook*, **1**, 30–33.

ISHAAYA, I. and CASIDA, J.E. (1974). Dietary TH6040 alters composition and enzyme activity of housefly larval cuticle. *Pesticide Biochemistry and Physiology*, **4**, 484–490.

ISHAAYA, I. and YABLONSKI, S. (1987). Toxicity of two benzoylphenyl ureas against insecticide resistant mealworms. In *Chitin and Benzoylphenyl Ureas* (J.E. Wright and A. Retnakaran, A., eds), pp. 131–140. Dr. W. Junk Publishers, Dordrecht.

ISHAAYA, I., BLUMBERG, D. and YAROM, I. (1989). Buprofezin – a novel IGR for controlling whiteflies and scale insects. *Mededelingen van de Faculteit Landbouwwetenschappen Rijksuniversiteit Gent*, **54**, 1003–1008.

ISHAAYA, I., MENDELSON, Z. and MELAMED-MADJAR, V. (1988). Effect of buprofezin on embryogenesis and progeny formation of sweetpotato whitefly (Homoptera: Aleyrodidae). *Journal of Economic Entomology*, **81**, 781–784.

ISHAAYA, I., YABLONSKI, S. and ASCHER, K.R.S. (1987). Toxicological and biochemical aspects of novel acylureas on resistant and susceptible strains of *Tribolium castaneum*. In *Proceedings of the Fourth International Working Conference on Stored-Product Protection* (E. Donahaye and S. Navarro, eds), pp. 613–622. Caspit, Jerusalem.

IZAWA, Y., UCHIDA, M., SUGIMOTO, T. and ASAI, T. (1985). Inhibition of chitin biosynthesis by buprofezin analogs in relations to their activity controlling *Nilaparvata lugens* Stal. *Pesticide Biochemistry and Physiology*, **24**, 343–347.

KANNO, H., IKEDA, K., ASAI, T. and MAEKAWA, S. (1981). 2-tert-butylimino-3-isopropyl-5-phenyl-perhydro-1,3,5-thiadiazin-4-one (NNI 750), a new insecticide. *1981 British Crop Protection Conference – Pests and Diseases*, pp. 56–69.

MASNER, P., ANGST, M. and DORN, S. (1987). Fenoxycarb, an insect growth regulator with juvenile hormone activity: a candidate for *Heliothis virescens* (F) control on cotton. *Pesticide Science*, **18**, 89–94.

MAYER, R.T., CHEN, A.C. and DELOACH, J.R. (1981). Chitin synthesis inhibiting insect growth regulator do not inhibit chitin synthesis. *Experientia*, **37**, 337–338.

MITSUI, T., NOBUSAWA, C. and FUKAMI, J. (1984). Mode of inhibition of chitin synthesis by diflubenzuron in the cabbage armyworm. *Mamestra brassicae* L. *Journal of Pesticide Science*, **9**, 19–26.

NAGATA, T. (1986). Timing of buprofezin application for control of the brown planthopper, *Nilaparvata lugens* Stal. (Homoptera: Delphacidae). *Applied Entomology and Zoology*, **21**, 357–362.

PELEG, B.A. (1982). Effect of a new insect growth regulator, RO 13-5223, on scale insects. *Phytoparasitica*, **10**, 27–31.

PELEG, B.A. (1988). Effect of a new phenoxy juvenile hormone analog on California red scale (Homoptera: Diaspididae), Florida wax scale (Homoptera: Coccidae) and the ectoparasite *Aphitis holoxanthus* DeBache (Hymenoptera: Aphelinidae). *Journal of Economic Entomology*, **81**, 88–92.

POST, L.C., de JONG, B.J. and VINCENT, W.R. (1974). 1-(2,6-Disubstituted benzoyl)-3-phenylurea insecticides: inhibitors of chitin synthesis. *Pesticide Biochemistry and Physiology*, **4**, 473–483.

SBRAGIA, R., BISABRI-ERSHADI, B. and RIGTERINK, R.H.

(1983). XRD-473, a new acylurea insecticide effective against *Heliothis*. *Tenth International Congress of Plant Protection*, Brighton, pp. 417–424.

WILSON, D. and ANEMA, B.P. (1988). Development of buprofezin for control of whitefly *Trialeurodes vaporariorum* and *Bemisia tabaci* on glasshouse crops in the Netherlands and the UK. *1981 British Crop Protection Conference – Pests and Diseases*, pp. 175–180.

YAROM, I., BLUMBERG, D. and ISHAAYA, I. (1988). Effect of buprofezin on California red scale (Homoptera: Diaspididae) and Mediterranean black scale (Homoptera: Coccidae). *Journal of Economic Entomology*, **81**, 1581–1585.

YASUI, M., FUKADA, M. and MAEKAWA, S. (1985). Effect of buprofezin on different developmental stages of the greenhouse whitefly, *Trialeurodes vaporariorum* (Westwood) (Homoptera: Aleyrodidae). *Applied Entomology and Zoology*, **20**, 340–347.

YASUI, M., FUKADA, M. and MAEKAWA, S. (1987). Effect of buprofezin on reproduction of the greenhouse whitefly, *Trialeurodes vaporariorum* (Westwood) (Homoptera: Aleyrodidae). *Applied Entomology and Zoology*, **22**, 266–271.

14
Cyromazine – An Insecticide that Affects the Cuticle

STUART E. REYNOLDS[1] and ANDREW C. KOTZE[2]

[1]*School of Biological Sciences, University of Bath, Claverton Down, Bath BA2 7AY, UK*
[2]*NWS Agriculture and Fisheries, P.M.B. 10, Rydalmere, NSW 2116, Australia*

Cyromazine

Cyromazine (2-cyclopropylamino-4,6-diamino-*s*-triazine; CGA 72662) is an insecticidal diaminotriazine used principally as a larvicide against Dipteran pests such as house fly and face fly (Hall and Foehse, 1980) and sheep blow fly (Friedel and McConnell, 1985). The insecticide has insect growth regulator (IGR)-like effects on target insects; death follows treatment only after a period of growth. The mode of action of cyromazine is not known.

A number of other diamino- and triaminotriazines also have insecticidal properties. First identified as chemosterilants (Borkovec and De Milo, 1967), it seems likely that these compounds are in fact larvicides (De Milo *et al.*, 1981). One of the most active of these compounds is 2,4-diamino-6-(2-furyl)-*s*-triazine (AI3-22641). The symptoms of poisoning with this compound are very like those associated with cyromazine (Pessah *et al.*, 1985; S.E. Reynolds, unpublished), and it seems likely that it shares cyromazine's mode of action.

Insecticides: Mechanism of Action and Resistance
© 1992   Intercept Ltd, P.O. Box 716, Andover, Hants SP10 1YG, UK

Actions of Cyromazine

Although cyromazine is particularly toxic to fly larvae, it also kills the Lepidopteran caterpillar *Manduca sexta* at somewhat larger doses (Hughes *et al.*, 1989; Reynolds and Blakey, 1989). The pathology of cyromazine poisoning in *Manduca* is very similar to that seen in the housefly *Musca domestica* (Price and Stubbs, 1984), and we have used the much larger *Manduca* as a model in which to investigate the insecticide's actions. In this paper we review the evidence that cyromazine may attack the cuticle.

Following transfer to cyromazine-treated diet, tobacco hornworm larvae show impaired growth, eventually developing cuticular lesions prior to death, which usually ensures after a period of days. The caterpillars literally burst.

Unlike the case with acylurea insecticides (Reynolds, 1987), death is not especially associated with subsequent moulting. Although insects treated with lower doses of cyromazine, which allow them to survive long enough to moult, sometimes show morphological abnormalities (especially in the pupal stage) (Reynolds and Blakey, 1989) these are quite different from the larval-pupal intermediate kind of malformation indicative of endocrine (juvenile hormone) disturbance (Sláma, Romaňuk and Šorm, 1974).

The toxicity of the insecticide to *Manduca* larvae depends on how long they are exposed to it. When the insecticide is given only in the fifth (and final) larval instar, the LC_{50} in diet is about 20 ppm. When newly hatched larvae are fed continuously on cyromazine-treated diet, however, the LC_{50} falls to less than 5 ppm (Reynolds and Blakey, 1989). This may be compared to a reported LC_{50} for housefly larvae of 0.1–0.3 ppm in diet (Miller *et al.*, 1981).

One of the earliest signs of poisoning in both caterpillars (Reynolds and Blakey, 1989) and fly larvae (Price and Stubbs, 1984) is an abnormally elongated body shape (Figure 1). In *Manduca*, this symptom is seen only a few hours after transfer of the insects to a treated diet, and is closely correlated with the declining growth rate. At this stage the poisoned insects look and feel turgid, as though they are suffering from abnormally elevated internal hydrostatic pressure. The insects are restricted in their ability to move, as though internal pressure prevented them flexing their bodies. Direct measurement using a pressure transducer attached to a cannula (Figure 2) confirms that the basal internal pressure within the haemocoel is almost doubled; active attempts to move lead to very large pressure transients in treated insects, but not in controls (Reynolds and Blakey, 1989).

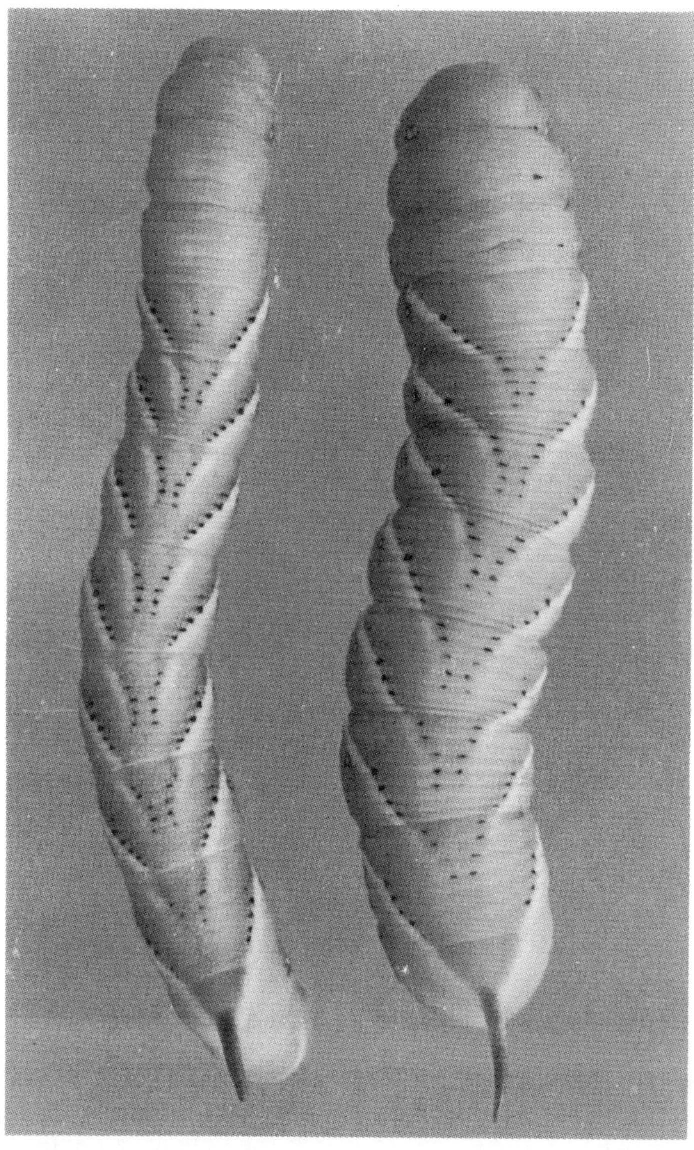

Figure 1 Characteristic long and thin appearance of cyromazine-treated *Manduca* caterpillar (*right*) compared with control (*left*). Cyromazine treatment (50 ppm in artificial diet) was begun at the beginning (day 0) of the fifth stage. Photo taken on day 2 (after Reynolds and Blakey, 1989.)

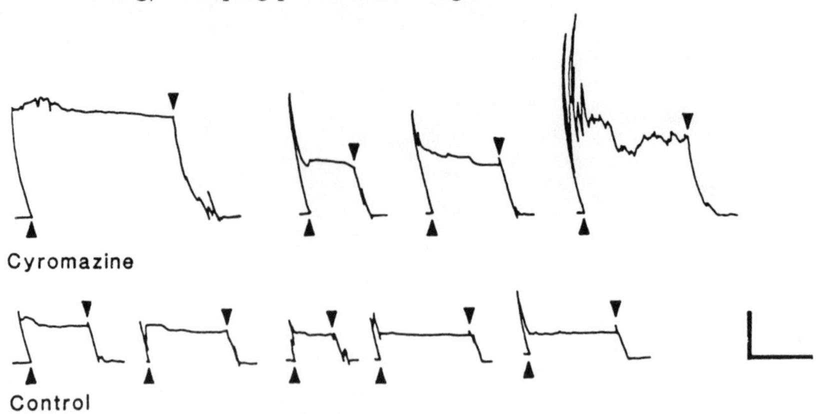

Cyromazine

Control

Figure 2   Cyromazine causes increased internal pressure. Examples of pressure records from actively moving day 1 fifth stage *Manduca* caterpillars:(*top*) cyromazine-treated (50 ppm in artificial diet given on day 0); (*bottom*) controls. Vertical deflections represent positive internal pressure. Arrowheads represent insertion (arrow points up) and withdrawal (arrow points down) of the hypodermic needle attached to the transducer. Scale bars represent 20mm Hg pressure (vertical) and 20 s (horizontal). (From Reynolds and Blakey, 1989.)

The Effect of Cyromazine on Cuticle Mechanics

The symptoms described above appear to be the consequence of an abnormally inextensible cuticle, which causes the observed elevated internal pressure by its failure to expand as the insect grows. This was first shown by using a simple creep test in which loops of cuticle from affected insects extended less rapidly under maintained tension than did loops from control insects (Reynolds and Blakey, 1989).

More discriminating measurements of cuticle mechanical properties can be made using an Instron testing machine which extends the cuticle loop at a constant rate while measuring the force required to do so (Kotze and Reynolds, 1990). This reveals (Figure 3; Table 1) that in affected insects cuticle stiffness (i.e. the ratio of applied force to extension) is increased by a factor of almost 10, while the force necessary to break the cuticle (strength) is increased about threefold. On the other hand, the extent to which the cuticle can extend before breaking is reduced by about half. These changes are not due to increased thickness of the affected cuticle, which is actually thinner than the control by about 20%. Since the altered mechanical properties of the cuticle of cyromazine-treated insects are maintained after the cuticle is separated from its epidermis, the changes must be intrinsic to the cuticle itself (Table 1).

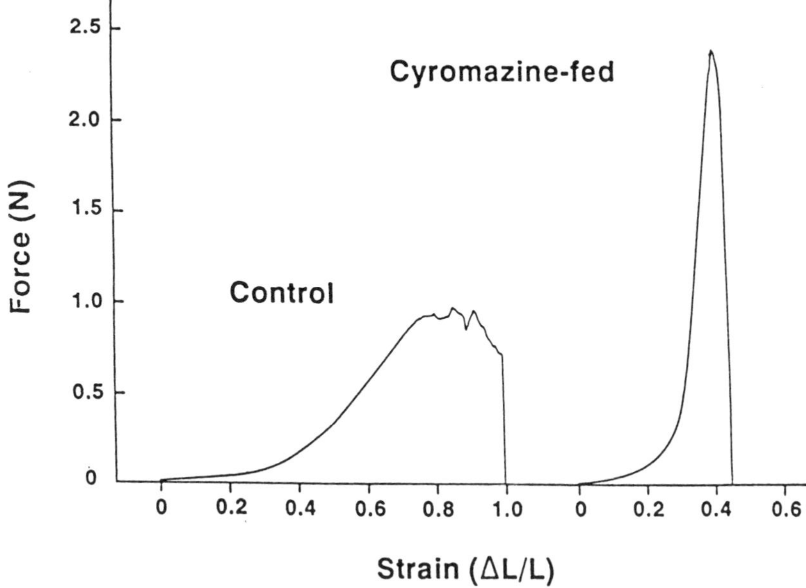

Figure 3   Cyromazine causes increased cuticle stiffness and strength. Force-strain curves for loops of abdominal cuticle (with epidermis intact) from day 1 fifth stage *Manduca* caterpillars fed for 24 hours with either control or cyromazine-treated (50 ppm) diet. (From Kotze and Reynolds, 1990.)

These alterations in the mechanical properties of the cuticle are exhibited very early in the poisoning process, with stiffness and strength clearly affected by 3 hours after exposure to cyromazine (Kotze and Reynolds, 1990). This corresponds very closely to the time at which growth and altered body shape are first seen (Reynolds and Blakey, 1989).

Increased cuticle stiffness is probably the cause of the altered body shape. The cuticle is folded into ridges which run around the circumference of the body wall, but not along its length (Figure 1). The unfolding of these ridges results in a more pronounced 'toe' in the initial (low force) region of the force-strain curve when testing in the axial than the circumferential dimension (the stiffness and strength of the cuticle in circumferential and axial directions are however similar in the linear region of the force-strain curve at higher forces). The shallow 'toe' region of the force-strain curve is little affected by cyromazine, whereas the slope of the linear region is greatly increased. Therefore, increased tension in the cuticle of insecticide-affected insects results in a greater unfolding relative to stretching, and thus elongation of the body.

Table 1   Effect of cyromazine on the mechanical properties of *Manduca* cuticle (data from Kotze and Reynolds, 1990).

Treatment	*cuticle mechanical properties*		
	stiffness ($\Delta N/\Delta$ strain)	force at yield point (N)	strain at yield point
Epidermis intact			
control	1.94 ± 0.18	0.83 ± 0.05	0.86 ± 0.06
cyromazine	18.53 ± 1.33	2.29 ± 0.08	0.44 ± 0.02
Epidermis removed			
control	5.75 ± 0.62	1.30 ± 0.07	0.36 ± 0.03
cyromazine	29.86 ± 1.21	2.66 ± 0.14	0.14 ± 0.01

Mechanical tests were on circumferential loops of abdominal cuticle taken from day 1 fifth instar stage *Manduca* caterpillars given 50 ppm cyromazine in diet for 24 hours.

Means + S.E., N = 10.

Increased cuticle stiffness may also be the cause of reduced growth. The rate of weight increase in tobacco hornworm caterpillars is exceptionally fast. Fifth instar larvae grow at a rate of about 0.1g h^{-1} throughout the feeding period (on the first day this corresponds to an hourly increase in weight of about 5%). Such a rapid increase in body size can only be sustained by a corresponding rate of increase in the linear dimensions of the body wall (about 1.7% h^{-1}). If the cuticle extends less readily than usual, this will limit the rate at which growth can occur. If this is the case, then it would be expected that the affected insects would display residual tension in the body wall. The high internal pressure observed in treated caterpillars (see above) confirms this.

The increased strength and stiffness of the cuticle in cyromazine-treated insects are consistent with an increase in the extent of intermolecular interactions between the various components of the cuticle. Such interactions could take the form of chemical cross-links between cuticle macromolecules (such as occur during sclerotization), or might be the result of an altered cuticle microenvironment. We have attempted to discover the nature of these interactions by biochemical analysis of affect cuticle.

Cuticle Chemistry

Manduca larvae were given 50 ppm cyromazine in diet for 24 hours (a dose which produces severe effects on the cuticle), before their cuticle was taken for analysis. In some cases the incorporation of radiolabelled precursors was measured. Full details are given in Kotze and Reynolds (1991).

Decreased cuticle hydration would be expected to cause an increase in stiffness. However, our measurements of cuticle water content actually show a slight but significant increase in the water content of cuticle samples from cyromazine-treated insects (Table 2). This small change cannot therefore explain the altered mechanical properties of the cuticle.

Table 2 Effect of cyromazine on cuticle composition and on incorporation into cuticle of biosynthetic precursors (data from Kotze and Reynolds, 1991).

	Control	*Cyromazine*	*N*
Water content	70.06 ± 0.6	73.20 ± 0.8	10
Chitin synthesis Incorporation of $[^{14}C]$-N-acetylglucosamine			
in vivo (nCi/larva)	22.80 ± 2.3	21.70 ± 1.0	10
in vitro (nCi/proleg)	1.76 ± 0.13	1.52 ± 0.21	5
Protein content SDS-soluble protein			
(mg/g larva)	0.98 ± 0.03	0.96 ± 0.03	10
Protein synthesis Incorporation of $[2\text{-}^{3}H]$-glycine			
(nCi/g larva)	145 ± 22	134 ± 13	7

Cyromazine given to day 0 fifth instar larvae at 50 ppm. Refer to original paper for exact conditions.

Means $\pm$ S.E.

Cyromazine treatment does not affect the synthesis of chitin. The incorporation of the labelled precursor [^{14}C]-N-acetylglucosamine was unchanged either *in vivo* or *in vitro* (Table 2). This is in accord with the findings of Hughes *et al.* (1989) that chitin content in *Manduca* cuticle is unaltered by cyromazine treatment.

The cuticle's content of SDS-soluble protein in cyromazine-treated insects is unchanged relative to the larval size (Table 2). The relative proportions of the principal SDS-soluble proteins identifiable on SDS-polyacrylamide gels in treated and untreated cuticle is indistinguishable (Figure 4A). Cuticle protein synthesis is apparently unaffected since the amount of [^{3}H]-glycine incorporated into SDS-soluble cuticle proteins (Table 2), and the electrophoretic pattern of the newly synthesised proteins (Figure 4B), are unchanged by exposure to cyromazine.

These findings indicate that the major macromolecular components of the cuticle and its water content are unchanged in cyromazine-affected insects, although we cannot exclude the possibility that cuticle proteins insoluble in SDS (a minor component) may be specifically affected (methods for extracting these proteins are destructive and so we have not attempted to analyse them). The chemical nature of the change(s) within the cuticle that lead to increased stiffness and strength thus remains unknown.

Accumulation within the Cuticle of Cyromazine

One possibility that we have considered is that it is cyromazine itself which is directly responsible for the altered mechanical properties of the cuticle. Cyromazine is selectively accumulated in the cuticle following exposure to the insecticide (Hughes *et al.*, 1989; Kotze and Reynolds, 1991). This raises the possibility that cyromazine itself might act as a cross-linking agent. Any cross-links involving cyromazine must be non-covalent and relatively weak, since 99% of the [^{14}C]-cyromazine in the cuticle can be extracted in 0.5M KCl solution, with more than 99% of this being in the form of chemically unchanged cyromazine (analysis by thin layer chromatography).

It is possible that cyromazine forms ionic cross-links between cuticle proteins. The insecticide molecule's positively charged amino groups might be expected to associate reversibly with the negatively charged carboxyl side chains of the predominately acidic cuticle proteins. Such ionic interactions might constitute temporary cross-links between cuticle proteins that would increase the strength and stiffness of the cuticle matrix. Sequestration by association with

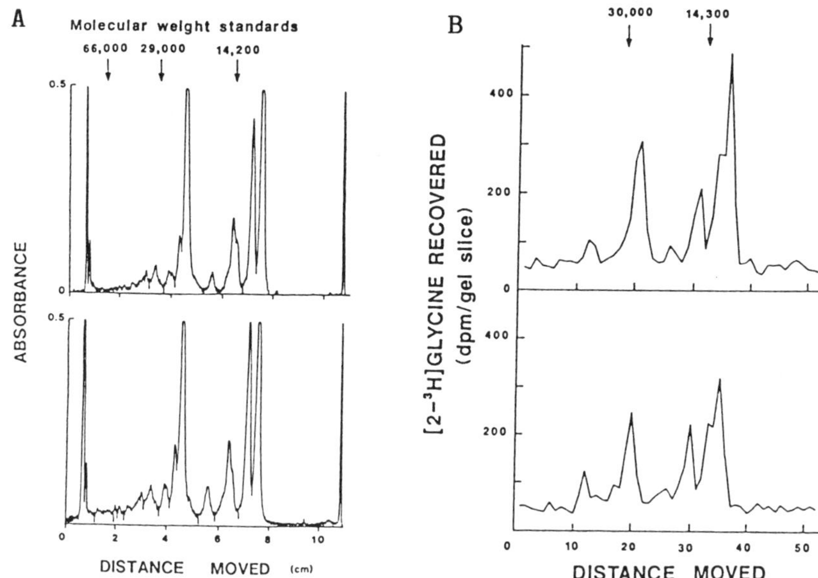

Figure 4 Cyromazine does not affect SDS-soluble cuticle proteins (**A**). Densitometric profiles of SDS-polyacrylamide gels of proteins extracted in SDS buffer from the cuticles of (*top*) control and (*bottom*) cyromazine-treated (24 hours, 50 ppm in diet) fifth stage *Manduca* larvae. (**B**). Profiles of [2³H]-glycine-derived radioactivity along SDS polyacrylamide gels of SDS-soluble proteins from (*top*) control and (*bottom*) cyromazine-treated (50 ppm in diet) fifth stage *Manduca* larvae. The insects were allowed to feed on the experimental diet for 6 hours, and were then injected with 5 uCi [2-³H]-glycine. After a further 7 hours the insects were killed and samples taken from abdominal segments 3 and 4. Distance moved along gel is described by numbering slices from 0 (origin) to 52 (dye front). (After Kotze and Reynolds, 1991.)

cuticle proteins might also explain why the insecticide accumulates in the cuticle.

It must be admitted, however, that so far attempts to alter the mechanical properties of excised cuticle loops by soaking them in cyromazine solutions have not met with success. Moreover, it is unclear why only some diamino- and triaminotriazines are insecticidal. Melamine (2,4,6-triamino-*s*-triazine) for example, which might be expected to form weak ionic cross-links of the kind discussed above, is completely non-toxic to *Manduca* larvae at 100 ppm in diet (S.E. Reynolds, unpublished).

Acknowledgements

ACK was supported by Australian Wool Corporation Research Fellowship No US21N. [^{14}C]-cyromazine was a gift from CIBA-GEIGY AG, Basel, Switzerland. We thank Jackie Rawlings for insect care.

References

BORKOVEC, A.B. and DE MILO, A.B. (1967). Insect chemosterilants. V. Derivatives of melamine. *Journal of Medicinal Chemistry*, **10**, 457–461.

DE MILO, A.B., BORKOVEC, A.B., COHEN, C.F. and ROBBINS, W.E. (1981). Larvicidal effects of substituted diamino- and triamino-s-triazines. *Agricultural and Food Chemistry*, **29**, 82–84.

FRIEDEL, T. and McCONNELL, P.A. (1985). Cyromazine inhibits reproduction and larval development of the Australian sheep blowfly (Diptera: Calliphoridae). *Journal of Economic Entomology*, **78**, 868–873.

HALL, R.D. and FOEHSE, M.C. (1980). Laboratory and field tests of CGA-72662 for control of house fly and face fly in poultry, bovine or swine manure. *Journal of Economic Entomology*, **73**, 564–569.

HUGHES, P.B., DAUTERMAN, W.C. and MOTOYAMA, N. (1989). Inhibition of growth and development of tobacco hornworm (Lepidoptera: Sphingidae) larvae by cyromazine. *Journal of Economic Entomology*, **82**, 45–51.

KOTZE, A.C. and REYNOLDS, S.E. (1990). Mechanical properties of the cuticle of *Manduca sexta* larvae treated with cyromazine. *Pesticide Biochemistry and Physiology*, **38**, 267–272.

KOTZE, A.C. and REYNOLDS, S.E. (1991). An examination of cuticle chitin and protein in cyromazine-affected *Manduca sexta* larvae. *Pesticide Biochemistry and Physiology* (in press).

MILLER, R.W., CORLEY, C., COHEN, C.F., ROBBINS, W.E. and MARKS, E.P. (1981). CGA-19255 and CGA-72662: Efficacy against flies and possible mode of action and metabolism. *The Southwest Entomologist*, **6**, 272–278.

PESSAH, I.N., MENZER, R.E. and BORKOVEC, A.B. (1985). Characterization of the biological activity of 2,4-diamino-6-(2-furyl)-s-triazine in the house fly (Diptera: Muscidae). *Annals of the*

Entomological Society of America, **78**, 873–880.

PRICE, N.R. and STUBBS, M.R. (1984). Some effects of CGA-72662 on larval development in the housefly, *Musca domestica* (L.). *International Journal of Invertebrate Reproduction and Development*, **7**, 119–126.

REYNOLDS, S.E. (1987). The cuticle, growth and moulting in insects: the essential background to the action of acylurea insecticides. *Pesticide Science*, **20**, 131–146.

REYNOLDS, S.E. and BLAKEY, J.K. (1989). Cyromazine causes decreased cuticle extensibility in larvae of the tobacco hornworm, *Manduca sexta. Pesticide Biochemistry and Physiology*, **35**, 251–258.

SLÁMA, K., ROMAŇUK, M. and ŠORM, F. (1974). *Insect hormones and bianalogues*. Springer Verlag, Vienna, Austria.

15
Novel Sulphur-containing Compounds with Juvenile Hormone Activity

ISTVÁN UJVÁRY, GYÖRGY MATOLSCY and LÁSZLÓ VARJAS

Plant Protection Institute, Hungarian Academy of Science, H-1022 Budapest, Herman Ottó út 15, Hungary

Introduction

Juvenile hormones (JHs) are homologous sesquiterpenoid derivatives involved in the regulation of gene expression during morphogenesis and reproductive development of insects (Riddiford, 1985; Palli *et al.*, 1990). The determination of the structure of the natural JHs provided novel lead compounds for the research on safer and selective insecticides resulting in several commercial juvenoids from about 15,000 of analogues synthesised during the past 25 years (for reviews, see Henrick, 1982; Wimmer and Romanňuk, 1989). Most of these compounds were designed, often using the so-called bioisosteric principle, by various structural modifications of the terpenoid chain (Figure 1). Intriguingly, however, several types of compounds with non-terpenoid like structures have also shown JH activity (Figures 2 and 3). Recently, we have found that certain nonjuvenoid-like bisisothiocyanate derivatives of α-amino acids ornithine and lysine and some related compounds show characteristic JH effect on *Manduca sexta* (Figure 3) (Ujváry *et al.*, 1989). In related studies we attempted to modify the structure of known juvenoids using the proinsecticide principle. Here we report preliminary results of these efforts aimed at finding novel JH analogues.

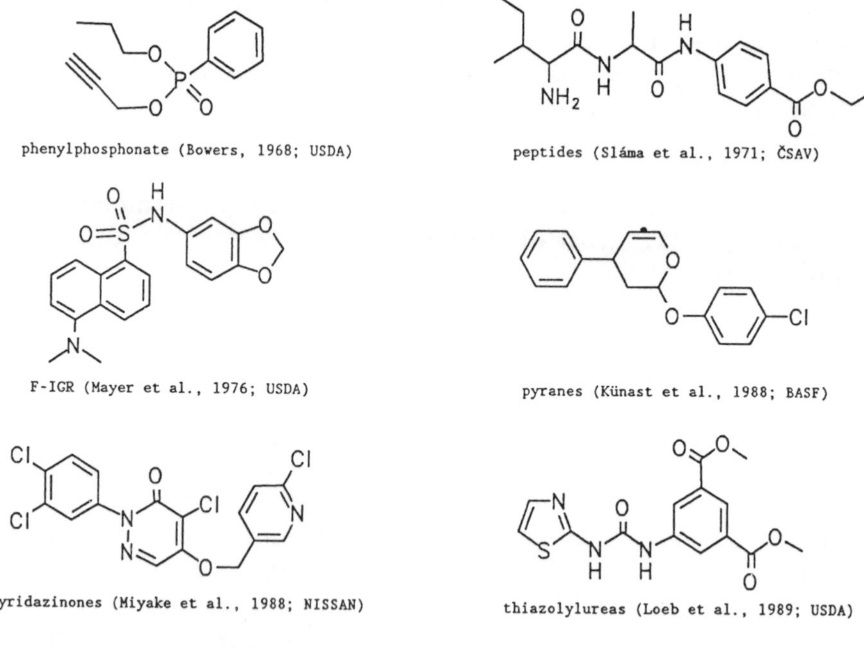

Figure 1   Typical bioisosteric modifications in juvenoids.

Figure 2   Structures of representative non-terpenoid juvenoids.

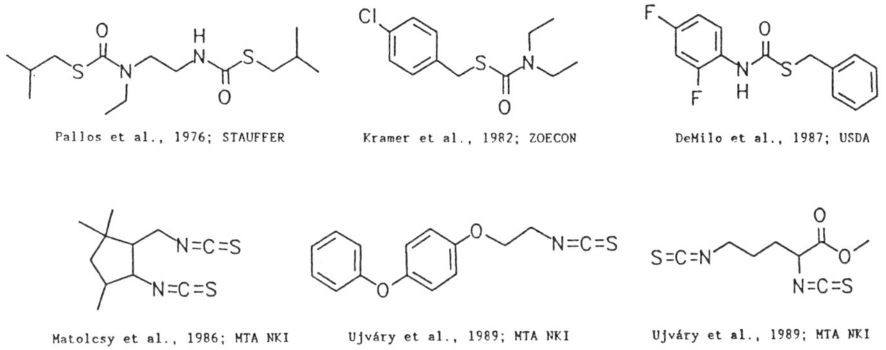

Figure 3 Structures of selected thiolcarbamate and isothiocyanate insect growth regulators

Projuvenoidal Carbamate Derivatives

The biological activity and selectivity of insecticides is basically determined by the structure of the molecule, but the intrinsic activity is usually modulated by penetration, transport and metabolism. Derivatization of an insecticide with appropriate detachable functional moieties can favourably influence the physiochemical properties of the parent molecule, thus altering its fate in biological systems. Fukuto and his co-workers have developed techniques for making proinsecticidal derivatives for carbamate neurotoxicants with improved toxicological properties (Fahmy and Fukuto, 1983; Fukuto, 1983). The most successful method of derivatization of methylcarbamates is the *N*-sulphenylation which technique has recently been used for several commercial proinsecticides with improved selectivity (Drabek and Neumann, 1985; Prestwich, 1990).

For JH analogues, Sláma and Romaňuk advanced the juvenogen concept denoting precursors, namely esters, of juvenoid alcohols that undergo biochemical activation in the insect organism (Sláma and Romanňuk, 1976).

In order to obtain novel juvenoid derivatives with modified physiochemical and biological properties we adopted the above-mentioned proinsecticide principle, used earlier solely for various neurotoxic carbamates. For our studies, the highly active carbamate type JH mimic fenoxycarb (Dorn *et al.*, 1981) appeared as an attractive model compound to which derivatization techniques could be applied.

Results and Discussion

Chemistry

The symmetrical and asymmetrical sulphur-containing carbamates derivatives (**1–5**) were synthesized as shown in Figure 4. The adaptation of known methods (Brown and Kohn, 1974; Fahmy *et al.*, 1978) allowed the synthesis of the sulphenyl and sulphinyl derivatives **1, 4, 5** and **2** respectively. The sulphonyl biscarbamate **3** was more conveniently prepared by heating the appropriate amine and sulfamide (McDermott and Spillane, 1984) to give the substituted sulfamide derivative which, in turn, was acylated with ethyl chloroformate affording the corresponding final product.

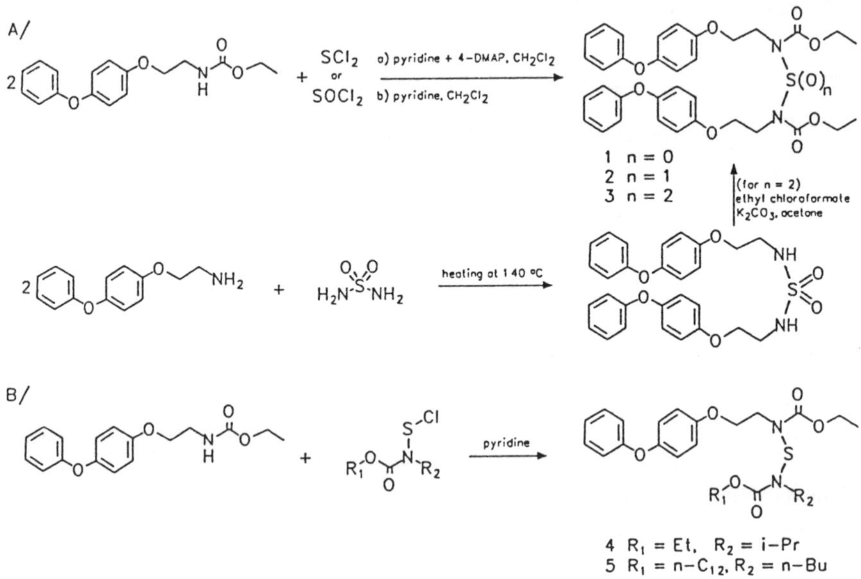

Figure 4 Synthesis of sulphur-containing carbamate derivatives

Determination of juvenile hormone activity

Inhibition of metamorphosis of the large white butterfly
Pieris brassicae *L. (Lepidoptera: Pieridae)*
The test compounds were dissolved in acetone and topically applied to 24-hour old last instar (L_5) larvae of large white butterflies reared on potted greenhouse-grown cabbage plants and maintained under laboratory conditions in an 18L:6D photoperiod at 25°C. The larval-pupal intermediates were classified according to a scoring system (0 = normal pupa; 4 = larval form with pupa-like organs on the head).

The morphogenetic activities of the compounds are expressed as ID_{50} values (50% inhibiting dose) (Table 1). The results show that the JH activity of the symmetrical sulphenyl and sulphinyl compounds (**1** and **2** respectively) and of the lipophilic asymmetrical derivative **5** improved seven- to eightfold with regard to fenoxycarb. The sulphonyl derivative **3**, however, is a poor JH mimic ($ID_{50} > 0.1$ µ/larva) indicating that the $N\text{-}SO_2\text{-}N$ system is too stable to be cleaved under physiological conditions.

Table 1 Morhogenetic activity of juvenoids on larvae of the large white butterfly, *Pieris brassicae* L.

Compound	ID_{50}[a] µg/specimen	Relative activity[b]
Fenoxycarb	0.021	1.0
1	0.0028	7.5
2	0.0026	8.1
3	>0.1	<0.2
4	0.041	0.5
5	0.0028	7.5

[a] The average inhibiting dose causing intermediate morphological disturbances (score = 2) according to the scoring system where: score 0 = normal pupa and score 4 = larval form with pupa-like organs on head. Each dose was tested on 25-50 of 24 hour-old last instar larvae, and the ID_{50} values were calculated by linear regression of log dose / average score relationship.

[b] The morphogenetic activity of fenoxycarb is taken as 1.0; the calculation is based on the ED_{50} values of the compounds.

Summary

Our studies demonstrate that application of the proinsecticide derivatization technique to the carbamate-type juvenoid fenoxycarb can result in a substantial increase in JH activity. The highest JH activity was observed for symmetrical biscarbamates with a sulphenyl or sulphinyl bridge and sulphenylated derivatives containing lipophilic carbamates as derivatizing agent. The increased activity of the derivatives can be attributed to their higher lipophilicity affecting translocation in the insect organism, thus facilitating its movement through membranes to the site of action. However, the structural modifications can also introduce the 'delay factor', as suggested by Fukuto for neurotoxic procarbamates, and protect the parent compound from fast metabolism.

Acknowledgements

This work was performed within the framework of the UNDP/ UNIDO sponsored projects HUN/82006 and HUN/86/006. The authors are grateful to Drs I. Bélai and F. Szurdoki for the synthesis of some of the compounds. Thanks are also due to Professor P. Sohár and Mrs K. Bauer for spectral data.

References

BOWERS, W.S. (1968). Juvenile hormone: Activity of natural and synthetic synergists. *Nature*, **161**, 895–897.

BROWN, M.S. and KOHN, G.K. (1974). *N*-Chlorothio carbamates. *US Patent* 3,843,689.

DEMILO, A.B., HAUGHT, S.B. and KELLY, T.J. (1987). Nonterpenoid *S*-benzyl thiocarbamates with juvenile hormone-like activity. In *Synthesis and Chemistry of Agrochemicals*. (D.R. Baker, J.G. Fenyes, W.K. Moberg and B. Cross, eds), ACS Symposium Series No **355**, pp. 260–272. American Chemical Society, Washington DC.

DORN, S., FRISCHKNECHT, M.L., MARTINEZ, V., ZUR-FLÜH, R. and FISCHER, U. (1981). A novel non-neurotoxic insecticide with a broad activity spectrum. *Zeitschrift für Pflanzenkrankheiten und Pflanzenschutz*, **88**, 269–275.

DRABEK, J. and NEUMANN, R. (1985). Proinsecticides. In *Insecticides* (D.H. Hutson and T.R. Roberts, eds), *Progress in Pesticide Biochemistry and Toxicology* Volume **5**, pp. 35–86. John Wiley and Sons, Chichester.

FAHMY, M.A., MALLIPUDI, N.M. and FUKUTO, T.R. (1978). Selective toxicity of *N,N'*-thiocarbamates. *Journal of Agricultural and Food Chemistry*, **26**, 550–557.

FAHMY, M.A. and FUKUTO, T.R. (1983). Rationale and chemistry of proinsecticidal methylcarbamates. In *Pesticide Chemistry: Human Welfare and the Environment* (J. Miyamoto and P.C. Kearney, eds), Volume **1**, pp. 193–200. Pergamon Press, Oxford.

FUKUTO, T.R. (1983). Structure-activity relationships in derivatives of anticholinesterase insecticides. In *Pesticide Chemistry: Human Welfare and the Environment* (J. Miyamoto and P.C. Kearney, eds), Volume **1**, pp. 203–212. Pergamon Press, Oxford.

HENRICK, C.A., STALL, G.B. and SIDDALL, J.B. (1973). Alkyl, 3,7,11-trimethyl-2,4-dodecadienoates a new class of potent insect regulators with juvenile hormone activity. *Journal of Agricultural and Food Chemistry*, **21**, 354–359.

HENRICK, C.A. (1982). Juvenile hormone analogs: Structure-activity relationships. In *Insecticide Mode of Action* (J.R. Coats, ed), pp. 315–402. Academic Press, New York.

KISIDA, H., NISHIDA, S., HATAKOSHI, M. and MATSUO, N. (1990). Pyriproxifen: A novel insect growth regulator. Poster No. 01A-21, presented at the Seventh International Congress of Pesticide Chemistry, Hamburg, August 5–10, 1990. *Book of Abstracts*, Volume **1**, p. 33.

KÜNAST, C., HIMMELE, W. and THEOBALD, H. (1988). Pyran derivatives: A new class of compounds inducing morphogenetic aberrations in the cotton stainer (*Dysdercus intermedius* Sign.). *Zeitschrift für Pflanzenkrankheiten und Pflanzenschutz*, **95**, 285–291.

LOEB, M.J., DEMILO, A.B., JONES, G. and JONES, D. (1989). Juvenile hormone-like effects of a thiazolylurea on lepidopteran larvae. *Archives of Insect Biochemistry and Physiology*, **12**, 79–88.

MATOLCSY, GY., UJVÁRY, I., RIDDIFORD, L.M. and HIRUMA, K. (1986). Alkylene-bis-isothiocyanates: Novel insect growth regulators. *Zeitschrift für Naturforschung*, **41c**, 1069–1072.

MAYERS, R.T., BRIDGES, A.C., COCKE, J., MEOLA, R. and OLSON, J.K. (1976). Fluorescent insect growth regulators (FIGRs): A new tool for insect physiologists. *Journal of Insect Physiology*, **22**, 515–520.

McDERMOTT, S.D. and SPILLANE, W.J. (1984). Synthesis and reactions of sulfamides. A review. *Organic Preparations and Procedures International*, **16**, 49–77.

MIYAKE, T., KUDO, M., UMEHJARA, Y., HIRATA, K., KAWAMURA, Y. and OGURA, T. (1988). NC-170, a new compound inhibiting the development of leafhoppers and planthoppers. *Proceedings of the Brighton Crop Protection Conference: Pest and Diseases.* **5-5**, pp. 535–542.

NIWA, A., IWAMURA, H., NAGAKAWA, Y. and FUJITA, T. (1990). Development of *N,O*-disubstituted hydroxylamines and *N,N*-disubstituted amines as insect juvenile hormone mimetics and the role of the nitrogenous function for activity. *Journal of Agricultural and Food Chemistry*, **38**, 514–520.

PALLI, S.R., OSIR, E.O., ENG., W.-S., BOEHM, M.F., ED-WARDS, M., KULCSÁR, P., UJVÁRY, I., HIRUMA, K., PRESTWICH, G.D. and RIDDIFORD, L.M. (1990). Juvenile hormone receptors in insect larval epidermis: Identification by photoaffinity labeling. *Proceedings of the National Academy of Sciences, USA*, **87**, 796–800.

PALLOS, F.M., LETCHWORTH, P.E. and MENN, J.J. (1976). Novel nonterpenoid insect growth regulators. *Journal of Agricultural and Food Chemistry*, **24**, 218–221.

PODUŠKA, K., ŠORM, F. and SLÁMA, K. (1971). Natural and synthetic materials with insect hormone activity. 9. Structure-juvenile activity relationships in simple peptides. *Zeitschrift für Naturforschung*, **26b**,719–722.

PRESTWICH, G.D. (1990). Pro-insecticides: Metabolically activated toxicants. In *Safer Insecticides: Development and Use* (E. Hodgson and R.J. Kuhr, eds), pp. 281–335. Marcel Dekker Inc., New York.

RIDDIFORD, L.M. (1985). Hormone action in the cellular level. In *Comprehensive Insect Physiology, Biochemistry and Pharmacology* (G.A. Kerkut and L.I. Gilbert, eds), Volume **8**, pp. 37–84. Pergamon Press, Oxford.

SLÁMA, K. and ROMAŇNUK, M. (1976). Juvenogens, biochemically activated juvenoid complexes. *Insect Biochemistry*, **6**, 579–586.

UJVÁRY, I., KIS-TAMÁS, A., VARJAS, L. and NOVÁK, L. (1983). Syntheses of silicon-containing juvenile hormone analogues. *Acta Chimica Hungarica*, **113**, 165–175.

UJVÁRY, I., MATOLCSY, GY., RIDDIFORD, L.M., HIRUMA, K. and HORWATH, K.L. (1989). Inhibition of spiracle and crochet formation and juvenile hormone activity of isothiocyanate derivatives in the tobacco hornworm, *Manduca sexta. Pesticide Biochemistry and Physiology*, **35**, 259–274.

WIMMER, Z. and ROMAŇNUK, M. (1989). Insect hormones and their bioanalogues. *Collection of Czechoslovak Chemical Communications*, **54**, 2302–2329.

16
Prenyl Imidazoles and Related Compounds Controlling Hormonal Development Processes of Insects

M. ETO AND K. KUWANO

Department of Agricultural Chemistry, Kyushu University, Fukuoka 812, Japan

Introduction

The growth and development of insects are under the control of several hormones. Ecdysteroids, or moulting hormones, secreted from the prothoracic gland are controlled by the peptide prothoracicotropic hormone and are responsible for cellular programming, co-operating with the sesquiterpenoid juvenile hormones (JH) secreted from the corpora allata. When the amount of JH secreted from the corpora allata is high, the epidermis is programmed for a larval moult. When the JH level declines, the cells are programmed for metamorphosis. JH is virtually absent in the pupae, but present in the adult in which it has some functions relating to ovarian development. Disturbing the normal hormone balance may cause crucial disorder in the growth and development of insects.

Organofluorine Compounds as Anti-JH Agents

We attempted to find anti-JH compounds which interfere with JH biosynthesis. Juvenile hormones are biosynthesized through the mevalonate pathways. Compactin inhibits hydroxymethylglutaryl-CoA reductase. Fluoromevalonolactone (FMev) blocks the JH-biosynthesis by inhibition of not only mevalonate kinase but also pyrophosphomevalonate decarboxylase *in vivo* (Quistad *et al.*, 1981: Nave *et al.*, 1985).

We prepared some derivatives of 3-fluoropropionic acid and related compounds including fluorohomomevalonolactone, hoping to block JH-biosynthesis in insects. However, none of them produced any distinctive anti-JH activity on the silkworm *Bombyx mori* larvae (Kuwano *et al.*, 1988). Although homomevalonate is a precursor of JH 0, I, II and iso-JH 0, its fluorinated lactone did not show anti-JH activity. We prepared both the (R)(-)- and (S)(+)- enantiomers of FMev with 80.4% and 87.4% e.e. respectively, and assayed their biological activities by topical application on fourth instar larvae of the silkworm (Shuto *et al.*, 1988). (S)-FMev was completely inactive, whereas only the (R)- enantiomer with the same configuration as natural mevalonate was active as an anti-JH agent. It induced precocious pupation with an ED_{50} of 100 µg/larva. At doses higher than 300 µg, lethal effects appeared. The inhibition of enzymes participating in the early part of the mevalonate pathways, which are common to the biosyntheses of many biologically important substances, may cause undesirable effects to non-target organisms.

Activities of Imidazole Derivatives

Blocking a final step in the biosynthetic pathways of JH is most preferable with regard to selective toxicity. Since cytochrome P 450 catalyzes epoxidation in the final step of JH-biosynthesis, combining the partial structures of terpenes and cytochrome P 450 inhibitors in a molecule is a rational strategy to design anti-JH compounds. We employed substituted imidazoles as a cytochrome P 450 inhibitive moiety.

Anti-JH activity inducing precocious pupation in *B. mori* larvae was found in imidazole derivatives containing combinations of a prenyl group and an aromatic group at 1- and 5-positions as shown in Figure 1. Complete inactivity was observed, however, in 1,2- and 1,4-disubstituted imidazoles. Thus, 1-citronellyl-5-phenylimidazole (KK-22), 1-benzyl-5-[(E)-2,6-dimethyl-1,5-heptadienyl]imidazole (KK-42) and some others were found to have potent anti-JH activity (Kuwano *et al.*, 1983, 1984, 1985; Kuwano and Eto, 1986). KK-42 is the most potent compound in the KK-compound series and was more than 100 times as active as (R)-FMev, inducing precocious pupation in fourth instar *B. mori* larvae upon topical application, with an ED_{50} value of less than 1 µg/larva. Precocious metamorphosis also occurred directly from the third instar larvae but few adults emerged from such precocious pupae.

	Structure	ED_{50}, µg/larva
KK-40	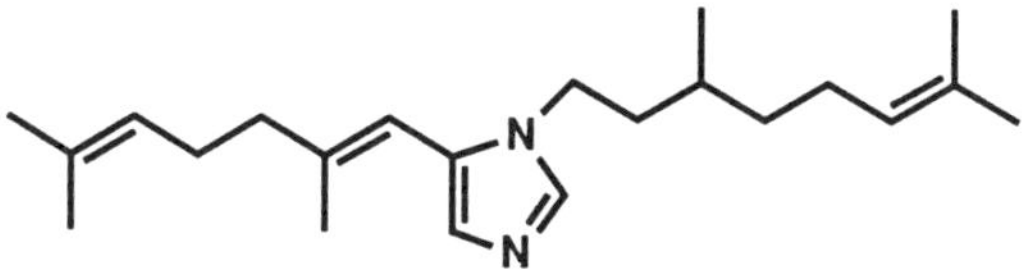	35.7
KK-22	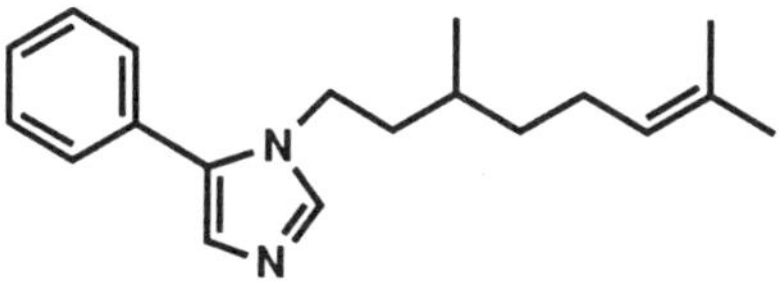	2.8
KK-42	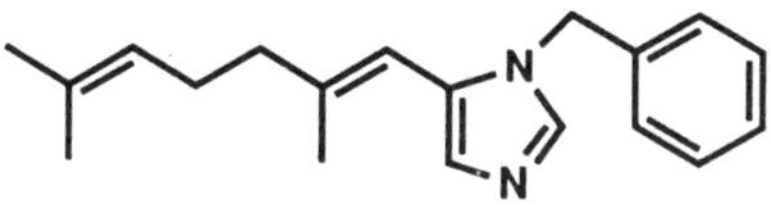	0.4
KK-110	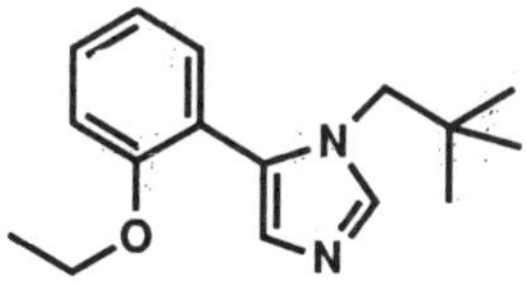	1.6
KK-135	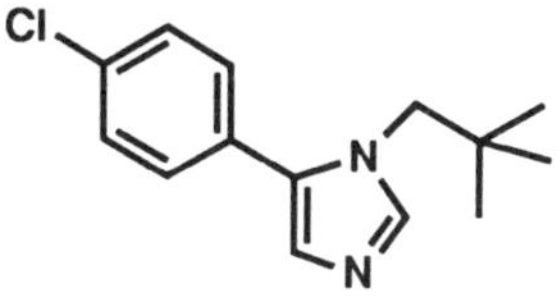	2.1
$(R)(-)$-FMev	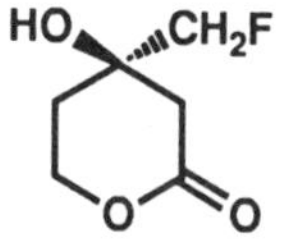	100
$(S)(+)$-FMev	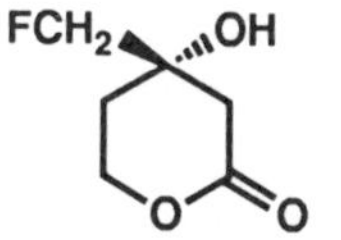	>1000 (no activity)

Figure 1 Induction of precocious pupation in fourth instar larvae of the silkworm (*Bombyx mori*).

1,5 Diphenyl imidazoles such as KK-40 resemble terpenoids more than do KK-22 and 42, but are much less active than the aromatic analogues. The aromatic ring appears to be recognized by some enzymes as identical with a part of the prenyl group. Proper replacement of the partial skeleton of bioactive terpenoids by an aromatic ring often increases activity. This effect may be due to the ring interacting with the target site and stabilizing the molecule against metabolic and chemical degradation. Moreover, the prenyl group can be replaced by a neoptyl group to form the active compound such as 1-neopentyl-5-(2-ethoxyphenyl)imidazole (KK-110) and its 4-chlorophenyl analogue(KK-135) (Kuwano *et al.*, 1988).

The imidazole derivatives terminated the pharate larval diapause in the silkmoth *Antheraea yamamai* eggs (Suzuki *et al.*, 1990; Kuwano *et al.*, 1991). The diapause occurs after the development of the embryo into the first instar larva in the egg. Although larval diapause in some insects has been said to be correlated with a high JH titre, the mechanism of the action of KK compounds is not yet known. The diapause of silkmoth eggs has not yet been broken artificially except by chilling the egg for a few months, a finding which has provided a convenient artificial hatching method. Silk derived from silkmoth cocoons is more valuable than that of *B. mori*.

Mode of Action of Imidazole Derivatives

The anti-JH effect of KK compounds of inducing the precocious pupation of silkworm larvae was counteracted by the simultaneous application of JH 1 or methoprene but not of JH III (Kuwano *et al.*, 1985). 2 µg per larva of KK-42 induced precocious pupation in 100% of the fourth instar larvae, while all of the larvae moulted into normal fifth instar when 10 µg of methoprene was topically applied. The 50% effective doses of methoprene and JH I suppressing the activity of KK-42 at a dose of 2 µg were 1.4 and 5.3 µg/larva respectively. This was also the case for the neopentyl imidazole derivatives such as KK-110 (Kuwano *et al.*, 1990). These results indicate that the precocious pupation induced by KK compounds is caused by JH deficiency.

Platt *et al.* (1990) found that the imidazole derivatives inhibit the *in vitro* biosynthesis of JH III in the cockroach *Dipoloptera punctata* corpora allata. 1-Isobutyl-5-(4-phenoxyphenyl) (KK-98) inhibited 50% the epoxidation of methyl farnesoate (MF) at 80 nM. However, the inhibitory activity is not always correlative with the activity to induce precocious pupation in *B. mori* larvae as shown in Figure 2;

KK-98, the most active MF epoxidase inhibitor in this series, was almost unable to induce precocious pupation, whereas KK-42, the most active precocious pupation inducer, was relatively poor in epoxidase inhibition. This discrepancy may be due to differences in the insect species used, transportation, and metabolism.

	Structure	JH synthesis, $IC_{50}(\mu M)$	$ED_{50,}$ $\mu g/larva$
KK-42		3.5	0.6
KK-83		1.5	1.9
KK-98		0.08	>80 (0)

Figure 2 Activities for the inhibition of JH III synthesis and induction of precicious pupation.

Precocious pupation can be induced by allatectomy at the time of the third ecdysis. The critical period (60 hours) for KK-42 induction of precocious pupation was longer than that for allatectomy (0 hours) in the N_4 race silkworm (Kadono-Okuda *et al.*, 1987a). On the other hand, KK compounds caused developmental disturbances in pupae (Kadono-Okuda *et al.*, 1987b). KK-42 delayed the pigmentation of compound eyes to correlate with adult emergence. Figure 3 shows that the ability of KK compounds to delay the eye pigmentation in pupal-adult development is well correlated with the activity to induce the precocious pupation. Ovarian growth was also suppressed by KK-42 treatment at pupation. These developmental disturbances in the pupal stage appear difficult to explain as anti-JH effects. The delayed effects of KK-42 on pupal-adult development, eye pigmentation, and ovarian growth were not counteracted by treatment with methoprene, but were reversed by 20-hydroxyecdysone.

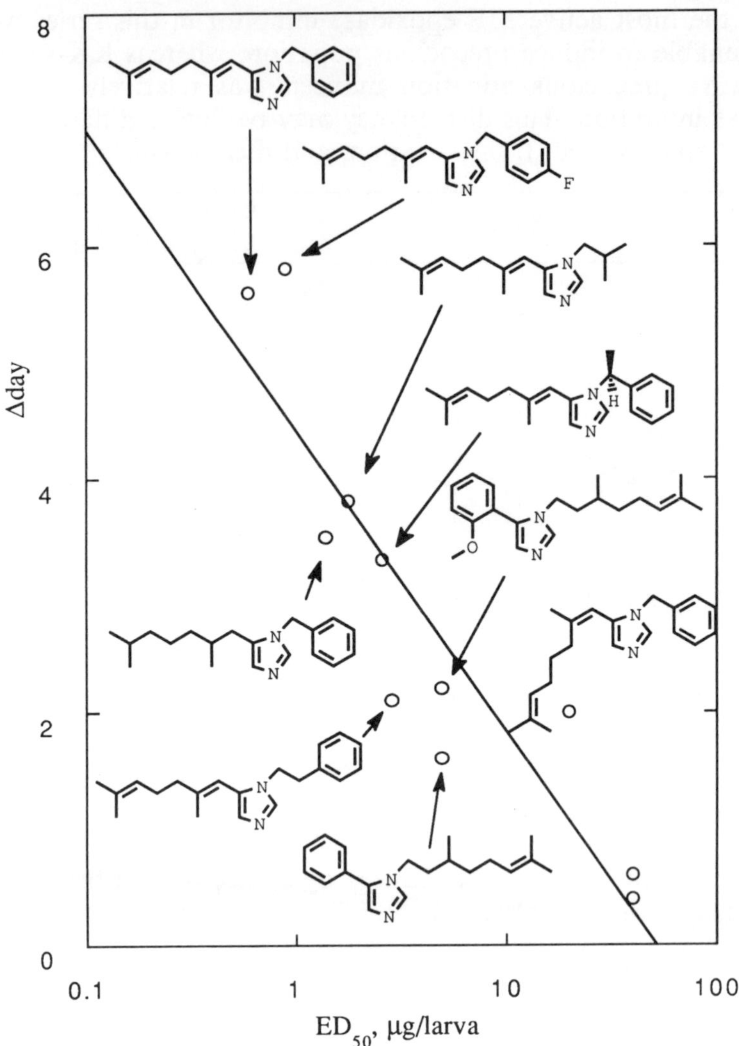

Figure 3 Correlation between activities to induce precocious pupation and to delay pupal-adult development of silkworm ($r = 0.907$). (ED_{50} = topical dose in μg per fourth instar larva which induces 50% precocious pupation; day = delay time (days) for eye pigmentation of pupae at a dose of 50 μg per pupa.)

KK-42 inhibited the increase of ecdysteroid titres in pupal haemolymph upon topical application (Kadono-Okuda *et al.*, 1987b). The ecdysteroid titre rapidly increases, then decreases in the first few days after larval-pupal ecdysis in normal pupae, whereas in the pupae treated with KK-42 such change in ecdysteroid titre occurs only after one week. A dose of 10 µg KK-42 per pupa suppressed completely the increase of ecdysteroid titre assayed on day 2, when the ecdysteroid titre attained the maximum level in the control. The 50% effective dose of KK-42 was estimated to be about 1 µg per pupa.

Similar suppression of ecdysteroid titre by KK-42 was also observed in larval stages (Kadono-Okuda *et al.*, 1987a). In the control, the ecdysteroid titre peak was found on day 3 of the fourth instar. In the fifth instar, a trace level was maintained for a few days before a sharp increase occurred after the onset of cocoon spinning. In KK-42 treated larvae, the ecdysteroids level remained low for one week before the titre increased toward cocoon spinning, resulting in precocious pupation.

In vitro experiments using prothoracic glands prepared from mature silkworm larvae, KK-42 inhibited ecdysone synthesis in the glands (Yamashita *et al.*, 1987). The 50% effective concentration was estimated to be about 0.5 nM. The inhibition of ecdysteroid biosynthesis by prothoracic glands was also confirmed in *Locusta migratoria* (Roussel *et al.*, 1987). Furthermore, 20-hydroxyecdysone rendered KK-42 ineffective in inducing precocious pupation (Kiuchi, personal communication).

Certain oxygenation steps catalyzed by cytochrome P 450 in ecdysteroid biosynthesis pathways from phytosteroids in insects may be inhibited by the imidazole derivatives. In conclusion, KK-compounds cause a variety of disturbances in the growth and development of insects by inhibiting the biosyntheses of both JH and ecdysteroids.

References

KADONO-OKUDA, K., KUWANO, E., ETO, M. and YAMASH-ITA, O. (1987a). Inhibitory action of an imidazole compound on ecdysone synthesis in prothoracic glands of the silkworm, *Bombyx mori. Development Growth and Differentiation*, **29**, 527–533.

KADONO-OKUDA, K., KUWANO, E., ETO, M. and YAMA-SHITA, O. (1987b). Anti-ecdysteroid action of some imidazole derivatives on pupal-adult development of the silkworm, *Bombyx mori. Applied Entomology and Zoology*, **22**, 370–379.

KUWANO, E. and ETO, M. (1986). Anti-juvenile hormone effects of an imidazole compound (KK-42) in different larval instars of *Bombyx mori. Agricultural and Biological Chemistry*, **50**, 2919–2920.

KUWANO, E., KIKUCHI, M. and ETO, M. (1988). Induction of

precocious metamorphosis by 1,5-disubstituted imidazoles with a non-terpene chain. *Agricultural and Biological Chemistry*, **52**, 1619–1620.

KUWANO, E., KIKUCHI, M. and ETO, M. (1990). Synthesis and insect growth regulatory activity of 1-neopentyl-5-substituted imidazoles. *Journal of the Faculty of Agriculture Kyushu University*, **35**, 35–41.

KUWANO, E., TAKEYA, R. and ETO, M. (1983). Terpenoid imidazoles: New anti-juvenile hormones. *Agricultural and Biological Chemistry*, **47**, 921–923.

KUWANO, E., TAKEYA, R. and ETO, M. (1984). Synthesis and anti-juvenile hormone activity of 1-citronellyl-5-substituted imidazoles. *Agricultural and Biological Chemistry*, **48**, 3115–3119.

KUWANO, E., TAKEYA, R. and ETO, M. (1985). Synthesis and anti-juvenile hormone activity of 1-substituted-5-[(E)-2,6-dimethyl-1,5-heptadienyl]imidazoles. *Agricultural and Biological Chemistry*, **49**, 483–486.

KUWANO, E., YAMASHITA, T., OTA, H. and ETO, M. (1988). Synthesis and biological activities of 3-fluoropropionic acid derivatives and related compounds. *Journal of the Faculty of Agriculture Kyushu University*, **32**, 147–156.

KUWANO, E., FUJISAWA, T., SUZUKI, K. and ETO, M. (1991). Termination of egg diapause by imidazoles in the silkmoth, *Antheraea yamamai*. *Agricultural and Biological Chemistry* (in press).

NAVE, J.F., D'ORCHYMONT, H., DUCEP, J.B., PIRIOU, F. and JUNG, M.J. (1985). Mechanism of the inhibition of cholesterol biosynthesis by 6-fluoromevalonate. *Biochemical Journal*, **227**, 247–254.

PLATT, G.E., KUWANO, E., FARNSWORTH, D.E. and FEYEREISEN, R. (1990). Structure/activity studies on 1,5-disubstituted imidazoles as inhibitors of juvenile hormone biosynthesis in isolated corpora allata of the cockroach *Diploptera punctata*. *Pesticide Biochemistry and Physiology*, **38**, 223–230.

QUISTAD, G.B., CERF, D.C., SCHOOLEY, D.A. and STAAL, G.B. (1981). Fluoromevalonate acts as an inhibitor of insect juvenile hormone biosynthesis. *Nature*, **289**, 176–177.

ROUSSEL, J.-P., KIKUCHI, M., AKAI, H. and KUWANO, E. (1987). Effets d'un dérivé de l'imidazole (KK-42) sur la biosynthèse de l'ecdysone par les glandes prothoraciques de Locusta migratoria, *in vitro*. *Competes Rendes Hebdomadaires des Séances de l'Academie des Science, Paris*, **305**, Série III, 141–144.

SHUTO, A., KUWANO, E. and ETO, M. (1988). Synthesis of two optical isomers of the insect and anti-juvenile hormone agent fluoromevalonolactone and their biological activities. *Agricultural and Biological Chemistry*, **52**, 915–919.

SUZUKI, K., MINAGAWA, T., KUMAGAI, T., HAYA, S., ENDO, Y., OSANAI, M. and KUWANO, E. (1990). Control mechanism of diapause of the pharate first-instar larvae of the silkmoth *Antheraea yamamai*. *Journal of Insect Physiology*, **36** 855–860.

YAMASHITA, O., KADONO-OKUDA, K., KUWANO, E., and ETO, M. (1987). An imidazole compound as a potent anti-ecdysteroid in an insect. *Agricultural and Biological Chemistry*, **51**, 2295–2297.

17
Effects of Broad-spectrum Antiviral Agents on Hemimetabolous Insects

I. GELBIČ and J. ŠULA

Czechoslovak Academy of Sciences, Institute of Entomology, Branišovská 31, 370 05 České Budějovice, Czechoslovakia

Introduction

In 1975 Holý prepared a new type of nucleoside analogue – (R,S)-9-(2,3-dihydroxypropyl)adenine (DHPA) – that has since been recognized as a broad-spectrum antiviral agent (DeClercq *et al.*, 1978; DeClercq and Holý, 1979). Later this compound was found to exhibit many biological activities. In mice (DeClercq *et al.*, 1981) the application of DHPA in drinking water caused a testicular aplasia accompanied by sterility. However, the effects were reversible, because the fertility was fully restored 5–6 weeks after the treatment. Sterilizing effects in insects were observed in adults of the bugs *Pyrrhocoris apterus* L. and *Dysdercus cingulatus* Fabr. (Sláma *et al.*, 1983) and in viviparous apterous females of *Aphis fabae* Scopoli (Gelbič *et al.*, 1984). Morphogenic abnormalities were described in the last larval instar of *Spodoptera littoralis* (Boisd.) by Gelbič and Holý (1985). Sterility induced in the bugs *P. apterus* and *D. cingulatus* was either partial or complete, depending on the dose of DHPA.

In this paper, a comparison of the effects of three different adenine analogues in some representatives of hemimetabolous insects is made, laying stress on their sterilizing effects in the development of ovaries and in changes of some enzyme activities.

Insecticides: Mechanism of Action and Resistance
© 1992   Intercept Ltd, P.O. Box 716, Andover, Hants SP10 1YG, UK

Material and Methods

Laboratory colonies of *P. apterus*, *D. cingulatus* and *A. fabae* were reared as described by Šula *et al.* (1989), Gelbič and Švec (1988) and Gelbič *et al.* (1984) respectively. The tested compounds, (R,S)-9-(2,3-dihydroxypropyl)adenine (DHPA), D-eritadenine and 9-(phosphonylmethoxyethyl)adenine (PMEA) (Figure 1), were prepared in the Institute of Organic Chemistry and Biochemistry of the Czechóslovak Academy of Sciences by Holý according to a previously described procedure (Holý *et al.*, 1982).

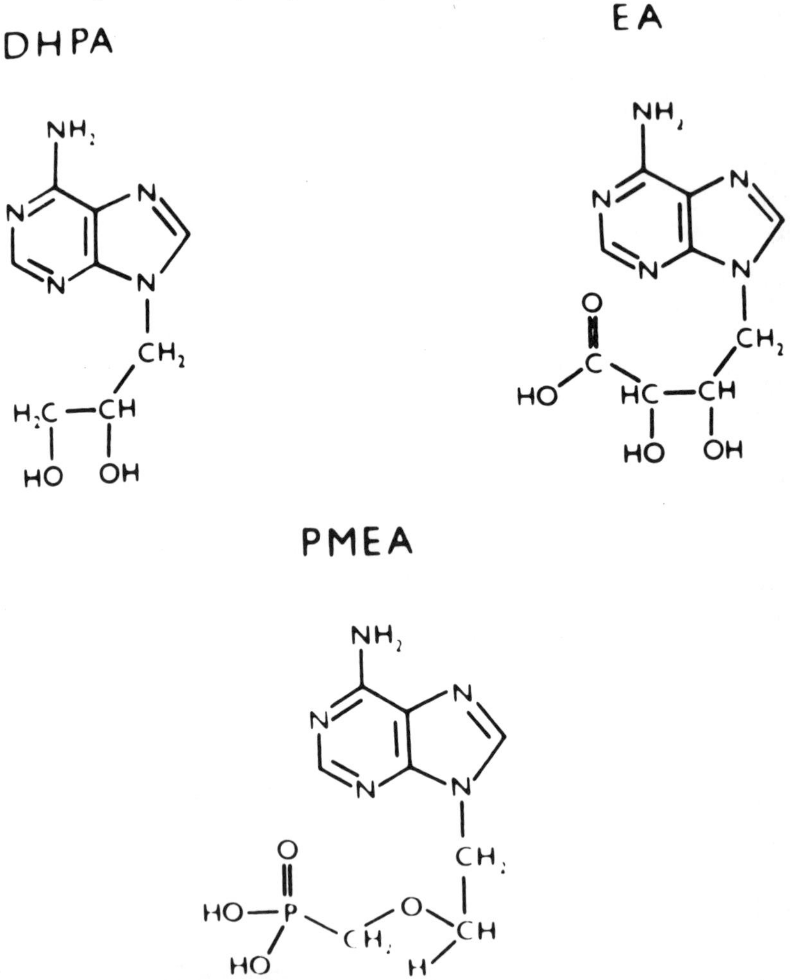

DHPA

EA

PMEA

Figure 1 Structure of tested compounds. DHPA = (R,S)-9-(2,3-dihydroxypropyl) adenine; EA = D-eritadenine; PMEA = 9-(phosphonylmethoxyethyl)adenine.

Experimental animals – *P. apterus* and *D. cingulatus* – were kept in Petri dishes (diameter 10cm). Aqueous solutions of the tested compounds were administered perorally to bugs in the following concentrations: 0.001%, 0.005%, 0.01%, 0.05%, 0.1% and 0.5%. In experiments with *A. fabae* DHPA was administered via their host plants (*Vicia faba*), which were grown in 0.1% and 0.01% DHPA solutions. The mortality, the number of eggs laid (or nymphs in viviparous aphids) and their hatchability were recorded. 2-, 4-, 6-, 8-, 10- and 12-days-old females of bugs were used for histological and electrophoretical analysis of ovaries. The dissected ovaries fixed in Bouin's solution (Dubosque-Brasil modification) were embedded in Paraplast after dehydration in ethanol. 5 µm sections were stained with Mayer's haematoxylin and additionally with eosine.

SDS-PAGE
Electrophoresis under denaturing conditions was performed according to Laemmli (1970) with 5–10% gradient slab gels 0.7mm thick.

Phosphatase activity
Acid phosphatase activity was determined using p-nitrophenylphosphate as the substrate (resulting concentration 5mM). The reaction proceeded in 0.05M acetate buffer pH 5.0, 4.2 and 3.5 for haemolymph, fat body and ovaries respectively. After 30 minutes the reaction was incubated at 30°C 0.1M NaOH and absorbance was read at 405 nm.

Alkaline phosphatase activity was also determined with p-nitrophenylphosphate as a substrate (5mM concentration) and at pH 8.4 for haemolymph and fat body and at 8.6 for ovaries. Ammediol buffer (0.05M) containing magnesium chloride (1mg 10^{-1}ml) was used. After 30 minutes reaction was incubated at 30°C 0.1M NaOH and the absorbance was read at 405 nm.

Protein determination
The quantity of proteins was measured after the method of Bradford according to Stoscheck (1990) using bovine serum albumin as a standard.

Results

Effect of nucleoside analogues on the reproduction of the hemimetabolous insects

P. apterus and *D. cingulatus* have two ovaries of the telotrophic type.

In both species the first oviposition occurs on day 7 or 8 after adult emergence. Further batches follow at 2- to 4-day intervals. The total number of eggs from the first three batches and their hatchability in control and treated females are given in Table 1.

Table 1   The effects of orally applied (RS)-DHPA, EA[*] and PMEA on the fecundity of *Pyrrhocoris apterus* L. and *Dysdercus cingulatus* Febr. females (single couples)

Compound	Concentration (%)	*Pyrrhocoris apterus*		*Dysdercus cingulatur*	
		Number of laid eggs/♀	Number of hatched eggs/♀	Number of laid eggs/♀	Number of hatched eggs/♀
Untreated control		171.1 ± 14.5[+]	148.8 ± 15.9	178.9 ± 11.4	161.8 ± 12.3
(RS)-DHPA	0.001	167.8 ± 12.5	149.3 ± 14.6	180.1 ± 12.4	160.3 ± 11.7
	0.01	141.4 ± 17.5	34.9 ± 7.4	161.5 ± 11.8	64.8 ± 14.6
	0.05	125.6 ± 14.8	0	116.8 ± 14.7	1.7 ± 2.1[x]
	0.1	60.7 ± 14.2	0	0	0
	0.5	0	0	0	0
EA	0.001	168.9 ± 13.6	151.6 ± 14.1	169.9 ± 17.6	161.1 ± 14.8
	0.005	87.8 ± 16.7	59.8 ± 9.7	72.3 ± 11.8	43.6 ± 9.7
	0.01	53.7 ± 10.6	32.1 ± 8.7	38.6 ± 9.4	7.6 ± 4.7
	0.05	33.7 ± 9.6	8.7 ± 4.7	0	0
	0.1	0	0	0	0
PMEA	0.001	163.8 ± 14.7	148.6 ± 17.1	180.1 ± 18.3	159.9 ± 16.4
	0.0005	157.6 ± 11.1	134.6 ± 13.3	165.1 ± 11.6	143.3 ± 13.6
	0.001	122.1 ± 10.9	101.8 ± 11.9	115.7 ± 14.1	98.1 ± 11.8
	0.005	98.1 ± 10.6	65.1 ± 8.9	47.8 ± 12.3	24.3 ± 11.4
	0.01	47.5 ± 6.9	7.8 ± 4.5	0	0
	0.05	0	0	0	0

[+] – Number of eggs from first three batches (in each group 10 females were used)
[x] – statistically significant at 5% level
[*] – D-eritadenine

The stage of ovarian development reached in *P. apterus* and *D. cingulatus* adults treated with nucleoside analogues was dose-dependent. Females treated continuously from their emergence by a given compound were affected in two ways: either the development of ovaries was blocked and no eggs were laid or at lower analogue doses the number of eggs and their hatchability was reduced (Table 1).

Changes in the ovarian development were noticed only after the application of higher doses of tested compounds. In *P. apterus* developmental changes in ovaries were found after application of DHPA, D-eritadenine and PMEA at concentrations of 0.5%, 0.1% and 0.05% respectively. Females of *D.cingulatus* were more sensitive to the tested compounds as the same effects were caused at lower doses – 0.1% DHPA, 0.05% D-eritadenine and 0.01% PMEA. At these concentrations practically no ovarian development was observed. The ovaries of continuously treated 21-day-old females of *D. cingulatus* were almost the same size as those of 1-day-old control females. In histological sections many histopathological changes were observed – vacuolization, mitotic inhibition *in germarium* and *in vitellarium*, numerous necroses, etc.

Irregularities in the development of the ovaries were also accompanied by changes in the electrophoretic pattern of proteins. After PAGE it was found that vitellogenin synthesis was inhibited in treated females. Nevertheless, the sterility of *P. apterus* and *D. cingulatus* caused by DHPA is reversible. Even after fourteen or twenty-one days of continuous exposure of females to the action of this compound, when the treatment was stopped eggs begin to grow for several days and after eight to ten days females lay eggs. In the first batch the number of eggs is much smaller in comparison with that of control females. In subsequent batches the number of eggs laid approaches that of control values (Table 2). In the case of D-eritadenine and PMEA, the reversibility of their effects was not so pronounced.

Females of *P. apterus* and *D. cingulatus* after treatment by DHPA at concentrations of 0.1% and 0.05% respectively, laid significantly fewer eggs whose hatchability was practically zero. This decrease of hatchability is caused by numerous irregularities in the embryonic development. When lower doses of DHPA were applied the hatchability was higher (Table 1). After application of DHPA, D-eritadenine and PMEA at concentrations of 0.01%, 0.001% and 0.0001% respectively, the eggs produced were as hatchable as those of the control females. The sterilization effect was also found in *A. fabae*. After DHPA application through the host plant, a partial or complete sterility was observed. The results are summarized in Table 3.

Effects of nucleoid analogues on enzymes

Using electrophoresis, six different esterases were found in the ovaries of *P. apterus* females. DHPA at 0.5% concentration increased the activity of one esterase band and slightly decreased the activity of two esterases with the greatest mobility. The activities

Table 2 Restoration of the fecundity of *Pyrrhocoris apterus* L. females after treatment with 0.5% (RS)-DHPA. Each group consisted of 10 females.

Length of exposure (days)	Batch No.	Number of laid eggs/ ♀	Number of hatched eggs/ ♀
Untreated control	I	46.7 ± 9.7	38.7 ± 9.1
	II	56.8 ± 7.8	47.8 ± 8.6
	III	60.1 ± 8.1	54.7 ± 6.4
7	I	14.1 ± 5.7	0
	II	21.9 ± 7.9	0
	III	48.9 ± 8.7	27.8 ± 7.7
14	I	11.1 ± 6.1	0
	II	27.8 ± 7.9	0
	III	39.8 ± 9.8	19.8 ± 4.3
21	I	10.8 ± 4.6	0
	II	17.1 ± 7.6	0
	III	26.7 ± 7.7	9.7 ± 4.3

of other esterases remained unchanged. This phenomenon was observed on day 4 after emergence, and it was especially conspicuous at six and eight days after emergence.

Acid and alkaline phosphatase activity
Practically no differences in acid phosphatase activity between the control and treated animals were found in the fat body. In the case of 0.5% DHPA-treated females acid phosphatase activity in the ovaries significantly differed from control and 0.1% DHPA-treated females in the former females, acid phosphatase activity remained low (roughly 12 nmol of the product/min.mg) while in control females after day 5 of imaginal life the activity reached the value of 60 nmol at day 8 and later lowered to 50 nmol. In 0.1% DHPA-treated females 95 nmol of the product was formed at day 10. Raised activity of acid phosphatase was noticed also in the haemolymph. This activity was lowest in control females and highest in females treated with 0.5% DHPA, showing a fivefold increase on days 7 and 9.

In the case of alkaline phosphatase, the haemolymph and ovarian enzymes showed no differences in comparison with control females. In the fat body a shift was noticed in the temporal pattern of the enzymatic activity (Figure 2).

Table 3 Basic data on the effects of (RS)-DHPA on viviparous females *Aphis fabae* Scopoli reared at 20 $\pm$ 0.5°C, 70-80$ R.H. and photoperiod 18 hours on the bean *Vicia faba*.

Type of experiments	Controls	Variant I		Variant II		Variant III	
Concentration		1 mg/ml^{-1}	0.1mg/ml^{-1}	1 mg/ml^{-1}	0.1mg/ml^{-1}	1 mg/ml^{-1}	0.1mg/ml^{-1}
Pre-reproductive period[+]	11.2 $\pm$ 0.6	10.4 $\pm$ 0.5	10.3 $\pm$ 0.6	10.3 $\pm$ 0.8^{x}	10.4 $\pm$ 0.7	11.1 $\pm$ 0.7^{x}	9.9 $\pm$ 0.8
Reproductive period[+]	18.7 $\pm$ 1.3	6.2 $\pm$ 1.1	10.5 $\pm$ 4.1		14.1 $\pm$ 1.9		12.1 $\pm$ 3.1
Post-reproductive period[+]	5.1 $\pm$ 1.2	3.8 $\pm$ 2.5	3.8 $\pm$ 1.9	14.9 $\pm$ 4.1	2.7 $\pm$ 1.8	9.7 $\pm$ 2.9	3.2 $\pm$ 1.8
Fecundity (no of nymphs/♀	78.4 $\pm$ 8.,7	24.9 $\pm$ 12.8	59.9 $\pm$ 27.1	12^{++}	69.1 $\pm$ 12.5	0	37.1 $\pm$ 6.3

[+] $=$ days; $^{\text{x}}=$ only up to imaginal ecdysis; $^{++}=$ produced only 1 ♀

The following variants of experiments were carried out:
I   Oral application to adults (last instar nymphs were placed on treated plants immediately before imaginal ecdysis).
II Oral application to nymphs (reared on treated plants from birth to imaginal ecdysis; the adults were transferred on the untreated plants).
III Oral application throughout life (from the birth of the nymph to the death of the adult aphid).

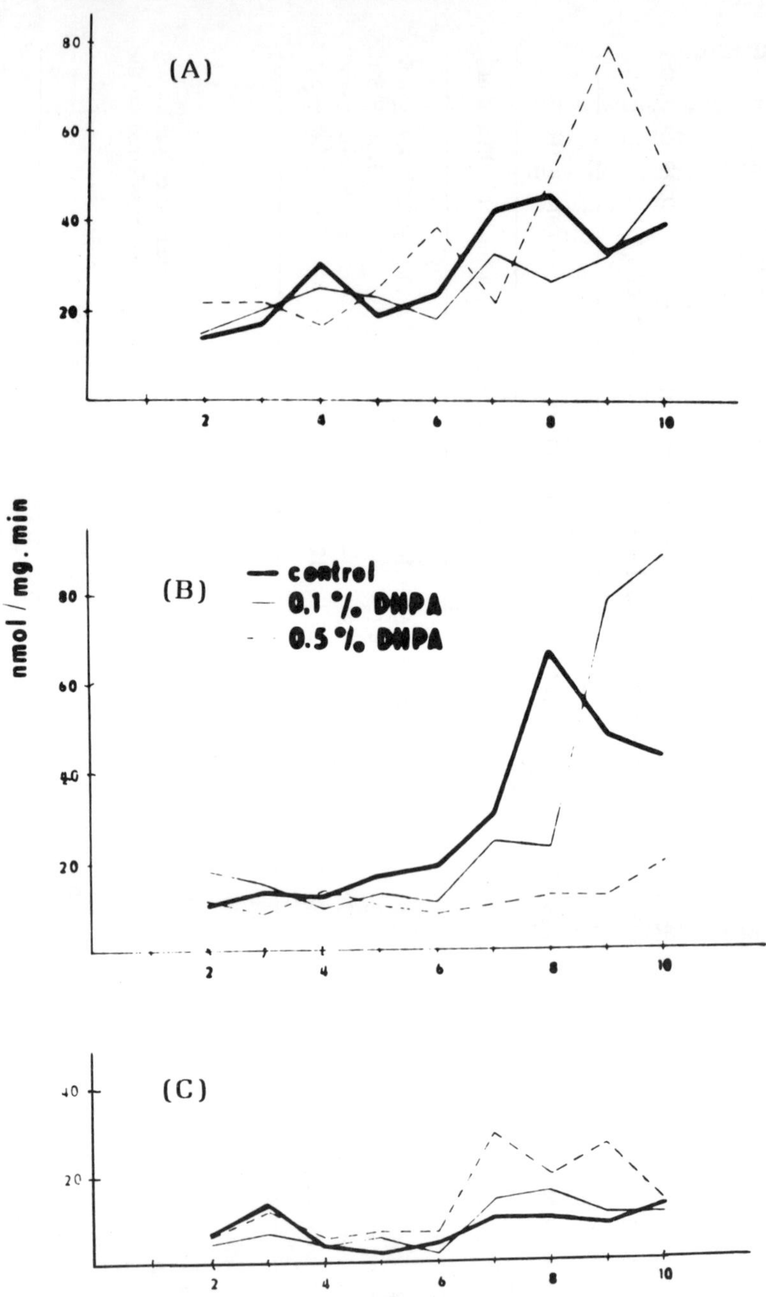

Figure 2   The temporal pattern of phosphatase activity in *Pyrrhocoris apterus* L. females. A = alkaline phosphatase in fat body; B = acid phosphatase in ovaries; C = acid phosphatase in haemolymph.

Discussion

Our results show that some nucleoside analogues with antiviral activity can affect insect fecundity and egg hatchability. Of the analysed species, *D. cingulatus* was found to be the most sensitive to the action of adenine analogues. Continual oral application of DHPA aqueous solution at 0.1% concentration or D-eritadenine and PMEA at concentrations of 0.05% and 0.01% respectively, inhibited oviposition, whereas doses five times higher were needed to prevent oviposition in the related species, *P. apterus*. Sterility evoked by DHPA in both species of bugs is fully reversible, perhaps because pathological changes observed in the ovaries of treated females were not sufficiently drastic to prevent the restoration of the functional state and the formation of ripe eggs. On the other hand, the same compound caused an irreversible sterility in nymphs of *A. fabae*. The mechanism of DHPA action in this insect seems, therefore, to be different. On the contrary, neither fecundity nor hatchability was affected in the holometabolous species *Spodoptera littoralis* Boisd., *S. frugiperda* Smith, *Mocus latipes* Gn., *Hylobius abietis* L., *Pachneus lithus* Germ., and *Leptinotarsa decemlineata* Stal., not even when a 1% solution of DHPA was used (Gelbič, 1989). It is presumed that doses necessary for sterilization are higher than those causing the inhibition of metamorphosis.

As the sterility evoked by some nucleoside analogues is reversible in some species and irreversible in others, it is probable that the sterility is caused by pathological changes in the ovaries and also (in the case of reversibility) by the inhibition of some proteosynthetic and enzymatic processes connected with egg formation. This conclusion is supported by Šula *et al.* (1987) who stated that DHPa inhibited the vitellogenin proteosynthesis. Votruba and Holý (1980) found that DHPA inhibited the enzyme S-adenosyl-L-homocysteine hydrolase. This enzyme is involved in the regulation of S-adenosyl-L-methionine mediated methylations.

The general features of sterility induced by nucleoside analogues are in many respects identical with those induced by the application of classical chemosterilants like tepa, metepa, hempa, etc. (Masner and Landa, 1971; Hussein and Steffan, 1976), but the mechanisms of their actions are different. The action of classical chemosterilants is based on alkylation (LaBrecque and Smith, 1968) whereas nucleoside analogues sterilize apparently by inhibiting the natural processes of alkylation (Sláma *et al.*, 1983). Nucleoside analogues also inhibit vitellogenin proteosynthesis (Šula *et al.*, 1987). Classical chemosterilants, on the other hand, inhibit only the nutritive phase of egg formation and vitellogenins accumulate in the haemolymph (Gelbič and Socha, 1976). Hempa was found to inhibit all esterase activity in the haemolymph, ovaries and fat body of *P. apterus*. DHPA only slightly inhibited some esterases in this bug species but induced higher activity of another esterase (Gelbič and Šula, 1990).

Similarly, as in the case of esterases, we did not observe any inhibition of acid and alkaline phosphatases. Němec and Sláma (1989) found inhibition of acid and alkaline phosphatase activity in the Malpighian tubules and in intestinal epithelium by DHPA and D-eritadenine. The differences between these results and ours can be explained by the fact that Němec and Sláma (1989) determined the phosphatase activity only at day 3 of imaginal life and so did not observe possible temporal shifts of this activity. The second, and perhaps more important, reason is that different organs for this analysis were used.

Presented data as well as those quoted by DeClercq *et al.* (1981), Jelínek *et al.* (1981), Gelbič *et al.* (1984), Beneš *et al.* (1984) and Gelbič and Švec (1988) show that some nucleoside analogues affect those organs with a development dependent on intensive cell proliferation, e.g. embryogenesis, oogenesis and the growth of the plant root system. As these compounds are adenine analogues, the possibility that they interfere with the synthesis of postsynthetic modifications of nucleic acids cannot be excluded. More research is needed to elucidate the mode of action of these nucleoside analogues and to find more efficient compounds which could be applicable in insect pest control.

References

BENEŠ, K., HOLÝ, A. and MELICHAR, O. (1984). The effect of 9-(S)-(dihydroxypropyl)adenine in seedling roots of *Vicia faba* L. in comparison with adenine, adenosine and some cytokinins. *Biologia plantarum*, **26**, 144–150.

DECLERCQ, E., DESCAMPS, J., DE SOMER, P., and HOLÝ, A. (1978). (S)-9-(2,3-dihydroxypropyl)adenine – an aliphatic nucleoside analog with broad-spectrum antiviral activity. *Science*, **200**, 563–565.

DECLERCQ, E. and HOLÝ, A. (1979). Antiviral activity of aliphatic nucleoside analogues: structure-function relationships. *Journal of Medical Chemistry*, **22**, 510–513.

DECLERCQ, E., LEYTEN, R., SOBIS, H., MATOUŠEK, J., HOLÝ, A. and DE SOMER, P. (1981). Inhibitory effect of a broad spectrum antiviral agent, (S)-9-(2,3-dihydroxypropyl)adenine, on spermatogenesis in mice. *Toxicology and Applied Pharmacology*, **59**, 441–461.

GELBIČ, I. (1989). The effects of selected antiviral agents on insects. In *Regulation of Insect Reproduction IV* (M. Tonner, T. Soldán, and B. Bennettová, eds), pp. 267–278. Academia, Prague.

GELBIČ, I. and HOLÝ, A. (1985). Effects of antiviral agent (R,S)-9-(2,3-dihydroxypropyl)adenine on the larval development of *Spodoptera littoralis* (Lepidoptera). *Acta Entomologica Bohemoslovaca*, **28**, 22–27.

GELBIČ, I. and SOCHA, R. (1976). The effects of the chemosterilant metepa on the development of ovaries of *Dixippus morosus* (Phasmoidea). *Acta Entomologica Bohemoslovaca*, **73**, 224–233.

GELBIČ, I. and ŠULA, J. (1990). Ovicidal and biochemical effects of hempa and nucleoside analogue on *Pyrrhocoris apterus* (L.) (Het., Pyrrhocoridae). *Journal of Applied Entomology*, **109**, 401–409.

GELBIČ, I. and ŠVEC, P. (1988). Changes in the fecundity and embryonic development of *Dysdercus cingulatus* caused by the effect of the antiviral agent (R,S)-9-(2,3-dihydroxypropyl)adenine. *Acta Entomologica Bohemoslovaca*, **85**, 327–334.

GELBIČ, I., TONNER, M. and HOLÝ, A. (1984). Aphid sterility induced by antiviral agent (R,S)-9-(2,3-dihydroxypropyl)adenine. *Acta Entomologica Bohemoslovaca*, **81**, 46–53.

HOLÝ, A. (1975). Aliphatic analogues of nucleosides, nucleotides, and oligonucleotides. *Collection of Czechoslovak Chemical Communications*, **49**, 187–214.

HOLÝ, A., VOTRUBA, I. and DECLERCQ, E. (1982). Synthesis and antiviral activity of stereoisomeric eritadenines. *Collection of Czechoslovak Chemical Communications*, **47**, 1392–1407.

HUSSEIN, E.M. and STEFFAN, A.W. (1976). Reproductive capacity of the bean aphid, *Aphis fabae* Scop. (Homoptera: Aphididae) as affected by Metepa. *Zeitschrift für angewandte Entomologie*, **80**, 69–82.

JELÍ NEK, R., HOLÝ, A. and VOTRUBA, I. (1981). Embroyotoxicity of 9-(S)-9-(2,3-dihydroxypropyl)adenine. *Teratology*, **24**, 267–275.

LABRECQUE, C.C. and SMITH, C.N. (1968). *Principles of insect chemosterilization*. Appleton Century Crofts, New York.

LAEMMLI, U.K. (1970). Cleavage of structural proteins during the assembly of the head of bacteriophag T4. *Nature*, **227**, 680–685.

MASNER, P. and LANDA, V. (1971). The formation of compound egg chambers in a bug (Hemiptera) sterilized with 6-azauridine. *Canadian Entomologist*, **103**, 1063–1078.

NĚMEC V. and SLÁMA, V. (1989). Inhibition of phosphatase by open-chain nucleoside analogues in insects. *Experientia*, **45**, 148–150.

SLÁMA, K., HOLÝ, A. and VOTRUBA, I. (1983). Insect sterility induced by a broad-spectrum antiviral agent (S)-9-(2,3-dihydroxypropyl)adenine. *Entomologia Experimentalis et Applicata*, **33**, 9–14.

STOSCHECK, C.M. (1990). Quantitation of proteins. In *Methods in Enzymology*, Volume **182m**, pp. 50–68. Academic Press Inc., San Diego.

ŠULA J., GELBIČ, I. and SOCHA, R. (1987). The effects of (R,S)-9-(2,3-dihydroxypropyl)adenine on the reproduction and protein spectrum in haemolymph and ovaries of *Pyrrhocoris apterus* (Heteroptera, Pyrrhocoridae). *Acta Entomologica Bohemoslovaca*, **84**, 1–9.

VOTRUBA, I. and HOLÝ, A. (1980). Inhibition of S-adenosyl-L-homocysteine hydrolase by the aliphatic nucleoside analogue 9-(S)-(2,3-dihydroxypropyl)adenine. *Collection of Czechoslovak Chemical Communications*, **45**, 3039–3044.

18
The Effects of Trichlorfon and Dichlorvos on the Moulting of *Periplaneta americana* L

U. WUTTIG and H. PENZLIN

Friedrich Schiller University, Institute of General Zoology and Animal Physiology, Ebertstraße 1, O-6900 Jena, Germany

We studied the effects of the organophosphorous insecticides trichlorfon (O,O-dimethyl-(1-n-hydroxy-2,2,2-trichloroethylphosphonate) (TCF) and dichlorvos (O,O-dimethyl-2,2-dichlorovinylphosphonate) (DDVP) on the larval adult development of the American cockroach, *Periplaneta americana* L. Both insecticides are widely used as contact and stomach insecticides. Their chemical structures are given in Figure 1. TCF and DDVP primarily interfere with cholinergic synaptic transmission by inhibiting acetylcholinesterase (AChE), (EC 3.1.1.7), (Metcalf, 1955; Gyrd-Hansen and Kraul, 1984; Eldefrawi, 1985; Farage-Elawar *et al.*, 1988). This inhibition is accomplished by the phosphorylation of a serine hydroxyl residing in the catalytic site of AChE and results in the accumulation of acetylcholine (ACh) in the synaptic cleft, and repetitive stimulation of the acetylcholine receptors with resulting acute cholinergic poisoning (Eldefrawi, 1985; Lund, 1985). In addition to the primary action, further effects of TCF and DDVP were described, namely the inhibition of the pseudocholinesterase (EC 3.1.1.8), (Rhyhänen *et al.*, 1984), the late toxic polyneuropathy and thereby the inhibition of the neuropathy target esterase (Johnson, 1982).

Until now, there is no evidence for the effects of TCF or DDVP on the moulting regulation in insects. Known is only the inhibition of

Insecticides: Mechanism of Action and Resistance

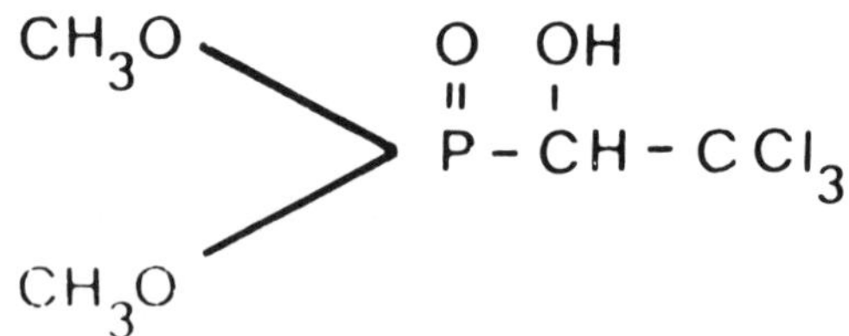

Figure 1 The chemical structures of TCF and DDVP.

moult in *Oziotelphusa senex senex* after treatment with the organophosphorous insecticides malathion and methylparathion

(Reddy, 1985).

Starting from this, we studied the effects of TCF and DDVP on the moult of *Periplaneta americana* in more detail.

Materials and Methods

Larvae of the last instar obtained from a colony maintained in our institute were used. In order to obtain homogeneous material, the animals were selected freshly moulted from the mass cultures and held in groups of ten animals in containers. All individuals were reared under constant and controlled environmental conditions. The administration of TCF and DDVP was performed by a single injection of 5 μg into the abdomen fifteen days after the last moult. At this time normally the activation of the prothoracic gland begins. In the case of TCF, 0.5 μg, 1 μg or 1.3 μg were dissolved in 5 μl Ringer solution injected. In control experiments, pure Ringer solution was administrated. DDVP was dissolved in acetone and diluted in Ringer solution. The volumes (5 μl) injected contained 2, 8 or 11 ng DDVP. The concentrations of insecticides administrated in this work lie under the highest sublethal concentrations which were ascertained in preceding studies.

The effectiveness of TCF and DDVP in retarding moulting was expressed as the retardation ratio (RR), calculated to the following formula

$$RR = \frac{t_c}{t_s}$$

where t_c is the average time (in days) between insecticide administration and the next moult, and t_s is the average time in control larvae treated only with Ringer or acetone solutions.

Results

Under the described conditions the mean duration of the last larval instar injected with Ringer solution was thirty days (n = 160).

Under the influence of the tested insecticides no animal died during the experiments. The larvae did not show any visible changes in their activity and they moulted normally.

Some remarks on the results revealed with TCF. TCF at concentrations of 1 μg and 1.3 μg per larva caused an elongation of the larval instar. Figure 2 gives the results with 1 μg TCF injected into the abdomen of the larvae on day 15. The TCF-treated larvae, (n = 32) showed a significantly (p < 0.01) prolonged moulting cycle compared to that of controls (n = 25). The moult occurs, on average, eight days later. While 100% of the control animals have moulted, only 28% of the TCF-treated have done so (RR = 1.5).

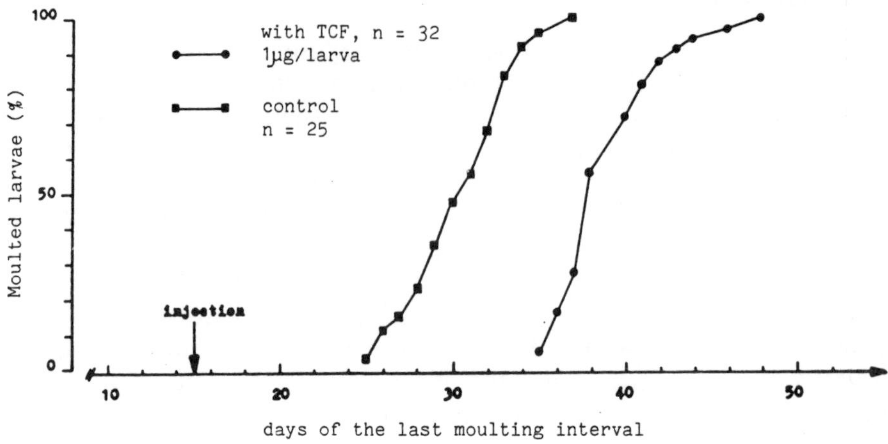

Figure 2   The effect of injected TCF (1 µg/larva) on the duration of the last larval instar of *Periplaneta americana*. Injections were made on day 15.

After injection of 1.3 µg TCF the moulting cycle was also elongated significantly (p < 0.001). The moult occurs on average ten days later (Figure 3). On day 35 100% of the controls (n = 27) had moulted, but only 29% of the TCF-treated larvae (RR = 1.6) (n = 13). Figure 4 summarizes the results obtained with TCF. It is shown that no significant difference in the moulting cycle has been found between control and experimental animals treated with 0.5 µg TCF (RR = 1.2).

Injections of dichlorvos resulted in similar effects to those following administration of TCF (Figure 5). 11 ng DDVP injected into the larvae (n = 44) caused a delay of the next moult of about seven days. This result differed significantly from that of controls (p < 0.01) (n = 43) (RR = 1.4). Larvae (n = 15) treated with 8 ng DDVP moulted, on average, four days later than the controls (n = 15). This is a significant difference (p < 0.01) (RR = 1.3). In contrast, no significant difference in the duration of the instar has been observed between control (n = 15) and experimental (n = 20) animals (RR = 1.13) after administration of 2 ng DDVP.

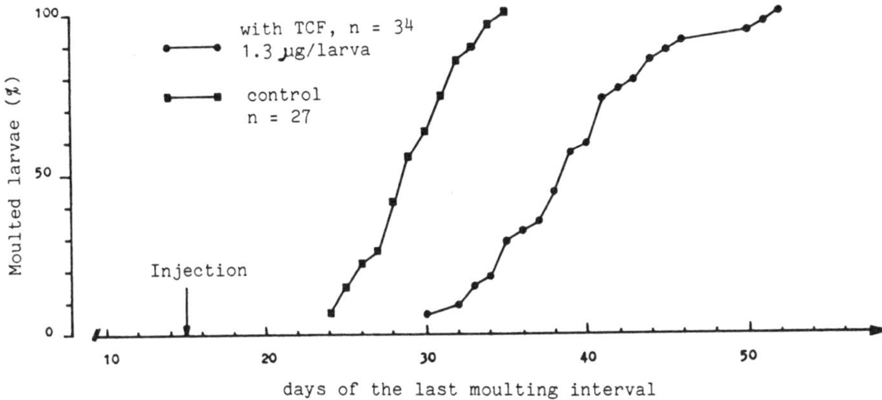

Figure 3 The effect of injected TCF (1.3µg/larva) on the duration of the last larval instar of *Periplaneta americana*. Injections were made on day 15

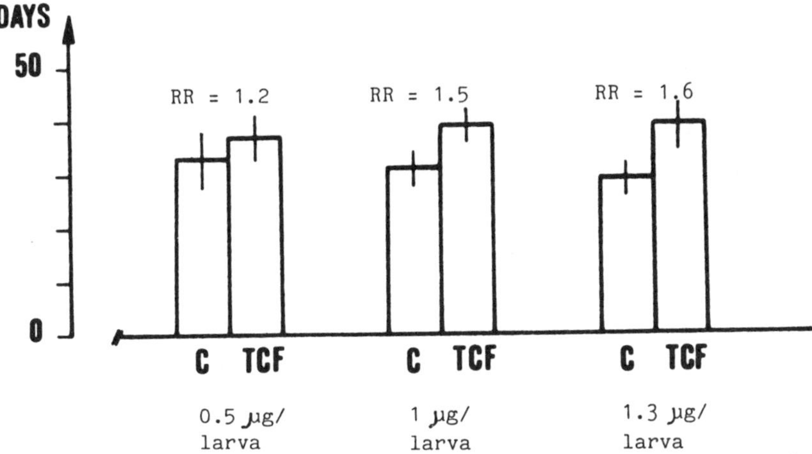

Figure 4 Effects of injected TCF on the duration of the last larval instar of *Periplaneta americana*. The injections were made on day 15 of moulting interval. Vertical bars represent ± S.E.M.; C = control; RR = retardation ratio

The effects of TCF and DDVP on the duration of the moulting cycle similar to those described here are also known from studies by Richter and Birkenbeil (1989), with extracts of *Ajuga reptans*. Such results are close to those obtained also in *Locusta migratoria* (Sieber and Rembold, 1983) and *Manduca sexta* (Schluter *et al.*, 1985) following azadirachtin treatment. An elongation of moulting interval has also been noted in *Oziotelphusa senex senex* after administration of the organophosphorous insecticides malathion and methylparathion (Reddy, 1985).

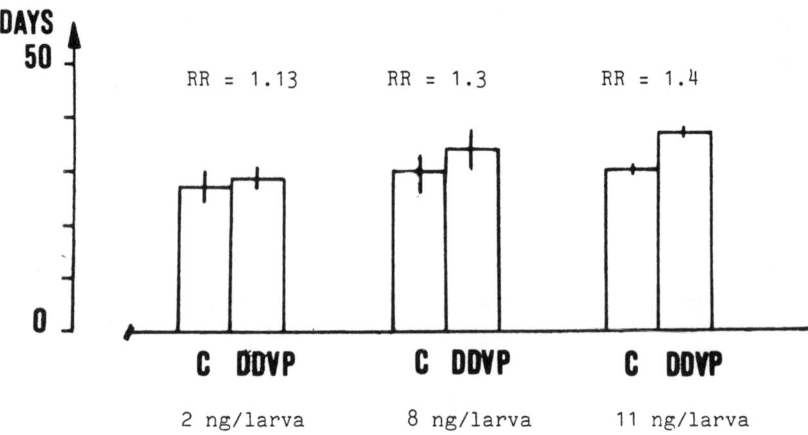

Figure 5   Effects of injected DDVP on the duration of the last larval instar of *Periplaneta americana*. The injections were made on day 15 of moulting interval. C = control; RR = retardation ratio.

The insecticide effects on the duration of the instar in our experiments suggests an impairment of the endocrines. Therefore we examined the effect of insecticide administration on the ecdysteroid production by the prothoracic gland *in vitro*. Several days after the last moult, the prothoracic glands of cockroach larvae were dissected in Ringer solution and collected in culture medium. The ecdysteroid titre of this medium was measured by radioimmunoassay (RIA) described by Eibisch *et al.* (1980). Each sample was measured twice. All results were expressed in nanogram of ecdysone equivalents, because the determination in this experiment by RIA was based on the cross reactivity of the prepared antiserum with ecdysone, 20-hydroxyecdysone and other ecdysteroids. Ecdysteroid measurements have been performed on control larvae and experimental individuals treated with 1.3 µg TCF or 11 ng DDVP.

The ecdysteroid titre during the last larval instar of *Periplaneta americana* is well known and it was corroborated in our experiments. Figure 6 shows the effects of the injected TCF on the *in vitro* production of ecdysone equivalents of single prothoracic gland on various days of the last larval instar. The first ecdysteroid peak on day 23 is unchanged under the influence of TCF in comparison to the control larvae. On this day the ecdysteroid titre reaches 3.9 and 5.2 ng ecdysone equivalents per gland respectively. A characteristic second peak of ecdysteroid titre occurs normally at the twenty-eighth day of the instar. The titre reaches 6.5 ± 1.8 ng ecdysone equivalents per gland. In TCF-treated larvae the second ecdysteroid peak is delayed and occurs six days later at the thirty-fourth day of the instar. On this day all control larvae had moulted into imagines.

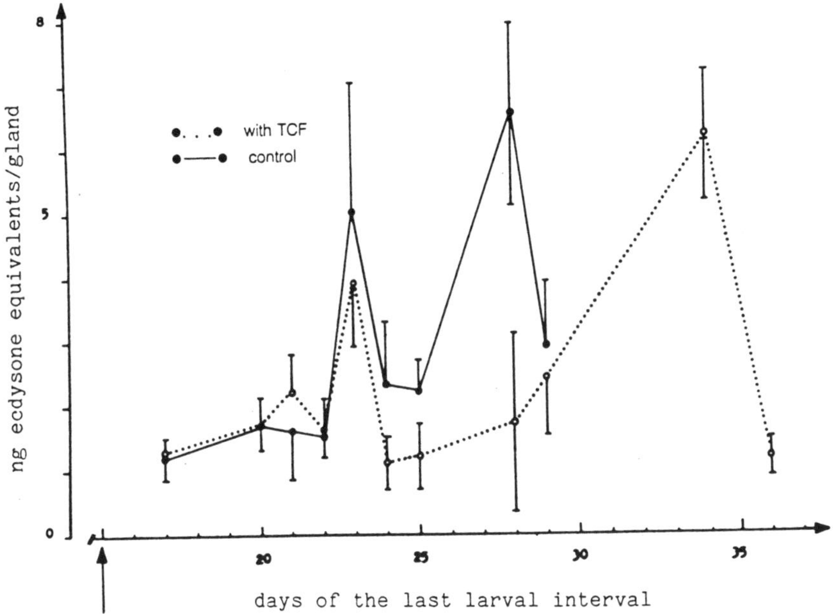

Figure 6   The effect of injected TCF (1.3 µg/larva) on the *invitro* production of ecdysone equivalents of single prothoracic gland on various days of the last larval instar of *Perioplaneta americana*. The second ecdysteroid increase of TCF-treated larvae on day 28 is significantly reduced ($p < 0.01$) in comparison to the control ($n = 19$).

Also after administration of DDVP the first peak of ecdysteroid titre remained unimpaired (Figure 7). Again, in control animals the second peak occurs at the twenty-eighth day of the instar. In larvae treated with DDVP the second ecdysteroid peak occurs seven days later at the thirty-fifth day of the instar. On this day all control larvae had moulted already into imagines.

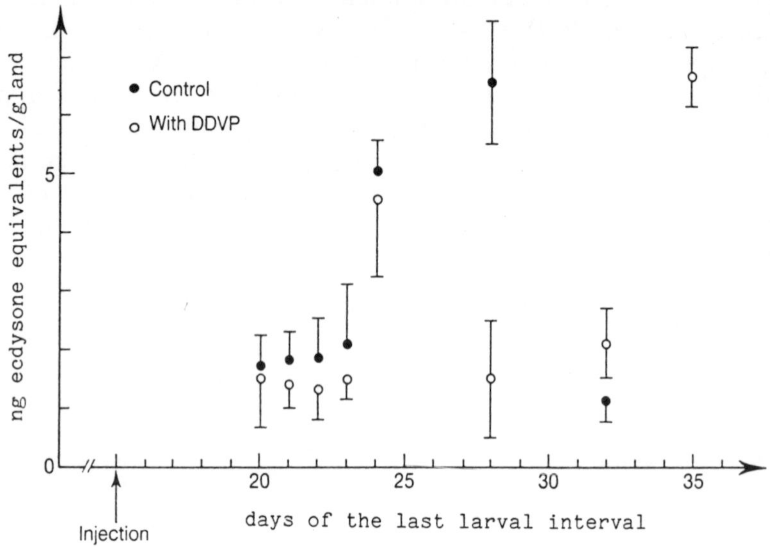

Figure 7   The effect of injected DDVP (11 ng/larva) on the *in vitro* production of ecdysone equivalents of single prothoracic gland on various days of the last larval instar of *Periplaneta americana*. The ecdysone equivalents production was studied in comparison to the delay in moulting. Data are the mean $\pm$ SE values. The second ecdysteroid increase of DDVP-treated larvae (n = 10) on day 28 is significantly reduced (p < 0.01). The ecdysteroid production on the day 35 was studied only in the DDVP-treated larvae because all control larvae had moulted already into imagines.

Our results indicate that the delays in the larval-adult development after TCF or DDVP administration correspond with the delays in the ecdysteroid productions. The maximal ecdysteroid titres were found two days before moulting as shown by Shaaya (1978) and Eibisch *et al.* (1980). Furthermore, our results clearly show that in *Periplaneta americana* only the second peak of moulting hormone titre is delayed after administration of TCF and DDVP. This second peak is supposed to be induced hormonally whereas the first is probably induced neuronally (Richter, 1983). This first peak remains unaffected. The observed delays in moulting as well as in appearance of the second ecdysteroid peak could be attributed to three mechanisms.

1. TCF and DDVP act directly on the prothoracic gland and retard

the release of ecdysteroids.

2. TCF and DDVP affect factors involved in the activation of the moulting gland like prothoracicotropic hormone (PTTH) or juvenile hormone (Gruetzmacher *et al.*, 1984; Bollenbacher *et al.*, 1987; Rountree *et al.*, 1987). It seems also possible that TCF and DDVP affect the hormone binding proteins (King and Tobe, 1988; Whitehead, 1989; Wiśniewski, 1989).

3. TCF and DDVP affect moulting via impairment of the cholinergic system. This hypothesis is based on the results given by Lester and Gilbert (1985, 1986, 1987). They observed an acetylcholine accumulation and increasing acetylcholinesterase activity in the larval *Manduca* brain of fifth instar and concluded that the cholinergic system is involved in the PTTH release.

In summary, the fact that TCF as well as DDVP are believed to be inhibitors of the AChE support the assumption that both insecticides affect moulting via impairment of the cholinergic system.

References

BOLLENBACHER, W.E., GRANGER, N.A., KATAHIRA, E.J. and O'BRIEN, M.A. (1987). Developmental endocrinology of larval moulting in the tobacco hornworm, *Manduca sexta*. *Journal of Experimental Biology*, **128**, 175–181.

EIBISCH, H., GERSCH, M., ECKERT, M. AND BÖHM, G.A. (1980). Investigation of the ecdysteroid production in prothoracic glands and of the ecdysteroid level in the haemolymph of *Periplaneta americana* larvae by means of radioimmunoassay (RIA) with an anti-ecdysterone serum. *Zoologische Jahrbücher Physiologie*, **84**, 153–163.

ELDEFRAWI, A.T. (1985). Acetylcholinesterase and anticholinesterase. In *Comprehensive Insect Physiology, Biochemistry, and Pharmacology* (G.A. Kerkut and L.I. Gilbert, eds), Volume **12**, pp. 115–130. Pergamon Press, Oxford.

FARAGE-ELAWAR, M., EHRICH, M.F., JORTNER, B.S. and MISRA, H.P. (1988). Effects of multiple oral doses of two carbamate insecticides on esterase level in young and adult chickens. *Pesticide Biochemistry and Physiology*, **32**, 262–268.

GRUETZMACHER, M.C., GILBERT, L.I. and BOLLENBACHER, W.E. (1984). Indirect stimulation of the prothoracic gland of *Manduca sexta* by juvenile hormone: Evidence for a fat body stimulatory factor. *Journal of Insect Physiology*, **30**, 771–778.

GYRD-HANSEN, N. and KRAUL, I. (1984). Obidoxime reactivation of organophosphate-inhibited cholinesterase activity in pigs. *Acta Veterinary Scan*, **25**, 86–95.

JOHNSON, M.K. (1982). The target for initiation of delayed neurotoxicity of organophosphorus esters: Biochemical and toxicological applications. In *Reviews in Biochemical Toxicology* (Hodgson, Bevel and Philpot, eds), Volume **4**, pp. 141–210. Elsevier, Amsterdam.

KING, L.E. and TOBE, S.S. (1988). The identification of an enantio-selective JH III binding protein from the haemolymph of the cockroach, *Diploptera punctata. Insect Biochemistry*, **18**, 793–797.

LESTER, D.S. and GILBERT, L.I. (1985). Choline acetyltransferase activity in the larval brain of *Manduca sexta. Insect Biochemistry*, **15**, 685–694.

LESTER, D.S. and GILBERT, L.I. (1986). Developmental changes in choline uptake and acetylcholine metabolism in the larval brain of the tobacco hornworm, *Manduca sexta, Developmental Brain Research*, **26**, 201–209.

LESTER, D.S. and GILBERT, L.I. (1987). Characterization of acetylcholinesterase activity in the larval brain of *Manduca sexta. Insect Biochemistry*, **17**, 99–109.

LUND, A.E. (1985). Insecticides: Effects on the nervous system. In *Comprehensive Insect Physiology, Biochemistry, and Pharmacology* (G.A. Kerkut and L.I. Gilbert, eds), Volume **12**, pp. 9–51. Pergamon Press, Oxford.

METCALF, R.L. (1955). *Organic Pesticides*. Interscience, New York.

REDDY, P.S. (1985). Moult-inhibition in the crab *Oziotelphusa senex senex* following exposure to malathion and methylparathion. *Bulletin Environmental Contam. Toxicology*, **35**, 92–97.

RICHTER, K. (1983). Experimentelle Untersuchungen der nervösen Steuerung der Prothorakaldrüse bei *Periplaneta americana. Zoologische Jahrbücher Physiologie*, **87**, 65–83.

RICHTER, K. and BIRKENBEIL, H. (1989). The effect of extracts from Ajuga reptans on moult regulation in the cockroach, *Periplaneta americana.* In *Insecticides-Mechanisms of Action and Resistance, Tagungsbericht, Nummer 274*, Akademie der Landwirtschaftswissenschaften der DDR.

RHYHÄNEN, R., HERRANEN, J., KORHONEN, K., PENTILLA, J., POLVILAMPI, M. and PUHAKAINEN, E. (1984). Relationship between serum lipids, lipoproteins and pseudocholinesterase during organophosphate poisoning in rabbits. *Int.*

Journal Biochemistry, **16**, 687–690.

ROUNTREE, D.B., COMBEST, W.L. and GILBERT, L.I. (1987). Protein phosphorylation in the prothoracic glands as a cellular model for juvenile hormone- prothoracicotropic hormone interactions. *Insect Biochemistry*, **17**, 934–948.

SCHLUTER, W., BIDMON, H.J. and GREWE, S. (1985). Azadirachtin affects growth and endocrine events in larvae of the tobacco hornworm, *Manduca sexta*. *Journal Insect Physiology*, **31**, 773–777.

SHAAYA, E. (1978). Ecdysone and juvenile hormone activity in the larvae of the cockroach *Periplaneta americana*. *Insect Biochemistry*, **8**, 193–195.

SIEBER, K.P. and REMBOLD, H. (1983). The effects of azadirachtin on the endocrine control of moulting in *Locusta migratoria*. *Journal Insect Physiology*, **29**, 523–527.

WHITEHEAD, D.L. (1989). Ecdysteroid carrier proteins. In *Ecdysone* (J. Koolman, ed). pp. 232–244. Georg Thieme Verlag, Thieme Med. Publ., Stuggart.

WIŚNIEWSKI, J.R. (1989). Identification of juvenile-hormone-binding proteins on blotted electropherograms using tritiated juvenile hormones. *Experientia*, **45**, 1124–1128.

Part 3
Targets and Insecticide Actions in the Insect Nervous System and in Fundamental Life Processes

19
Proctolin – Structure/Biological Function Relationship Studies

D. KONOPINSKA[1], G. ROSINSKI[2], W. SOBOTKA[3], AND A. PLECH[4]

[1]*Institute of Chemistry, University Wroclaw, ul. Joliot-Curie 14, 50-383 Wroclaw, Poland*
[2]*Department of Animal Physiology, A. Mickiewicz University, 60-021 Poznan, Poland*
[3]*Institute of Organic Industry, ul. Annopol 6, 03-236 Warsaw, Poland*
[4]*Silesian Academy of Medicine, Department of Pharmacology, Zabrze, Poland*

Introduction

Proctolin, the first structurally characterized myotropic insect neuroregulator, a pentapeptide L-arginyl-L-tyrosyl-L-leucyl-L-prolyl-L-threonine (Arg-Tyr-Leu-Pro-Thr), was first isolated by Brown and Starratt (1975) from whole body extracts of the American cockroach (*Periplaneta americana*). Proctolin has been shown to occur in species from six insect orders (Brown, 1977a; Holman and Cook, 1979a, 1979b; Orchard *et al.*, 1989) and other invertebrates (Brown, 1977b; Kravitz *et al.*, 1984; Adams and O'Shea, 1983). On the basis of myotropic properties, seen in the stimulation of contractions of smooth skeletal muscles (O'Shea and Adams, 1981; Adams and O'Shea, 1983; Starratt and Steele,1984), heart muscles of insects (Miller and Sullivan, 1981; Watson *et al.*, 1983) proctolin was considered to be an insect neurotransmitter. Recently, with the accumulation of more data concerning its biological properties, Orchard *et al.* (1989) expressed their opinion that proctolin fulfils a

Insecticides: Mechanism of Action and Resistance
© 1992   Intercept Ltd, P.O. Box 716, Andover, Hants SP10 1YG, UK

neuromodulatory function in insects.

Proctolin has been the subject of numerous studies to define the site of its occurrence in insects, estimate its stability both *in vivo* and *in vitro*, evaluate its possible insecticidal activity and biological effects on vertebrates, and to elucidate its structure-function relationship. The last of the mentioned research areas initiated a permanent demand for the synthesis of proctolin and its analogues.

Initially, proctolin was detected in the cockroach (*P. americana*) hindgut and then in other parts of the insect body such as the nervous tissues, digestive tract and the brain of insects and other invertebrates (Brown, 1977b, Adams and O'Shea, 1983; Kravitz *et al.*, 1984). Proctolin stability in insects was studied both *in vitro* and *in vivo*. Rapid *in-vivo* degradation of 3^H- or ^{14}C-labelled Tyr residue in position 2 of proctolin (Starratt and Steele, 1984; Quistad *et al.*, 1984) yielded the constituent amino acids (at least Tyr), tetrapeptide Tyr-Leu-Pro-Thr and dipeptide Arg-Tyr.

When topically applied, proctolin does not penetrate the larval cuticle (*Manduca sexta*) (Quistad *et al.*, 1984) thus eliminating the possibility of its practical use for insect control.

Further investigations into the biological effects of proctolin, particularly in vertebrates, shed some light on its physiological properties. After intracerebronventricular administration in rats increases in blood pressure and heartbeat frequency were observed (Schultz *et al.*, 1981; Konopinska *et al.*, 1989).

In vitro studies on the influence of proctolin on the impaired biological properties of blood cells derived from leukemic patients indicated a positive effect in the restoration of phagocytic activity of human granulocytes (Konopinska *et al.*, 1986a) and a stimulatory effect on the transformation rate of cultured lymphocytes (Kazanowska *et al.*, 1989).

Structural/Biological Function Relationship Studies on Proctolin

The structure-myotropic function examinations in insects resulted in the elaboration of several methods to synthesize proctolin and its analogues (Starratt and Brown, 1977, 1979; Sullivan and Newcomb, 1982; Konopinska *et al.*, 1986a, 1986b, 1988a, 1988b; Bartosz-Bechowski *et al.*, 1990). Structural modifications included the subsequent replacement of the proctolin native amino acid residue in positions 1 to 5 of the peptide chain as well as the synthesis of analogues with a shortened or elongated proctolin chain (Starratt and Brown, 1979; Sullivan and Newcomb, 1982). Comprehensive studies on biological properties have focused on contractions of hindgut muscles of *P. americana* (Starratt and Brown, 1979; Sullivan and Newcomb, 1982). Hence, the significance of the particular amino acid residues in the proctolin molecule, with respect to its myotropic properties, has been assessed. Lack of biological activity within physiological concentrations (10^{-10}-10^{-7}M) was observed in the majority of proctolin analogues except [Phe(p-OMe)^2]-proctolin, which was about three times more active than proctolin itself (Starratt and Brown, 1979).

The subject of our studies on the structure/biological function relationship of proctolin was a series of its analogues modified in positions 1 to 3 of the peptide chain and N-des-Arg tetrapeptide proctolin derivatives. The biological evaluations in our studies (Konopinska *et al.*, 1986b, 1988a, 1988b, 1990, 1991; Bartosz-Bechowski *et al.*, 1990) were performed on two insect species, *P. americana* and *Tenebrio molitor*, using the cardiostimulating *in-vitro* test according to Miller (1979) or Rosinski and Gäde (1988).

The N-terminal substitution and addition analogues [Arg^0]-, [Lys^1]-proctolin, maintained their activity within the insect hindgut (Sullivan and Newcomb, 1982) and on heart action (Konopinska *et al.*, 1988a, 1990). Replacement of the N-terminal Arg by the other basic amino acid (*Table 1*) residues such as L-ornitine (Orn), L-histidine (His), p-amino-L-phenylalanine [Phe(p-Gn)], ε-guanidino-ε-amino-L-n-caproic acid (Gac), α-guanidino-δ-amino-L-n-valeric acid (Gav) and α-guanidino-β-phenyl(p-amino)-L-propionic acid (Gap) led, within physiological concentrations, to the creation of inactive analogues which confirms the importance of the Arg residues for the myotropic function of proctolin (Konopinska *et al.*, 1990).

The role of the aromatic amino acid residue in position 2 of the proctolin chain has been more widely discussed in the literature (Starratt and Brown, 1979; Sullivan and Newcomb, 1982). Thus, replacement of Tyr by Phe, D-Tyr (Starratt and Brown, 1979), His or L-tryptophan (Trp) (Sullivan and Newcomb, 1982) resulted in decreasing stimulation of proctodeum muscle contraction in *P. americana* (Starratt and Brown, 1979; Sullivan and Newcomb, 1982). A surprising result was that substitution of p-methoxy-L-phenyla-lanine [Phe(p-OMe)] for Tyr increased the activity by almost three times, suggesting that a free hydroxyl group of Tyr is not essential for proctolin myotropic action (Starratt and Brown, 1979). All these observations supported the hypothesis that proctolin myotropic activity depends on the amino acid residue in position 2 of the peptide skeleton, and on the presence of the oxygen atom in a *para* position of the phenyl ring (Sullivan and Newcomb, 1982).

Table 1 Cardiostimulatory effect on insects of proctolin analogues modified at position 1 of the peptide chain according to Konopinska *et al.* (1988b, 1990)

Peptide X^1-Tyr-Leu-Pro-Thr	Biological activity relative to proctolin (%)	
	P. americana at 10^{-9}M	*T. molitor* at 10^{-8}M
X^1 =		
Arg (proctolin)	100	100
Lys	110	50
homo-Arg	inactive	110
Orn, Phe(p-NH$_2$), His		
L-citrulline (Cit),		
γ-amino-buyric (GABA),		
Phe(p-Gn), Gav, Gac,		
Gap and Lys-Arg	inactive	inactive

To verify that assumption, a synthesis of further proctolin analogues was carried out (Konopinska *et al.*, 1986b, 1988a, 1988b; Bartosz-Bechowski *et al.*, 1990) which included replacement of Tyr moiety by the following amino acid residues: Phe(p-NH$_2$), p-N,N-dimethylamino-L-phenylalanine [Phe(p-NMe$_2$)], p-nitro-L-phenylalanine [Phe(p-NO$_2$)], Phe(p-OMe), (Konopinska *et al.*, 1986b), p-ethoxyl-L-phenylalanine [Phe(p-OEt)], p-guanidino-L-phenylalanine [Phe(p-Gn)], β-amino-γ-phenyl-p-hydroxyl-L-butyric acid [Afb(p-OH)], β-amino-γ-phenyl-p-amino-L-butyric acid [Afb(p-NH$_2$)], β-amino-γ-phenyl-p-nitro-L-butyric acid [Afb(p-NO$_2$)] (Bartosz-Bechowski *et al.*, 1990), L-Dopa (Konopinska *et al.*, 1988b), 3'-amino-L-tyrosine [Tyr(3'-NH$_2$)], 3'-nitro-L-tyrosine [Tyr(3'-NO$_2$)] (Bartosz-Bechowski *et al.*, 1990) and β-cyclohexyl(methoxyl)-L-alanine [Cha(4-OMe)], an amino acid residue with a cyclohexyl system instead of the phenyl ring (Konopinska *et al.*, 1986b) (Table 2).

When evaluated for cardioexcitatory activity in *P. americana* and *T. molitor* heart, bioassayed according to Miller (1979) or Rosinski and Gäde (1988), the majority of the analogues increased the heartbeat frequency of both insect species in a comparable manner with, or even stronger than, proctolin itself. [L-Dopa^2]-proctolin revealed at physiological concentrations a particularly high activity

Table 2 Proctolin analogues modified at position 2 of the peptide chain and their cardioexcitatory effect in insects according to Konopinska *et al.*, (1988b, 1991) and Bartosz-Bechowski *et al.*, (1990)

Peptide	Biological activity relative to proctolin (%)	
	P. americana at 10^{-9}M	*T. molitor* at 10^{-8}M
Arg-X^2-Leu-Pro-Thr		
$X^2 =$		
Tyr (proctolin)	100	100
Phe(p-NH_2)	380	180
Phe(p-NMe_2)	100	320
Phe(p-NO_2)	370	170
Phe(p-OMe)	390	155
Phe(p-OEt)	100	0
Phe(p-Gn)	20	0
L-Dopa	500	50
Tyr (3′-NH_2)	110	100
Tyr (3′-NO_2)	120	0
Afb(p-OH), Afb(p-NH_2),		
Afb(p-NO_2)	inactive	inactive
X-Leu-Pro-Thr		
X =		
Tyr(3′-NH_2)	deaccelerate (10%)	deaccelerate (10%)
Phe(p-NH_2), Phe(p-NO_2),		
Phe(3′-COOEt, 4′-OMe)	inactive	inactive

(Konopinska *et al.*, 1988b) (Table 2) causing a higher chronotropic effect against *P. americana* heart than proctolin, whereas in *T. molitor* heartbeat frequency was considerably decreased. Similar species specificity was noticed for [Tyr(3'-NO_2)^2]-proctolin, which accelerated *P. americana* heart contractions but did not have any effect on the heartbeat frequency of *T. molitor*. On the other hand [Tyr(3'-NH_2)^2]-proctolin maintains proctolin activity against both insect species (Bartosz-Bechowski *et al.*, 1990). Lack of cardiotropic action in the case of [Cha(4-OMe)^2]-proctolin, deprived of the aromatic ring, in *P. americana* and *T. molitor* heart muscles (Konopinska *et al.*, 1986b) and also with [Afb(p-X)^2]-proctolin, where X = -OH, -NH_2 and -NO_2, was observed. Thus, the substitution of the Afb(p-X) residue for a Tyr proctolin peptide chain elongated by one $-CH_2-$ group led to the loss of myotropic activity in insects.

In conclusion, one can infer that proctolin myotropic effects depend on the presence of the *para* aromatic ring in position 2 with the polar groups of oxygen and nitrogen atoms. The occurrence of the free electron pairs in these substituents most likely facilitates interaction with the receptor site on both bioassayed insect species. Some other substituents such as hydroxyl, amino or nitro groups at the 3' position of the phenyl ring cause a distinct increase of the myotropic effect in the cockroach but lead to its impairment in the yellow mealworm. Further species-specific activity occurs with $[Phe(p\text{-}OEt)^2]$-proctolin which reveals a chronotropic activity against *P. americana* only. On the other hand, elongation in position 2 of the proctolin chain by a methylene $(-CH_2-)$ group as in the case of $[Afb(p\text{-}X)^2]$-proctolin, led to the loss of cardiotropic activity in both insects. Perhaps this was due to the change of the spatial arrangement of the biologically active proctolin conformation.

The role of L-leucine (Leu), the hydrophobic amino acid in position 3 of the proctolin peptide chain, has not yet been discussed in detail. Two known peptides $[D\text{-}Leu^3]$- and $[Ala^3]$-proctolin were inactive in hindgut stimulatory tests against *P. americana* (Starratt and Brown, 1979). Studies on the significance of Leu in position 3 of the proctolin skeleton for its myotropic function in insects may shed some light on the hydrophobic interaction of the proctolin molecule with its receptor site. Moreover, it should be pointed out that the Leu residue in proctolin is situated between the Tyr residue which plays an important role in creating the cardiotropic effect in insects (Konopinska *et al.*, 1986b, 1988b; Bartosz-Bechowski *et al.*, 1990) and the Pro-residue which is responsible for the stabilization of the biologically active proctolin conformation (Bertins and Nikiforovich, 1979). In our studies we performed a synthesis of a series of proctolin analogues modified at position 3 of the proctolin skeleton by replacement of the native Leu by the following amino acid residues: L-valine (Val), Thr, glicyne (Gly), Pro, 1-aminocyclopentane acid (Acp) and 1-aminocyclohexane acid (Ach) (Table 3) (Konopinska *et al.*, 1991).

Their biological activity was evaluated in cardioexcitatory tests (Rosinski and Gäde, 1988) on two insects, *P. americana* and *T. molitor*. Species specificity was observed since $[Val^3]$- and $[Gly^3]$-proctolin preserved 55% and 30% of proctolin's cardiotropic activity in and *T. molitor*, whereas the other analogues were inactive. In the case of *P. americana*, the $[Val^3]$-proctolin revealed

25% of the activity of the native peptide. The biological results show that the Leu residue in position 3 plays an essential role in the hydrophobic interaction of proctolin with its receptor site in *P. americana*.

Table 3 Cardioexcitatory effect on insects of proctolin analogues modified at position 3 of the peptide chain

Peptide Arg-Tyr-X^3-Pro-Thr	Biological activity relative to proctolin (%)	
	P. americana at 10^{-9}M	*T. molitor* at 10^{-8}M
X^3 =		
Leu (proctolin)	100	100
Val	25	55
Gly	10	30
Thr	inactive	inactive
Pro	inactive	inactive
Acp	inactive	inactive
Ach	inactive	inactive

Furthermore, we became interested in the des-Arg proctolin comprising the following tetrapeptides: Tyr(3'-NH$_2$)-Leu-Pro-Thr (Bartosz-Bechowski *et al.*, 1990), Phe(p-NH$_2$)-Leu-Pro-Thr, Phe(p-NO$_2$)-Leu-Pro-Thr and Phe(3'-Co$_2$Et,4'-OMe)-Leu-Pro-Thr (Konopinska *et al.*, 1991) (Table 2). These peptides were bioassayed in the cardiostimulatory test on *P. americana* and *T. molitor*. Almost all of the peptides were inactive, except Tyr(3'-NH$_2$)-Leu-Pro-Thr, which showed weak (about 10%) cardioacceleratory properties in both insects (Table 2). Results presented here show that the inhibitory effect of peptides is connected with their chemical structure.

Biological Effect of Proctolin in Rats

Recently, we became interested in the influence of proctolin and some of its analogues on the behaviour (*in vivo*) of rats. We evaluated their cardiovascular and antinociceptive effects after intracerebroventricular (icv) injection. Cardiovascular effects of

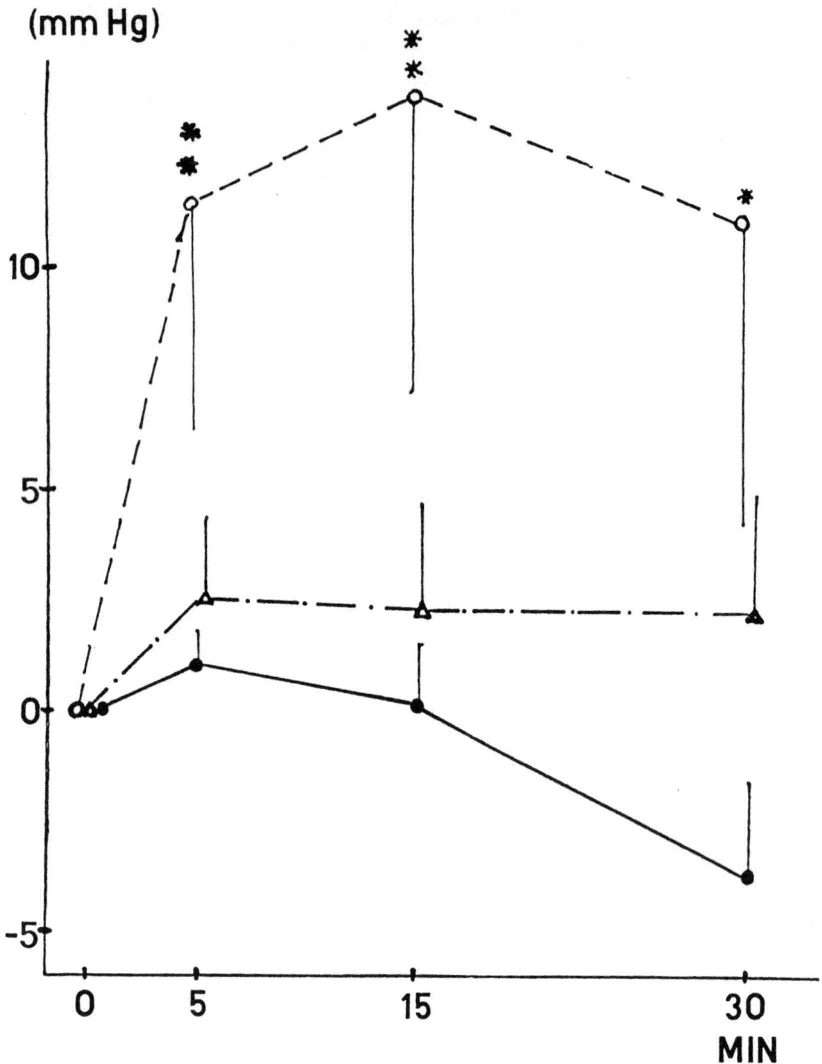

Figure 1 Changes of arterial blood pressure after intracerebroventricular (icv) injection of proctolin (PR) into rats. ●—●, control (0.9% NaCl, 10 μl (n = 7)); o- - -o, PR (100 nmol 10 μl^{-1}, icv (n = 8)); Δ----Δ PR 100 nmol and HCL. Naloxone 200 nmol in 10 ml icv. * $p < 0.05$; $\overset{*}{*}$ $p < 0.01$.

proctolin and some of its analogues modified in position 1 or 2 of the peptide chain such as [Lys^1]-, [His^1]-, [Cit^1]-, [Cha(4-OMe)^2]- and [Phe(p-NH$_2$)^2]-proctolin were evaluated (by icv application) at doses

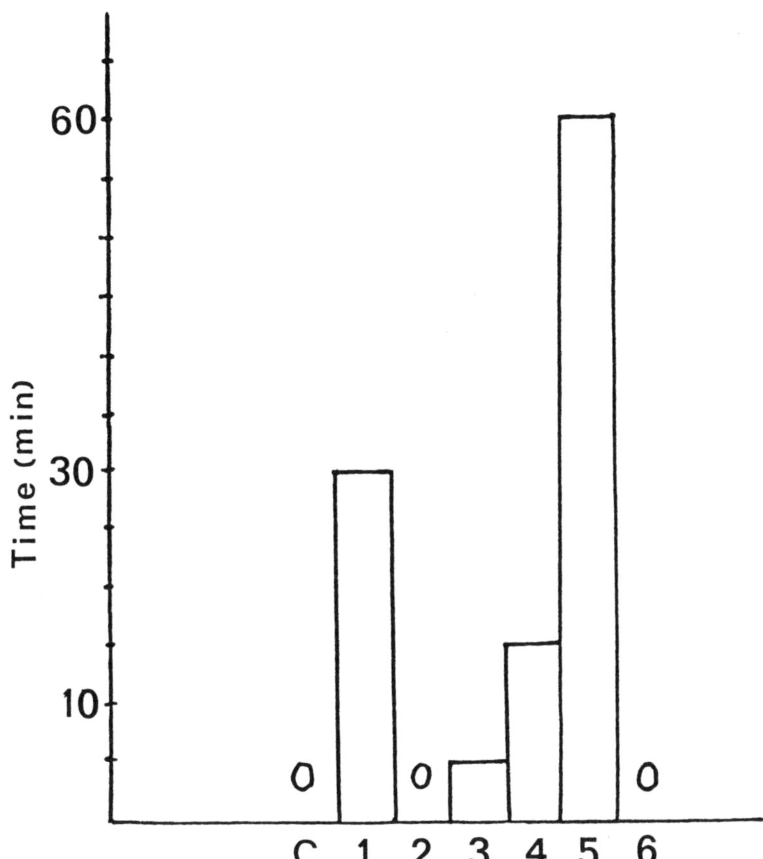

Figure 2 Duration of the significant antinociceptive effect of proctolin and some of its analogues in rats; proctolin injected intracerebroventricularly (icv) into the lateral brain at the dose of nmol. Statistical evaluation performed by Student's test. C, control (0.9% NaCl); 1, proctolin; 2, naloxone hydrochloride (Polfa) 1 mg Kg^{-1} ip + proctolin; 3, [Lys^1]-proctolin; 4, [Phe(p-OMe)^2]-proctolin; 5, [Cha(4-OMe)^2]-proctolin; 6, [L-Dopa^2]-proctolin.

of 100 and 500 nmol each. The arterial blood pressure, heartbeat frequency and respiratory rate of animals after icv administration of proctolin and its analogues were determined according to the Jedrusiak method (Jedrusiak *et al.*, 1991b). Among the tested peptides only proctolin, [Phe(p-NH$_2$)^2]- and [Cha(4-

OMe)^2]-proctolin (at a dose of 100 nmol) induced significant increases of arterial blood pressure, blocked by naloxone injection (Figure 1), whereas other peptides were inactive. These results suggest that the hypertensive effect of proctolin and its derivatives is due to their stimulation of the opioid receptor in the rat central nervous system.

Moreover, the investigated peptides have no influence on heartbeat frequency, except in the case of [Phe(p-NH$_2$)^2]- and [Cha(4-OMe)^2]-proctolin. Both of these peptides, in contrast to proctolin itself, induced a transient increase of the heartbeat frequency in the rat (Jedrusiak *et al.*, 1991a).

Antinociceptive action in rats was bioassayed with proctolin and its following derivatives: [Lys^1]-, [Phe(p-NH$_2$)^2]-, [Cha(4-OMe)^2]- and [L-Dopa^2]-proctolin at a dose of between 10 to 100 nmol icv, according to the method of O'Callaghan and Holtzman (1975). The investigated proctolin [Phe(p-NH$_2$)^2]- and [Cha(4-OMe)^2]-proctolin derivatives revealed antinociceptive activity, prevented by naloxone injection. [Cha(4-OMe)^2]-proctolin in particular showed strong activity causing a long-term antinociceptive effect last one hour (Figure 2).

Results presented here pointed out that cardiovascular as well as antinociceptive effects of proctolin and some of its analogues were connected with their interaction with opioid receptors in rats. Moreover, lack of cardiovascular and antinociceptive activity in the proctolin analogues modified at position 1 of the peptide chain, support our earlier hypothesis (Konopinska *et al.*, 1990) that the presence of the N-terminal Arg-residue in the proctolin peptide skeleton is essential for its biological activity.

Conclusions

The results reported here indicate that the insect neuropeptide proctolin and some of its analogues, which show biological activity in arthropods, also exert biological effects in mammals *in vivo*. The antinociceptive and cardiovascular effects observed in rats testify that proctolin and its derivatives modified at position 2 of the peptide chain have a neuromodulatory action in vertebrates as well as in invertebrates.

References

ADAMS, M.E. and O'SHEA, M. (1983). Peptide contransmitter at a neuromuscular junction. *Science*, **221**, 286–289.

BARTOSZ-BECHOWSKI, H., ROSINSKI, G., KONOPINSKA, D., SUJAK, P. and SOBOTKA, W. (1990). Further studies on proctolin analogs modified in the position 2 of the peptide chain and their influence on heartbeat frequency of insects. *International Journal of Peptide and Protein Research*, **36**, 450–457.

BERTINS, J.R. and NIKIFOROVICH, G.V. (1979). Theoretical conformational analysis of proctolin molecule. *Bioorganiszeskaya Chimya*, **5**, 1581–1583.

BROWN, B.E. (1977a). Occurrence of proctolin in 6 orders of insects. *Journal of Insect Physiology*, **23**, 861–864.

BROWN, B.E. (1977b). Proctolin – a peptide transmitter candidate in insects. *Life Sciences*, **17**, 1241–1252

BROWN, B.E. and STARRATT, A.N. (1975). Isolation of proctolin, a myotropic peptide from *Periplaneta americana*. *Journal of Insect Physiology*, **21**, 1879–1881.

HOLMAN, G.M. and COOK, B.J. (1979a). Analytical determination of proctolin by HPLC and its pharmacological action in the stable fly. *Comparative of Biochemistry and Physiology*, **62C**, 231–235.

HOLMAN, G.M. and COOK, B.J. (1979b). Evidence for proctolin and a second myotropic peptide in the cockroach *Leucophea maderaea, Insect Biochemistry*, **9**, 149–154.

JEDRUSIAK, J., PLECH, A., BRUS, R., SOBOTKA, W. and KONOPINSKA, D. (1991a). Studies on central effect of [Lys^1]-proctolin an analogue of proctolin. *Bulletin of the Polish Academy of Science*, in press.

JEDRUSIAK, J., PLECH, A., BRUS, R., KONOPINSKA, D. and SOBOTKA, W. (1991b). Cardiovascular effects on synthetic proctolin analogues in rats. Manuscript in preparation.

KAZANOWSKA, G., BOGUSLAWSKA-JAWORSKA, J. and KONOPINSKA, D. (1989). Effect of natural and synthetic peptides on the biological function of leukemic cells. In *Modern Trends in Human Leukemia VIII* (R. Neth, ed), pp. 223–225. Springer Verlag.

KONOPINSKA, D., KAZANOWSKA, B., SOBOTKA, W. and BOGUSLAWSKA-JAWORSKA, J. (1986a). An insect neuropeptide proctolin. A new synthesis and its novel property of restoration of phagocytosis by defective human PMN-leukocites. *Bulletin of the Polish Academy of Science*, **34**, 327–331.

KONOPINSKA, D., SOBOTKA, W., LESICKI, A., ROSINSKI, G. and SUJAK, P. (1986b). Synthesis of proctolin analogs modified in the position 2 of peptide chain and their cardioexcitatory effect on cockroach *Periplaneta americana* L. and yellow mealworm *Tenebrio molitor* L. *International Journal of Peptide and Protein Research*, **27**, 597–603.

KONOPINSKA, D., ROSINSKI, G., LESICKI, A., SUJAK, P., SOBOTKA, W. and BARTOSZ-BECHOWSKI, H. (1988a). New N-terminal modified proctolin analogs. Synthesis and their cardioexcitatory effect on insects. *International Journal of Peptide and Protein Research*, **31**, 463–467.

KONOPINSKA, D., ROSINSKI, G., LESICKI, A., SUJAK, P. and SOBOTKA, W. (1988b). [L-Dopa^2]-proctolin: synthesis and its high cardioexcitatory effect on *Periplaneta americana* L. *Bulletin of the Polish Academy of Science*, **36**, 17–20.

KONOPINSKA, D., ROSINSKI, G., LESICKI, A., SUJAK, P., BARTOSZ-BECHOWSKI, H., SOBOTKA, W., PLECH, A. and BRUS, R. (1989). Biological properties of proctolin and some of its synthetic analogs. In *Insecticide Mechanism of Action and Resistance* (Akademie der Landwertschaftswissenschaften der DDR, ed), *Tagungsberichte*, **274**, 105–109.

KONOPINSKA, D., BARTOSZ-BECHOWSKI, H., ROSINSKI, G., LESICKI, A., SUJAK, P. and SOBOTKA, W. (1990). Role of guanidine group at the N-terminal proctolin peptide chain in cardioexcitatory effects in insects. *International Journal of Peptide and Protein Research*, **35**, 12–17.

KONOPINSKA, D., BARTOSZ-BECHOWSKI, H., ROSINSKI, G. and SOBOTKA, W. (1991). Further proctolin analogs and their biological properties. *Polish Journal of Pharmacology*, in press.

KRAVITZ, E.A., BELTZ, B., GLASMAN, S.G., GOY, M., HARRIS-WARWIK, R., JOHNSON, M., LIVINGSTONE, M. and SCHWARTZ, T. (1984). The well modulated lobster: the role of serotonin, octopamine and proctolin in the lobster nervous system. *Pesticide Biochemistry*, **22**, 133–147.

MILLER, T.A. (1979). Nervous. vs. neurohormonal control of insect heartbeat. *Annals of Zoology*, **19**, 77–89.

MILLER, T.A. and SULLIVAN, R.E. (1981). Some effects of proctolin on the cardiac ganglion of the main lobster, *Homarus americanus*. *Journal of Neuroscience*, **12**, 626–639.

O'CALLAGHAN, J.P. and HOLTZMAN, S.G. (1975). Quantification of the analgesic activity of narcotic antagonists by a modified hot-plate procedure. *Journal of Pharmacology Experimental Therapy*, **192**, 497–505.

ORCHARD, I., BALANGER, J.H. and LANGE, A.B. (1989). Proctolin: a review with emphasis on insects. *Journal of Neurobiology*, **20**, 470–496.

O'SHEA, M. and ADAMS, M.E. (1981). Pentapeptide (proctolin) associated with an identified neuron. *Science*, **213**, 567–569.

QUISTAD, R.B., ADAMS, M.E., SCARBOROUGH, R.M., CARNEY, R.L. and SCHOOLEY, D.A. (1984). Metabolism of proctolin, a pentapeptide neurotransmitter in insects. *Life Sciences*, **34**, 569–576.

ROSINSKI, G. and GÄDE, G. (1988). Hyperglycaemic and myoactive factors in the corpora cardiaca of the mealworm *Tenebrio molitor*. *Journal of Insect Physiology*, **33**, 451–463.

SCHULTZ, H., SCHWARTZBERG, H. and PENZLIN, H. (1981). The insect neuropeptide proctolin can effect the CNS and smooth muscle of mammals. *Acta Biologica Germanica*, **40**, K1–K5.

STARRATT, A.N. and BROWN, B.E. (1977). Synthesis of proctolin, a pharmacologically active pentapeptide in insects. *Canadian Journal of Chemistry*, **55**, 4238–4242.

STARRATT, A.N. and BROWN, B.E. (1979). Analogs of the insect myotropic peptide proctolin. Synthesis and structure-activity studies.*Biochemistry and Biophysics Research Communication*, **90**, 1125–1130.

STARRATT, A.N. and STEELE, R.W. (1984). *In vivo* inactivation of the insect neuropeptide proctolin in *Periplaneta americana*. *Insect Chemistry*, **14**, 97–102.

SULLIVAN, R.E. and NEWCOMB, R.W. (1982). Structure-function analysis of an arthropod peptide hormone – proctolin and synthetic analogs compared on the cockroach hindgut receptor. *Peptides*, **3**, 337–344.

20
The GABA-Activated Cl⁻ Channel in Insects as Target for Insecticide Action – A Physiological Study

HARALD C. VON KEYSERLINGK[1] and R. JOHN WILLIS[2]

[1]*Schering AG Agrochemical Research, PO Box 65 01 11, D-1000 Berlin 65, Germany*
[2]*Schering Agrochemicals Ltd, Saffron Walden, Essex CB10 1XL, UK*

Introduction

For the past 200 years the majority of insect control technologies have been and still are based on molecules which interfere with the voltage activated Na^+ channel, the cholinergic synapse and the GABAergic synapse. The idea that one major inhibitory system in insects, the GABA activated Cl^- channel is an important target of insecticide action is based on the observation that cyclohexane and cyclodiene insecticides antagonize GABA-activated Cl^- currents (Matsumura and Ghiasuddin, 1983; Wafford *et al.*, 1989) and that avermectins agonize inhibitory Cl^- currents in most insect preparations (Tanaka and Matsumura, 1985; Abalis *et al.*, 1986; Bokisch and Walker, 1986; Hart *et al.*, 1986). This GABA-activated Cl^- channel in insect nerve and muscle cell membranes has since become a major focus of attention in both academic and industrial research institutions (Lummis, 1990; Sattelle, 1990). GABA receptors associated with Cl^- ion channels have been found in insect nerve and muscle cells (Table 1).

Neuronal GABA receptors have been localized in the

Insecticides: Mechanism of Action and Resistance
© 1992   Intercept Ltd, P.O. Box 716, Andover, Hants SP10 1YG, UK

Table 1 Established and putative localisations of GABA activated Cl⁻ channels in insects

---

neuron

dendritic, postsynaptic

soma   function   ??

terminal, presynaptic   ??

skeletal muscle

postsynaptic

extrasynaptic

other muscles   ??

---

postsynaptic, dendritic membranes in the neuropile, but also in the membranes of nerve cell somata where their function is not really clear. These nerve cell bodies do not receive inhibitory, GABA-emitting nerve terminals. In most cases these somata are not actively involved in electrical signalling at all but seem to be responsible for housekeeping support for the electrically active dendritic, axonal and terminal ramifications of the neuron.

Some but not all skeletal muscle fibres in insects are endowed with GABA-activated Cl⁻ channels, and some of these have been well characterized (Cull Candy, 1986; Scott and Duce, 1987; Bermudez and Beadle, 1988; Murphy and Wann, 1988; Fraser *et al.*, 1990).

Less information is available of GABA receptors in visceral and autonomous nerve-muscle systems, such as pulsating muscles regulating the flow of body fluids, ventilatory muscles regulating respiration, reproductive organs and the stomatogastric system.

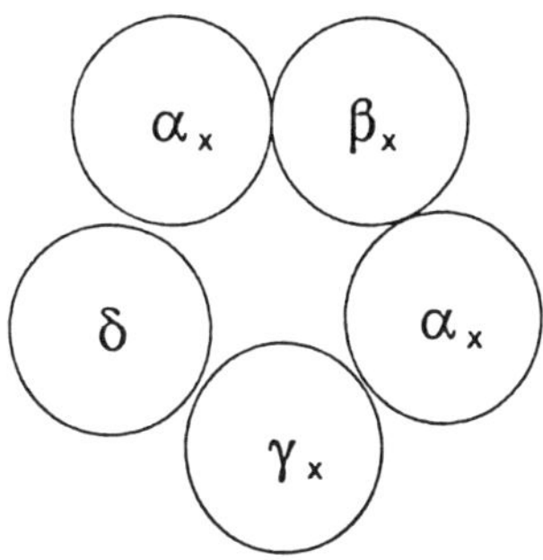

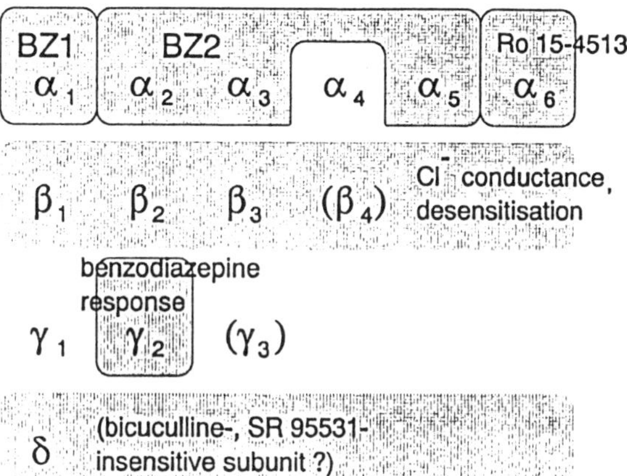

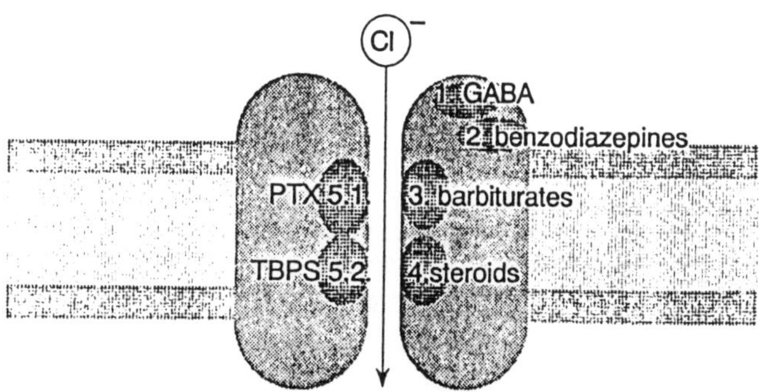

Figure 1 The GABA activated Cl⁻ ionophore complex.
(a) The hypothetical pentameric structure of the channel.
(b) Putative associations between protein structure and pharmacological profile or function.
(c) The five major ligand binding domaines.

GABA receptors with a unique pharmacological profile have recently been well characterized in the limulus heart (Benson, 1989).

The main physiological role of GABAergic innervation in central nervous and in peripheral neuromuscular systems seems to be the organization of simultaneous and consecutive excitatory events into coordinated output. Coordinated behaviour is possible only through a delicately tuned balance between excitatory and inhibitory synaptic events resulting in patterns of action potentials which encode meaningful messages for the receiving effector systems (Watson and Burrows, 1987; Egelhaaf *et al.*, 1990).

Molecular biologists have recently identified a bewildering variety of different $GABA_A$ receptor subunits in mammalian nervous systems based on varying degrees of sequence homologies, differing regional localizations and differing pharmacological profiles (Shivers *et al.*, 1989; Sieghart, 1989; Betz, 1990; von Blankenfeld *et al.*, 1990; Draguhn *et al.*, 1990; Olsen *et al.*, 1990; Verdoorn *et al.*, 1990; Sigel *et al.*, 1990; Lüddens *et al.*, 1991). Which of these twelve or fourteen different proteins associated with GABA-activated Cl⁻ channels in vertebrates match GABA receptor subunits from insect sources is currently a major focus of research activities (Figure 1).

The GABA Receptor Pharmacology

This GABA-activated Cl⁻ ionophore complex is characterized by at least five major separate binding domaines for agonists, antagonists, modulators and channel blockers (Figure 1 (c) and Table 2).

The success of every structure-activity-optimization endeavour depends entirely on the reliability and relevance of the activity data the chosen assay can feed back.

Which assays give the clearest information on the pharmacology of GABA receptors in insects? Nerve cell bodies dissociated from locust thoracic ganglia under continuous, rapid saline perfusion have proved to be a very stable and useful preparation (Figure 2b). GABA-activated Cl⁻ channels are found on every soma, as has previously been described (Lees *et al.*, 1987; Benson, 1988).

These nerve cell bodies are not from identified cells. Pharmacological differences found may be due to different receptor densities or receptor types in these different cells rather than to different properties of the investigated ligands. The experimental set-up allows us to record from these cells for many hours and thus to characterize voltage- activated and ligand-activated currents before each pharmacological experiment. In this way it is possible to do comparative pharmacological studies on a fairly homogeneous

Table 2 Putative ligand binding domaines at the GABA activated Cl⁻ ionophore complex

---

1.   agonists

GABA, muscimol, ZAPA

antagonists

bicuculline, SR 95531

2.   modulators: benzodiazepines

2.1.   BZ 1   zolpidem, oxoquazepam

2.2.   unspecific   flunitrazepam

2.3.   "peripheral"   chlorodiazepam

2.4.   inverse agonist   Ro 15-4513

3.   modulators at the channel site

3.   barbiturates   pentobarbital

4.   modulators at the channel site

4.   steroids

5.   channel blockers

5.1.   picrotoxinin

5.2.   TBPS

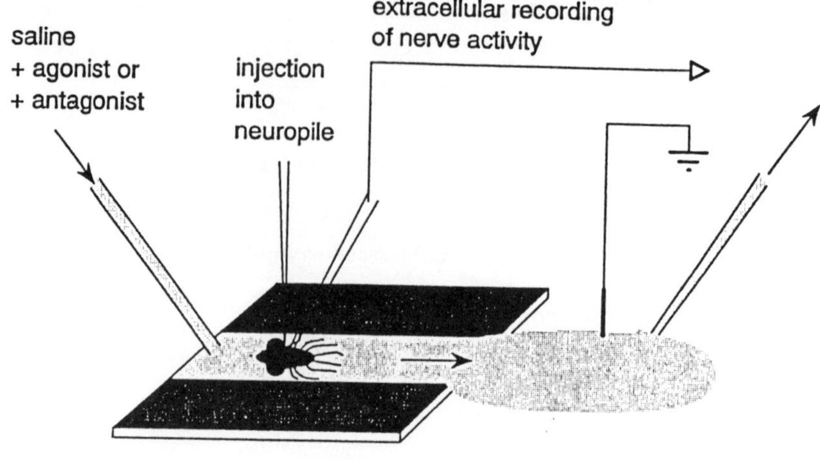

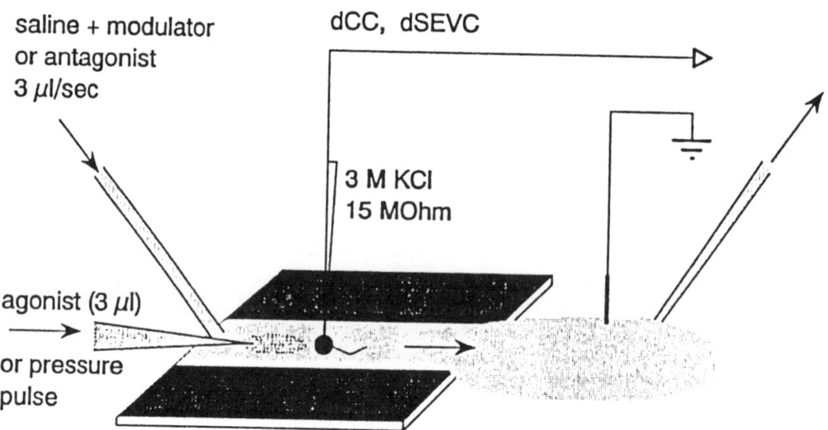

Figure 2   The experimental set up
(a) Recording of the spontaneous firing activity of the isolated synganglion of *Calliphora erythrocephala* wandering stage larvae. Neuropile injections of drugs are possible.
(b) Arrangement for single electrode voltage clamp (SEVC) experiments on isolated nerve cell bodies dissociated from locust metathoracic ganglia. The middle groove for the saline flow between the two parafilm sheeths covering either side of the microscopic cover slip is so narrow that one µl saline covers the nerve cell and bridges both edges.

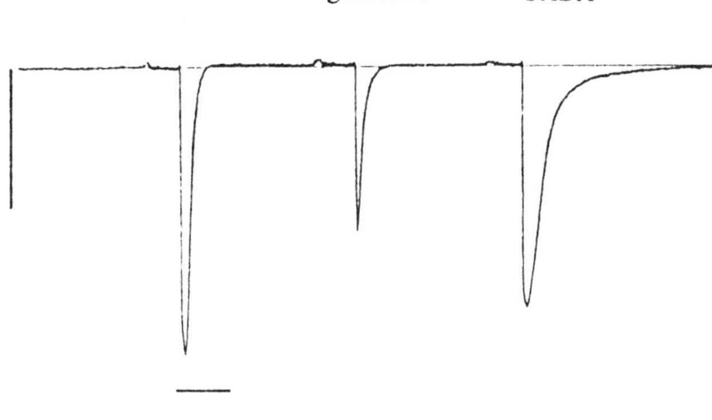

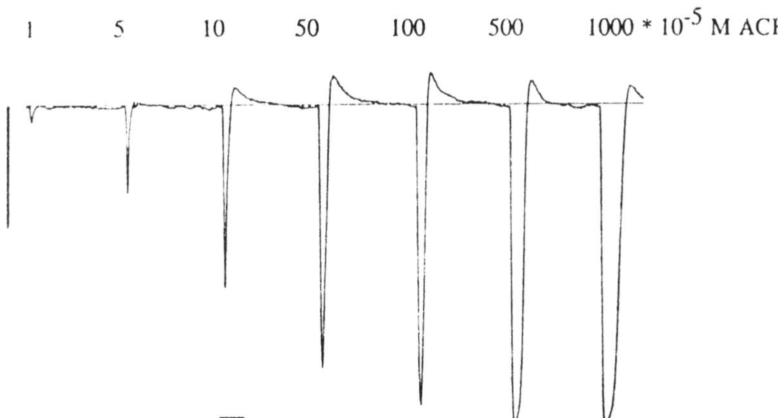

Figure 3   Inward currents recorded from isolated locust nerve cell bodies evoked by hand applied 3 µl pulses of transmitter solutions. Calibration: 1 nA, 10 sec.

(a)   Cell clamped at a holding potential (hp) of -80 mV. ACH, glutamate and GABA are dissolved in saline at 10^{-4} M.

(b)   Concentration response curve to ACH. Hp -100 mV (note biphasic response which occurs always at holding potentials more negative than -90 mV).

population of characterized cell types. All cells respond to acetylcholine, GABA and glutamate (Figure 3a). The hyperpolarizing responses to glutamate are based on inward currents which reverse at the same membrane potentials as the GABA-activated current does (Figure 4a). The reversal of both the GABA- and the

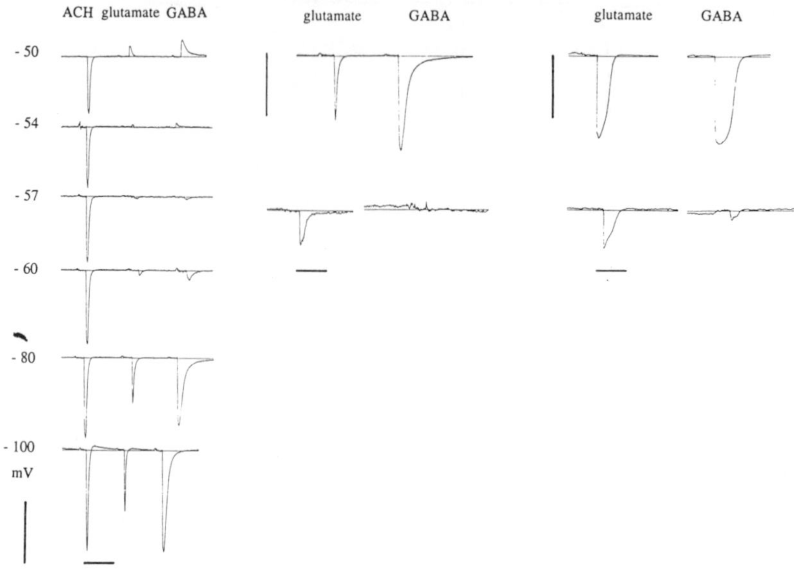

Figure 4 Comparison of currents evoked by 3 µl pulses of glutamate and GABA at 10^{-4} M. Calibration: 1 nA, 10 sec.

(a) Glutamate- and GABA-evoked currents reverse at precisely the same membrane potential (this cell -55 mV). Calibration: 0,5 nA at hp between -50 and -60 mV

(b) No cross desensitisation between GABA and glutamate after 10 minutes superfusion with GABA 5 * 10^{-5} M. Calibration: 0,5 nA, 5 sec.

(c) Superfusion with picrotoxinin, 10^{-5} M, for 10 minutes suppresses the GABA-activated current. Calibration: *upper trace* = 1 nA, 5 sec; *lower trace* = 0,5 nA, 5 sec.

glutamate-activated current are very sensitive to changes in the extracellular and the intracellular Cl^- concentration, thus suggesting that Cl^- ions are carrying the glutamate-activated current also. GABA- and glutamate-activated Cl^- currents do not cross desensitize (Figure 4b) and they are not blocked by the same compounds (Figure 4c) suggesting that they are funnelled through separate proteins.

Dose response curves to ACH, GABA and glutamate indicate that

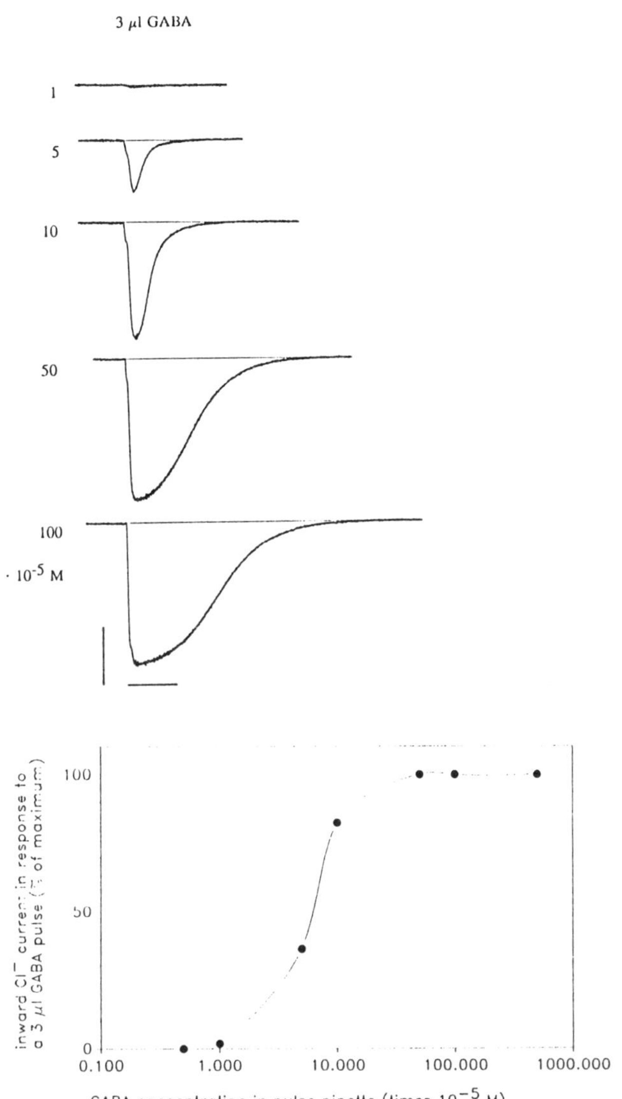

Figure 5 Quantification of the response to GABA. SEVC, hp -80 mV.
(a) Concentration response curve. Hand application of three µl
   GABA pulses. Calibration: 0,5 nA, 5 sec.

the cells differ in their sensitivities to these agonists and with respect
to the maximum peak current that can be elicited by each individual

0.05 - 4 s GABA 10^{-4}

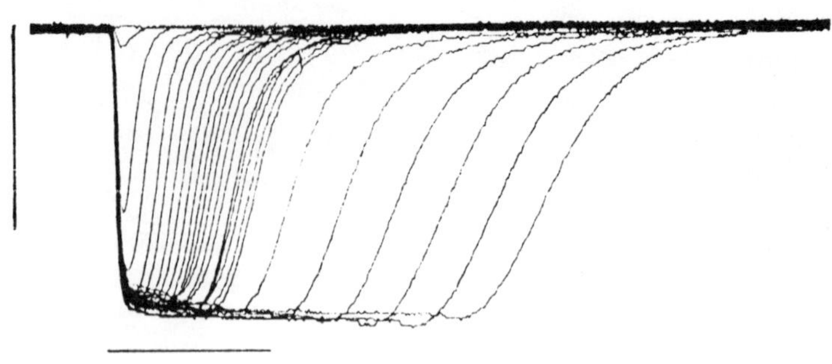

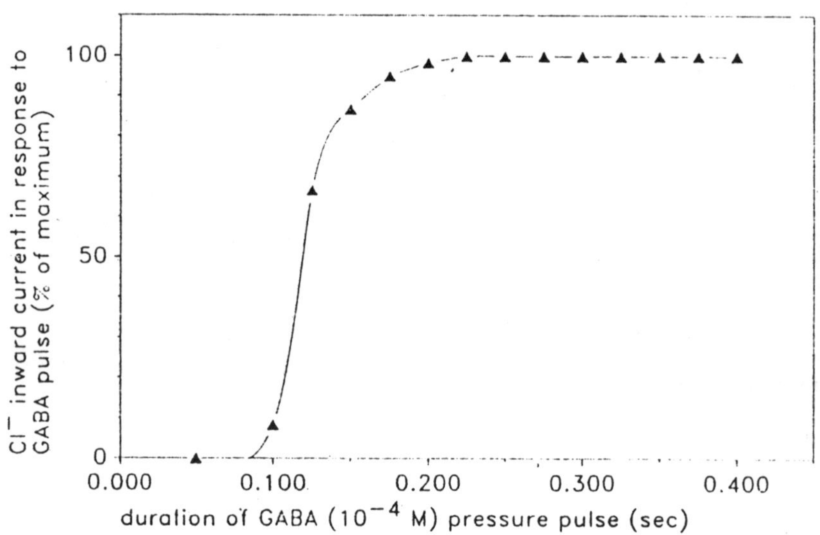

Figure 5(b) Pulse duration response response curve. Stimulator driven pressure pulses of GABA at 10^{-4} M. Calibration: 0,5 nA, 2 sec.

agonist. Typical maximum currents induced by GABA pulses are between -9.5 and -1.5 nA at -80 mV holding potential. 50% maximum current can be induced by GABA pulse concentrations between 10 and 5000 µM. In most cells the inward current elicited by pressure pulses of GABA, 10^{-4} M inactivates within seconds after the end of the GABA pulse.

The dose response curves obtained by either increasing GABA

concentrations (Figure 5a) or by increasing the pulse duration at a fixed GABA concentration ($5 \cdot 10^{-5}$ M) (Figure 5b) are very steep. The GABA-activated Cl⁻ currents are most sensitive to pharmacological modification in the range between 20% (potentiators) and 80% (attenuators) of their maxima.

Using this preparation as functional assay to study the pharmacology of the insect GABA-activated Cl⁻ channel, we found the following results.

The agonist site

Several GABA agonists can be identified and there is little variation in the relative agonist sensitivity among the different cells (Figure 6).

Muscimol, ZAPA and its seleno-analogue are the most potent agonists but isoguvacin THIP and thiomuscimol are active as well. Although several of these agonists are potent compared to the endogenous ligand GABA, 10^{-5} M is a relatively high concentration in general terms of drug activity.

As has been described before (Sattelle, 1990; Benson, 1988), the competitive and diagnostic mammalian $GABA_A$ antagonists bicuculline and SR 95 531 show no effect up to 10^{-4} M concentrations (Lees *et al.*, 1987). One interesting aspect has emerged, when it was shown conclusively that in the granular layer of the mammalian cerebellum – a region with a high density of GABA, muscimol and TBPS binding sites – no bicuculline and SR 95 531 binding can be detected, suggesting that in this part of the mammalian brain there is a $GABA_A$ receptor population whose pharmacological profile seems to match that of the insect GABA receptor (Olsen *et al.*, 1990).

The modulatory site

Four putative benzodiazepine binding domains have been identified in vertebrate tissues, three in different CNS regions and one in the periphery.

Ligand binding studies had shown that the most potent benzodiazepine in insects was chlorodiazepam, Ro 05-4864, the selective ligand for the peripheral benzodiazepine receptor in vertebrates which is not associated with a GABA-activated Cl⁻ channel complex (Sattelle, 1990; Lunt *et al.*, 1988). Again, it has been shown recently that Ro 05-4864 binding sites also occur in the piscine cerebellum (Eshleman and Murray, 1990).

Physiological data from locust somata on potentiation or attenuation of GABA-activated Cl⁻ currents by ligands at the

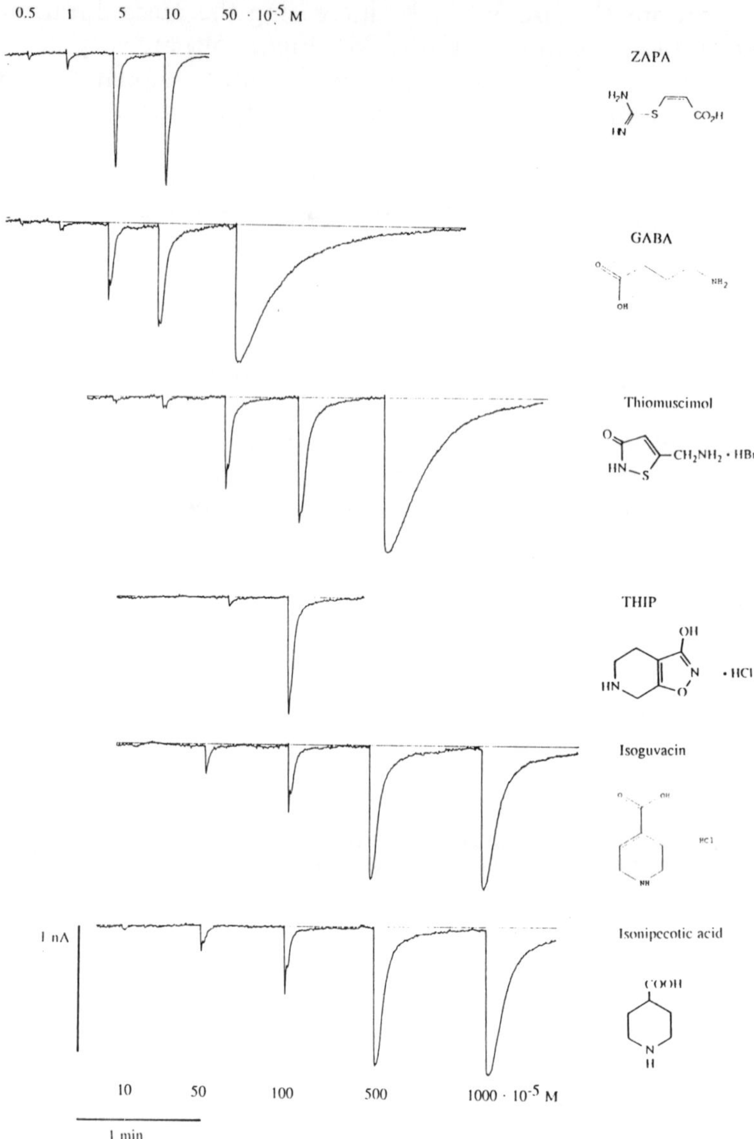

Figure 6 Concentration response curves to various GABA agonists. SEVC, hp -80 mV. Calibration: 1 nA, 1 minute.

benzodiazepine site turned out to be greatly variable between individual cells.

Among non-selective benzodiazepine ligands, diazepam is inactive

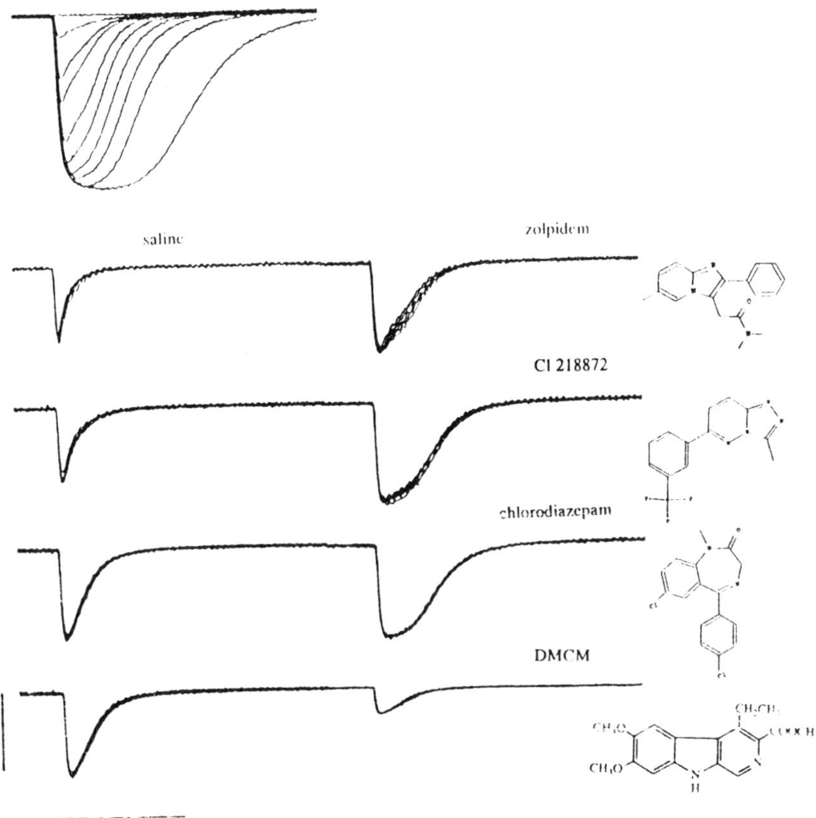

Figure 7 Modification of GABA-activated currents by ligands of benzodiazepine binding sites. Stimulator driven pressure pulses of GABA at 10^{-4} M. Pulse frequency: 1/min. SEVC, hp -80 mV. Calibration: 0,5 nA, 5 sec.

(a)   Superimposed traces of inward currents evoked by GABA pressure pulses of increasing duration (10, 100, 150, 200, 300, 400, 500, 750, 1000, 2000 msec).

(b–e)   Five superimposed traces. Left hand side: saline. Right hand side: 10–15 minutes after perfusion with the modulator, 10^{-5} M.

but flunitrazepam consistently potentiates the GABA response with the factor 1.4.

The selective BZ 1 ligands zolpidem, Cl 218872 and oxoquazepam gave capricious results, potentiating GABA responses strongly in

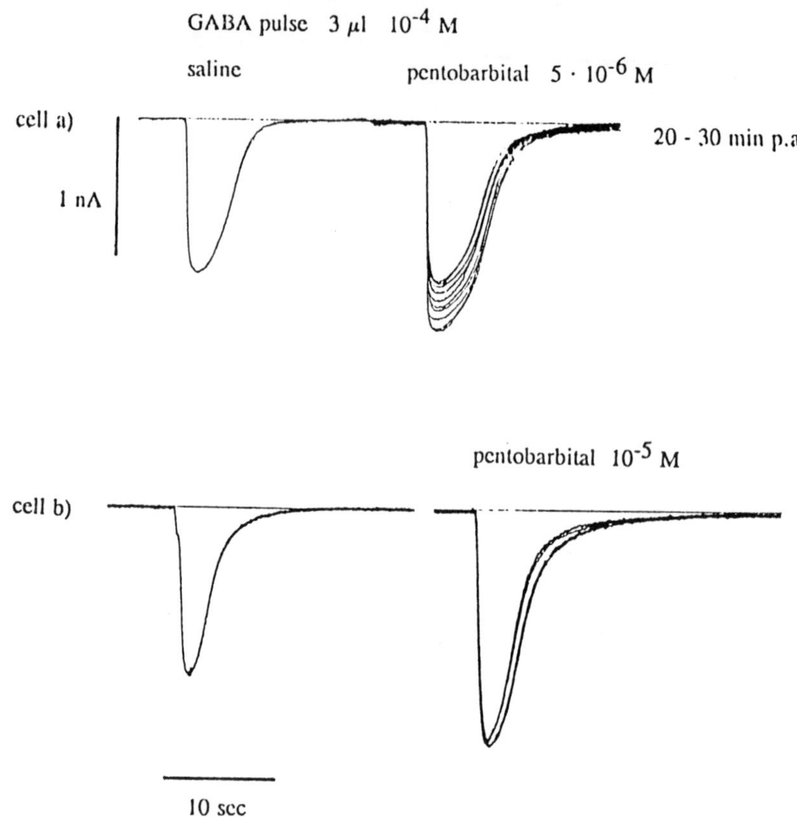

Figure 8   Modification of GABA-activated currents by pentobarbital. Stimulator driven pressure pulses of GABA at 10^{-4} M. Pulse frequency: 1/min. SEVC, hp -80 mV.

some cells but being inactive or even attenuating them in others. The most potent but most capricious ligand in this respect was zolpidem (Niddam *et al.*, 1987) (Figure 7b). Benzodiazepine type 1 binding sites are found in particularly high densities in the mammalian cerebellum (Eshleman and Murray, 1990).

The 'selective peripheral' BZ ligand, chlorodiazepam, potentiates the GABA responses to a significantly higher degree than flunitrazepam does although here, too, individual neurones differed substantially (Figure 7d).

The inverse agonist, DMCM, attenuates the GABA responses in locust somata (Figure 7e).

Thus, the benzodiazepine modulatory sites on the GABA receptor

Cl⁻ ionophore complex, which not long ago were believed to occur in vertebrates only as late specialization in evolution, seem to play a more important role in the regulation of insect activity and behaviour as has hitherto been recognized. These binding sites clearly warrant a more detailed examination. Subpopulations of benzodiazepine receptors with differing pharmacological profiles possibly exist within insect central nervous systems in a manner analogous to the vertebrate brain.

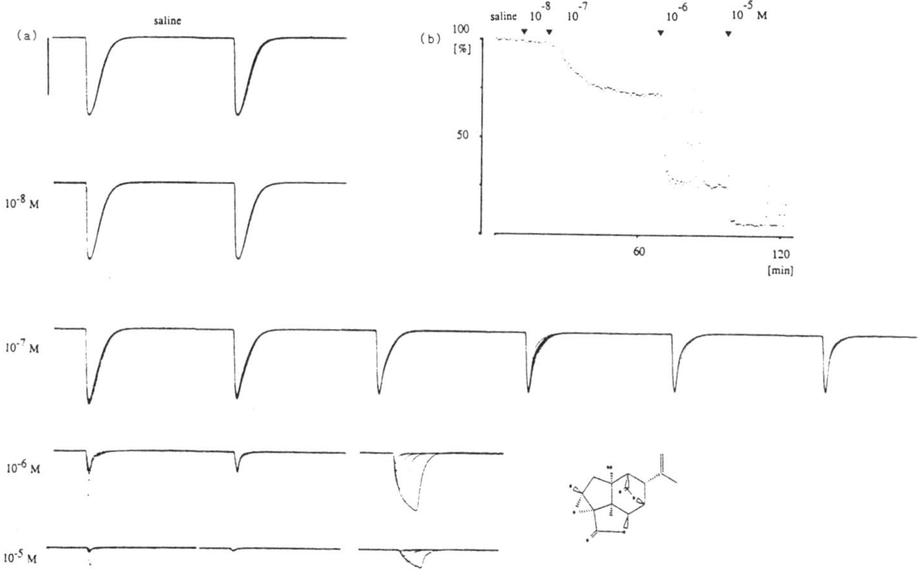

Figure 9 Block of GABA-activated currents by picrotoxinin. Stimulator driven pressure pulses of GABA at 10^{-4} M. Pulse frequency: 1/min. SEVC, hp -80 mV. Calibration: 0,5 nA, 5 sec.

(a) At least five traces superimposed. At 10^{-6} M and 10^{-6} M PTX, after reaching the plateau, the duration of the GABA pulse has been increased 2, 5, 10 and 20 times. Increasing GABA doses evoke increasing inward currents. This is not expected to happen with purely non-competitive channel blockers.

(b) The peak current amplitude of each GABA activated current as recorded during the entire experiment.

The channel modulation sites

It is well known that pentobarbital significantly potentiates the GABA-activated Cl⁻ current at concentrations above $5 \cdot 10^{-6}$ M (Figure 8) (Sattelle, 1990; Lees *et al.*, 1987). The pharmacology of this binding site, however, has not yet been studied in any detail in insects. This lack of information is even more apparent with regard to the steroid binding site.

The channel blocker sites

Among the non-competitive Cl⁻ channel blockers, picrotoxinin and the TBPS analogues belong to the most potent compounds producing $\approx$ 50% block at about $5 \cdot 10^{-7}$ M.

However, the non-competitive block induced by picrotoxinin, even at high concentrations, is never complete. Pulses of high GABA concentrations (10^{-3} M) are still capable of inducing a slow inward current that reverses at the same membrane potential as the transient, unmodified Cl⁻ current. This slow Cl⁻ current can also be activated by muscimol (10^{-4} M) in the picrotoxinin (10^{-5} M) blocked GABA receptor Cl⁻ channel complex (Figure 9). Thus, some interaction between the agonist site and the PTX site seems to occur at high agonist concentrations.

Summary

In summary, there are several agonists, several modulators and several non-competitive antagonists which are recognized by the GABA-activated Cl⁻ channel in locust somata. These molecules alter the functional behaviour of this protein complex in concentration ranges which are not far from the action of cyclohexane, cyclodiene- and avermectin-insecticides. So, apparently, on the cell membrane we are unable to distinguish the action of an insecticide from the action of one of these pharmacological tools.

Could then perhaps one of these ligands serve as lead structure for insecticide synthesis?

Insecticide Action

The principal property of an insecticide is the fact that these molecules invading the organism from outside through the integument, take over the control from the central nervous system and override the activities of most or all innervated cells by altering the messages encoded in the firing patterns of neuronal action potentials as received by the various effector systems.

To what extent are these pharmacological tools with significant receptor activity in insect nerve cell somata capable of disrupting the activities of intact ganglia or whole insects?

Figure 10 illustrates the barriers a xenobiotic molecule encounters in a contaminated insect.

Table 3 shows some of the most important properties an

insecticide must have before it can be developed into a product that we are able to offer to the public.

The performances of our potential lead structures against these backgrounds are summarized in Tables 4 and 5.

The whole organ activity data from isolated CNS of *Calliphora* larvae match surprisingly well with the receptor data from locust nerve cell bodies. No gross pharmacological differences can be detected between the GABA receptors in the locust neuronal soma membranes and the functional dendritic GABA receptors in isolated ganglia of the fly larvae. Only ZAPA and its analogues do not cross the blood brain barrier. They are active, however, when injected directly into the neuropile.

The most obvious pharmacological differences between the neuronal GABA receptors in the CNS and the peripheral GABA receptors in muscle membranes are the poor sensitivity of the peripheral GABA responses to Cl⁻ channel blockers such as picrotoxinin and TBPS, yet their relatively high sensitivity to agonists such as muscimol and ZAPA.

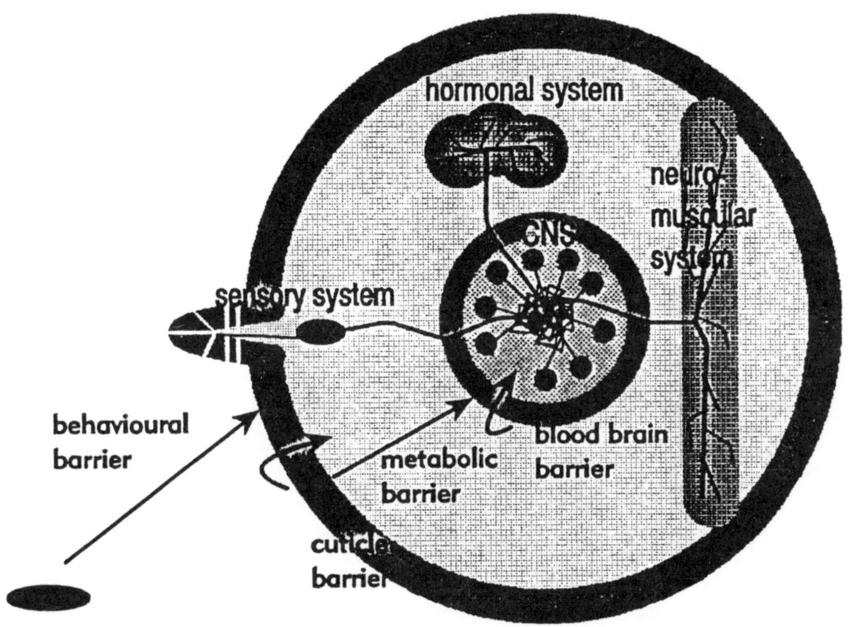

Figure 10 Diagram of the entry of a xenobiotic molecule into an insect and of the principle components of the insect nervous system.

The housefly injection data (Table 5) indicate that most of the ligands which show significant potency at the receptor level induce symptoms or knockdown conditions.

Both pharmacological modifications of GABA-activated Cl^- current, potentiation and attenuation or blockade, disrupt the voluntary neuromuscular system of insects to such an extent that complete paralysis may be induced for many hours or even days. This, however, is not necessarily associated with the death of the affected animal. These deeply intoxicated insects recover after hours or days, most of them without obvious residual lesions.

In other words: pharmacological modification of GABA receptor functioning is not lethal on its own. Drastic potentiation or blockade of GABA-activated Cl^- current for hours does not kill the affected isolated brain cell or the isolated ganglion.

Secondary processes elsewhere in the organism must be initiated and sustained before the entire animal irreversibly collapses. These secondary intoxication processes are not very well understood. They cannot reliably be monitored in the voluntary neuromuscular system. Some functions of the autonomous nervous system regulating pulsating and ventilatory movements, haemocoel pressure and visceral muscles such as movements of the intestine, excretory and reproductive organs may provide more accurate information about the actual status of intoxication and the distance or proximity to death.

At first sight our data suggests that agonists such as muscimol, THIP and ZAPA may be viewed as potential candidates as lead structures for insecticide synthesis.

Among the non-competitive antagonists, TBPS clearly stands out. Some benzodiazepine- and β carboline-type modulators can also be disruptive to the activities of whole ganglia and intact insects. Comparing the activity of these molecules in houseflies and lepidopterous larvae we find that none of these GABA agonists, antagonists and modulators shows any effect against the most important pest species, *Heliothis virescens* larvae, although there is evidence that GABA-activated Cl^- channels do exist in the central nervous system of this species.

This observation is puzzling and casts doubts on the routine of using locusts, cockroaches and flies as models to extrapolate activity data across the evolutionary tree.

When a lead structure is selected, clear aims for optimization have to be defined. In the case of muscimol-type agonist receptor, potency, irreversible lethal action and cuticle penetration as well as activity against lepidopterous larvae have to be increased

Table 3   Some preconditions or key hurdles for successful product development

potency	against the actual pest instar of the pest species after self contamination
activity	against resistant populations
no activity	against indifferent or "beneficial" arthropods in terrestrial and aquatic environments
no activity	against vertebrates
halflife	to be active but not accumulative in the field
no leaching	into water systems
cost efficacy	relative to available standards

substantially before competitive insecticidal activity can be expected.

None of the compounds discussed so far show insecticide activity after selfcontamination or topical application in the most important insect pest species.

This is a serious limitation since simultaneous optimization of both potency at the receptor protein and diffusion through the arthropod integument has proved to be a difficult task. The toxicokinetic behaviour of these ligands of the GABA-activated Cl^- channel does not permit the contamination of this vulnerable protein with a sufficiently high concentration for a sufficiently long time to induce irreversible lesions.

Since 1980 about 3,600 patents claiming insecticide activity have been filed. We have not been able to identify structures which have been derived purely from receptor pharmacology and have been optimized to potent insecticides. Professor Casidas trioxabicyclooctane-type molecules developed from TBPS probably are the only outstanding exception (Palmer *et al.*, 1990).

This does not necessarily imply that it is impossible to achieve good cuticle penetration and to retain or improve receptor potency

Tables 4a-c

GABA receptor pharmacology	locust soma	fly CNS		locust muscle
		EC_{20} (X 10^{-5} M)		
1.1. GABA agonists				
muscimol	1	--	1	10
Z-3-(amidinothio)propenoic acid	1	>>	50	2
GABA	2	>>	10000	10
Z-3-(amidinoseleno)propenoic acid	5	--	500	10
E-3-(amidinothio)propenoic acid	10	>>	100	100
thiomuscimol	20	--	10	1000
isoguvacine HCl	20	--	10	100
THIP	40	--	10	100
isonipecotic acid	50	>>	100	200
Z-3-(2-benzimidazolylthio)propenoic acid	>> 100	>>	50	5
1.2. GABA antagonists				
bicuculline methiiodide	>> 10	>>	10	>> 10
SR 95531	>> 10	>>	10	>> 10

GABA receptor pharmacology	locust soma	fly CNS		locust muscle
	potentiation at 10^{-5} M	EC_{20} (X 10^{-5} M)		
2.1. benzodiazepines, BZ1				
zolpidem	3.8	++	1	1
Cl 218872	2.2	++	100	>> 1
oxoquazepam	1.6	++	1	>> 1
2.2. benzodiazepines, mixed				
flunitrazepam	1.4	rythm	10	>> 10
2.3. benzodiazepines, "peripheral"				
Ro 05-4864 clorodiazepam	3.6	>>	50	>> 1
2.4. benzodiazepines, "atypical, 6"				
Ro 15-4513	1.3	>>	50	>> 1
2. "inverse agonists"				
DMCM	0.6	>>	50	>> 1

GABA receptor pharmacology	locust soma	fly CNS		locust muscle
	potentiation at 10^{-5} M	EC_{20} (X 10^{-5} M)		
3. channel site, barbiturates				
Pentobarbital	2.1	--	80	50
		EC_{20} (X 10^{-5} M)		
4. channel blockers				
Picrotoxinin	0.01	++	0.1	50
TBPS	0.05	++	0.1	>> 10
lindane	0.05	++	0.1	0.1

Tables 5a-c

GABA receptor pharmacology	Musca domestica adult female injection			topical		Heliothis virescens larvae III injection		topical	
			ED$_{50}$ (ug/insect)						
1.1. GABA agonists									
muscimol	S	K	0,5	>>	10	>>	10	>>	10
Z-3-(amidinothio)propenoic acid		>>	0,5	>>	10	>>	1		
GABA		>>	100	>>	10				
Z-3-(amidinoseleno)propenoic acid		>>	1	>>	10	>>	10		
E-3-(amidinothio)propenoic acid		>>	2	>>	10				
thiomuscimol	S	K	5	>>	10				
isoguvacine HCl	S	S	10	>>	10				
THIP	S	K	5	>>	10	>>	10		
isonipecotic acid		>>	10	>>	10				
Z-3-(2-benzimidazolylthio)propenoic acid		>>	1	>>	10				
1.2. GABA antagonists									
bicuculline methiiodide	S	S	10	>>	10	>>	10	>>	10
SR 95531		>>	10						

GABA receptor pharmacology	Musca domestica adult injection			topical	
		ED$_{50}$ (ug/insect)			
2.1. benzodiazepines, BZ1					
zolpidem		>>	2	>>	10
Cl 218872		>>	2,8		
oxoquazepam		>>	3,7		
2.2. benzodiazepines, mixed					
flunitrazepam	S	(K)	10	>>	10
2.3. benzodiazepines, "peripheral"					
Ro 05-4864 clorodiazepam	S	S	1,6	>>	10
2.4. benzodiazepines, "atypical, 6"					
Ro 15-4513		>>	0,5		
2. "inverse agonists"					
DMCM		>>	0,5		

GABA receptor pharmacology	Musca domestica adult injection			topical		Heliothis virescens larvae III injection		topical	
			ED$_{50}$ (ug/insect)						
3. channel site, barbiturates									
Pentobarbital	S	K	10	>>	10	>>	5	>>	50
4. channel blockers									
Picrotoxinin	C	K	1	>>	10	>>	1	>>	100
TBPS	C	T	0,22	>>	2	>>	1	>>	2
lindane	C	T	0,1	T	0,1	T	1	T	5

at the same time. It only points to the fact that these are separate optimization routes following different structure activity rules.

In general, however, we have experienced that insecticides have been discovered the other way round.

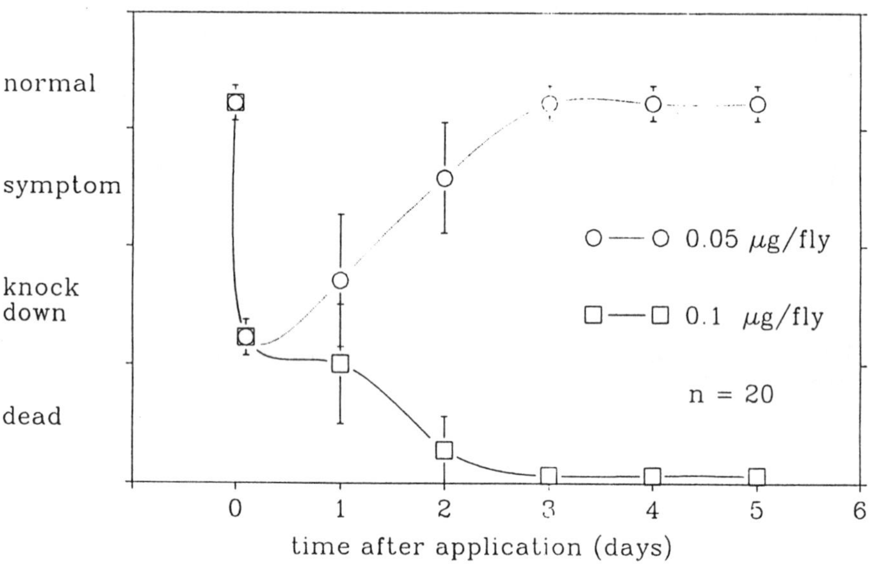

Figure 11   Structure, mammalian toxicity and insecticide activity after topical application against female houseflies of SN606011

The mode of action of a novel azole insecticide

A few members of a herbicide series had shown some activity in an insecticide screen. Analogue synthesis around these lead structures succeeded in shifting the activity away from herbicidal to insecticidal activity only.

What is the mode of action of this molecule? Topically applied to houseflies, it rapidly induces hyperexcitation and knockdown. This knockdown condition lasts for a day or two until, depending on the applied dose, either death or recovery occurs (Figure 11).

Recordings of the electrical activity of the flight muscle cells for several days during the entire intoxication process clearly show that the nerve cell activity is significantly increased soon after topical application of the compound. The patterns of these muscle potentials show that during these many hours of continuous spiking activity the intervals between successive action potentials are reduced (Figures 12a, 12b, 12c).

The isolated CNS preparation from *Calliphora* larvae responds to this compound with a gradual fusion of the cyclic waves of spontaneous firing activity resulting in uninterrupted firing for many hours at concentrations as low as $5 \cdot 10^{-8}$ M (Figure 13).

In the isolated locust nerve cell body this compound antagonizes the GABA-activated Cl⁻ current at 10^{-8} M concentrations (Figure 14).

Detailed competition studies have to be done before the precise binding domain on this GABA receptor-Cl⁻ ionophore complex can be identified with certainty.

How specific is the activity of this molecule and which other proteins are affected?

Table 6 summarizes the effects of SN 606 011 on the different cellular functions looked at so far. These data show clearly that up to concentrations of $5 \cdot 10^{-6}$ M, no other system is affected. At 10^{-5} M, however, we see some effects in the neuromuscular periphery. The GABA responses of the locust extensor tibia are attenuated but not blocked (Figure 15).

Conclusions

Circumstantial, pharmacological evidence suggests that the insect GABA receptors described so far may be linked closest to the those GABA receptor types which are found in high density in the vertebrate cerebellum (bicucullin and SR 59 531 insensitive; Ro 05-4868 sensitive, zolpidem sensitive, Ro 15-4513 sensitive).

It seems likely that GABA receptors of different subunit composition exist, both within the insect central nervous system and the periphery.

The apparent insensitivity of lepidopterous larvae to the GABA pharmacology suggests differing toxicokinetics and requires more detailed investigation.

The novel azole insecticide, SN 606 011, belongs to the most potent antagonists of GABA-activated Cl⁻ current in insects.

Table 6 Actions of SN 606011 in insect neuromuscular systems

---

neuron (dissociated locust nerve cell body)

responses to		10^{-x} M
GABA	block	8
ACH	> >	5
glutamate	> >	5
action potential	> >	5
membrane resistance	> >	5
resting potential	> >	5

---

muscle (locust extensor tibiae muscle)

responses to		
GABA	block	5
glutamate	> >	5
octopamin	> >	5
proctolin	> >	5
5-HT	> >	5
ACH	> >	5
membrane resistance	> >	5
resting potential	> >	5

heartbeat (cockroach)

frequency	> >	5

The ubiquitous presence of glutamate-activated Cl^- channels in insect nerve cell body membranes which are not found in the mammalian CNS raises questions about their function, their pharmacology and their vulnerability in toxicological terms.

Finally, we would like to share a few ideas and speculations.

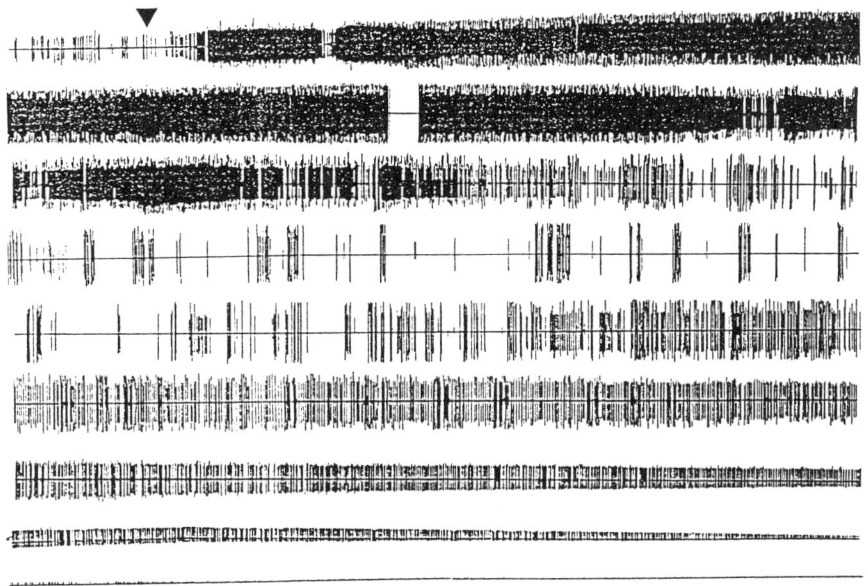

Figure 12(a) Activity of SN606011 against female houseflies. Simultaneous extracellular recording from two dorsolongitudinal flight muscles (4L display positive, 4R display negative). Continuous record for two days. arrow topical application of 0,07 ug SN606011. *Note* (1). Rapid onset of neurotoxic symptoms indicating rapid penetration through the cuticle. (2) Convulsive hyperexitation resulting in complete disruption of sensory and locomotor functions. (3) Almost perfect recovery and normal stimulus response behaviour. (4) Progressive decline of the amplitude of the muscle potentials associated with irreversible lesions resulting in the death of the insect. Calibration: 100 mV, 1hour.

• Blockade or uninterrupted firing of nerve action potentials for many hours can be survived without serious, long-lasting lesions. Which processes in the cascade of events during intoxication are lethal?

• Could it be that in insects the peripheral, voluntary neuromuscular system is a suitable target for paralysing posture and

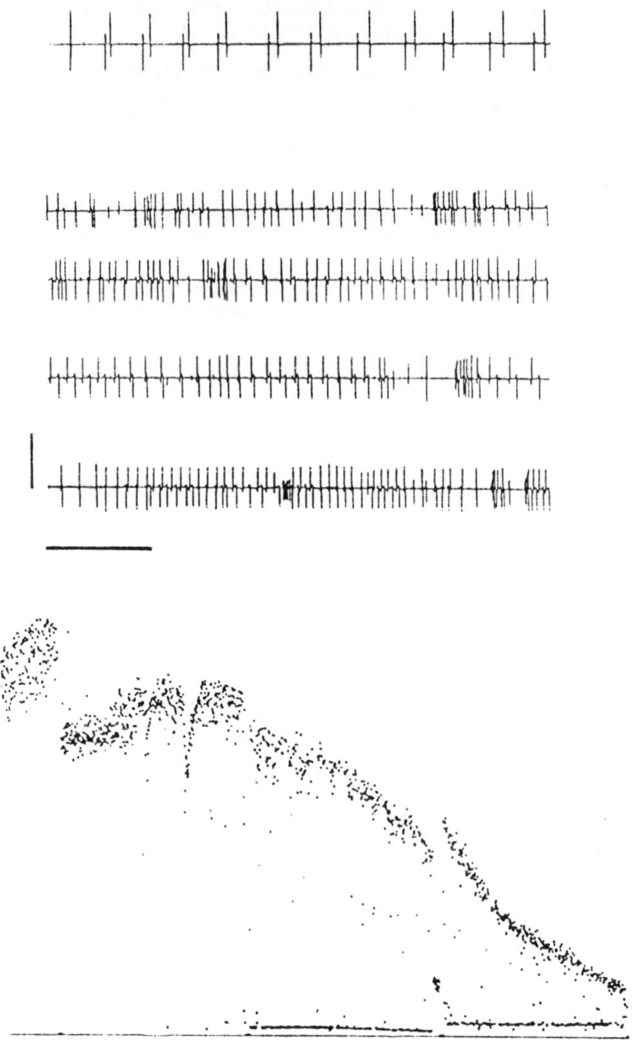

Figure 12(b) Firing pattern of these two separate motorunits. Upper trace: normal, regular spiking pattern. Lower traces: episodes from deep intoxication. Change in inter-spike interval: higher firing frequency and intermittent high frequency bursts. Calibration: 100mV, 1sec

(c) Four episodes out of record (a) played back to a digital spike analyser. Progressive decline of the amplitude of these muscle potentials during the late stages of intoxication. Calibration: 1 hour.

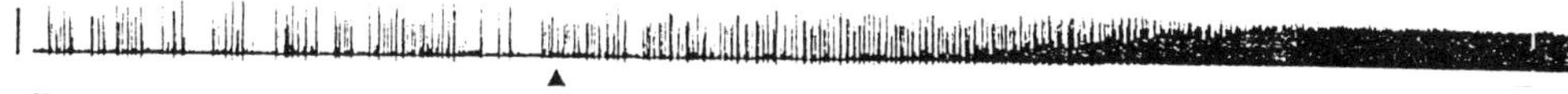

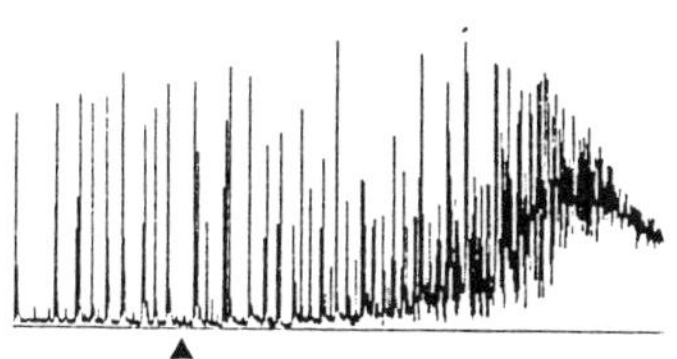

Figure 13   Activity of SN 606011 in the isolated CNS preparation of *Calliphora* larvae (see figure 2a).
(a) Time frequency histogramme. Arrow = start of perfusion with SN 606011, 10^{-7} M. Calibration: 500 spikes/sec, 5 minutes.
(b) Integral of total nerve activity (area under the curve). Arrow = start of perfusion with SN 606011, 5.10^{-8} M. Calibration: 5 minutes.

locomotion only but that it is much less suited as a target for the disruption of life itself? And could it be that the neurosecretory system is a more relevant target in toxicological terms? Would it be possible to interfere with these homeostatic functions directly, leaving the components of membrane excitability undisturbed?

● The selection of a lead structure with a high potential for success may not be trivial at all. Pharmacological potency alone certainly is not sufficient. Not only good neuropharmacology is required but enlightened insect toxicology as well.

● Could it be that one of the most important contributions of more detailed observations than screening data can provide lies in the precision with which error messages can be formulated? What is the key hurdle for the failure of this particular compound or idea? Every structure optimization procedure is a trial-and-error endeavour which is exclusively guided by the relevance of the activity data that we are able to feed back. Trial and error is good as long as one is able to correct the error fast. The tip of the electrode is one superb and indispensable guide to identify precisely the hurdle one may have bumped against.

This process of a first attempt, identifying and memorizing the error and re-trying a corrected version again is the operational principle which makes even simple nervous systems so tremendously powerful. Neurobiologists call such procedures orientation and learning. We should try to learn more from the operational functioning of nervous systems.

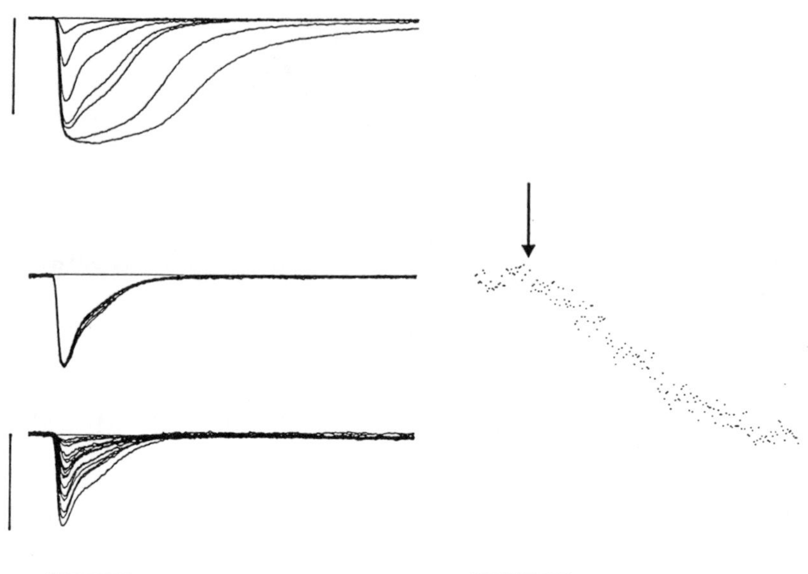

Figure 14 Activity of SN 606011 on isolated nerve cell bodies dissociated from locust metathoracic ganglia (see figure 2b). SEVC, hp -80 mV.

(a) *Upper trace*: superimposed records of GABA (10^{-4} M) activated Cl^- currents evoked by agonist pulses of increasing duration (10, 100, 200, 300, 400, 500, 1000, 2000 msec). Calibration: 1 nA, 10 sec. Middle trace: 10 superimposed records of GABA activated Cl^- currents during saline perfusion. *Lower trace*: 20 super-imposed records of GABA activated Cl^- currents after saline perfusion with SN 606011, 10^{-8} M.

(b) Continuous record of the peak current amplitudes of successive GABA induced responses. Arrow = start of perfusion with SN 606011, 10^{-8} M. Calibration: 30 minutes.

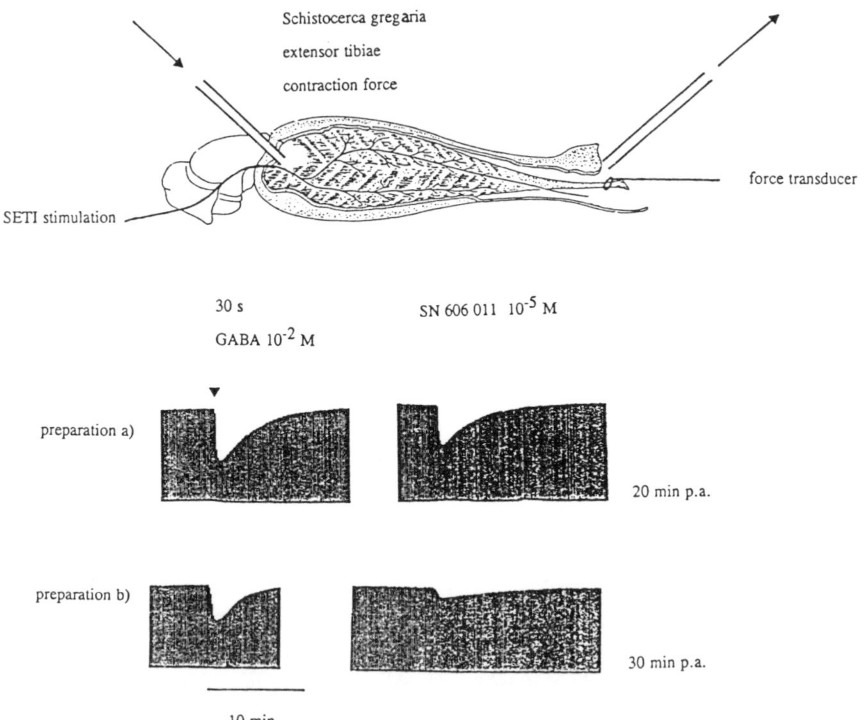

Figure 15 Activity of SN 606011 (10^{-5} M) on the GABA-evoked attenuation of the electrically induced contraction force in the locust extensor tibiae muscle. SETI (N3b) stimulation: 1,5 V, 0,8 msec, 50 Hz, 200 msec train duration, one train every 20 sec.

Acknowledgements

We would like to thank in particular J.D. Turner for his interest, suggestions and help with regard to the benzodiazepine pharmacology. We thank D. Funke, B. Herz and M. Pöhlmann for technical assistance, and I. Willecke for patient help with the manuscript.

References

ABALIS, I.M., ELDEFRAWI, A.T. and ELDEFRAWI, M.E. (1986). Actions of avermectin B_{1a} on the γ-aminobutyric acid $_A$ receptor and chloride channels in rat brain. *J. Biochem. Toxicol.*, **1**, 69–82.

BENSON, J.A. (1988). Transmitter receptors on insect neuronal somata: GABAergic and cholinergic pharmacology. In *Neurotox*

'88: *Molecular Basis of Drug and Pesticide Action* (G.G. Lunt, ed), pp. 193–206. Elsevier Science Publishing BV (Biomedical Division).

BENSON, J.A. (1989). A novel GABA receptor in the heart of a primitive arthropod, *Limulus polyphemus*. *J. Exp. Biol.*, **147**, 421–438.

BERMUDEZ, I. and BEADLE, D.J. (1988). Electrophysiological and ultrastructural properties of cockroach (*Periplaneta americana*) myosacs. *J. Insect Physiol.*, **34 (9)**, 881–889.

BETZ, H. (1990). Ligand-gated ion channels in the brain: the amino acid receptor superfamily. *Neuron*, **5**, 383–392.

BOKISCH, A.J. and WALKER, R.J. (1986). The action of avermectin (MK 936) on identified central neurones from *Helix* and its interaction with acetylcholine and gamma-aminobutyric acid (GABA) responses. *Comp. Biochem. Physiol.*, (C), **84**, 119–125.

CULL CANDY, S.G. (1986). Miniature and evoked inhibitory junctional currents and gamma aminobutyric-acid-activated current noise in locust *Schistocerca gregaria* muscle fibre. *J. Physiol.* (London), **374**, 179–200.

DRAGUHN, A., VERDOORN, T.A., EWERT, M., SEEBURG, P.H. and SAKMAN, B. (1990). Functional and molecular distinction between recombinant rat $GABA_A$ receptor subtypes by Zn^{2+}. *Neuron*, **5**, 781–788.

EGELHAAF, M., BORST, A. and PILZ, B. (1990). The role of GABA in detecting visual motion. *Brain Research*, **509**, 156–160.

ESHLEMAN, A.J. and MURRAY, T.F. (1990). Dependence on γ-aminobutyric acid of pyrethroid and 4'-chlorodiazepam modulation of the binding of t-[^{35}S]butylbicyclophosphorothionate in piscine brain. *Neuropharmacol.*, **29 (7)**, 641–648.

FRASER, S.P., DJAMGOZ, M.B.A., USHERWOOD, P.N.R., O'BRIEN, J., DARLISON, M.G. and BARNARD, E.A. (1990). Amino acid receptors from insect muscle: electrophysiological characterisation in *Xenopus* oocytes following expression by injection of messenger RNA. *Mol. Brain. Res.*, **8 (4)**, 331–342.

HART, R.J., LACEY, G. and WRAY, D.W. (1986). The postsynaptic actions of avermectin at locust thoracic ganglia. *J. Physiol.* (London), **376**, 43.

LEES, G., BEADLE, D.J., NEUMANN, R. and BENSON, J.A. (1987). Responses to GABA by isolated insect neuronal somata: pharmacology and modulation by a benzodiazepine and a barbiturate. *Brain Res.*, **401**, 267–278.

LÜDDENS, H. *et al.* (1991). *TiNS*.

LUMMIS, S.C.R. (1990). GABA receptors in insects. *Comp. Biochem. Physiol.*, **95C (1)**, 1–8.

LUNT, G.G., BROWN, M.C.S., RILEY, K. and RUTHERFORD, D.M. (1988). The biochemical characterisation of insect GABA receptors. In *Neurotox '88: Molecular Basis of Drug and Pesticide Action* (G.G. Lunt, ed), pp. 185–192. Elsevier Science Publishing BV (Biomedical Division).

MATSUMURA, F. and GHIASUDDIN, S.M. (1983). Evidence for similarities between cyclodiene type insecticides and picrotoxin in their action mechanisms. *J. Environ. Sci. Health*, **18 (1)**, 1–14.

MURPHY, V.F. and WANN, K.T. (1988). The action of GABA receptor agonists and antagonists on muscle membrane conductance in *Schistocerca gregaria*. *Br. J. Pharmacol.*, **95 (3)**, 713–722.

NIDDAM, R., DUBOIS, A., SCATTON, B., ARBILLA, S. and LANGER, S.Z. (1987). Autoradiographic localisation of [^{3}H]zolpidem binding sites in the rat CNS: comparison with the distribution of [^{3}H]flunitrazepam binding sites. *J. Neurochem.*, **49 (3)**, 890–899.

OLSEN, R.W., McCABE, R.T. and WALMSLEY, J.K. (1990). GABA$_A$ receptor subtypes: autoradiographic comparison of GABA, benzodiazepine and convulsant binding sites in the rat central nervous system. *J. Chem. Neuroanat.*, **3**, 59–76.

PALMER, C.J., SMITH, I.H., MOSS, M.D.V. and CASIDA, J.E. (1990). 1-[4-[(trimethylsilyl)ethynyl]phenyl]-2,6,7,-trioxabicyclo[2.2.2]octanes: a novel type of selective proinsecticide. *J. Agric. Food Chem.*, **38**, 1091–1093.

SATTELLE, D.B. (1990). GABA receptors of insects. *Advances in Insect Physiology*, **22**, 1–113.

SCOTT, R.H. and DUCE, I.R. (1987). Inhibition and gamma aminobutyric acid-induced conductance on locust *Schistocerca gregaria* extensor tibiae muscle fibres. *J. Insect Physiol.*, **33 (3)**, 183–190.

SHIVERS, B.D., KILLISCH, I., SPRENGEL, R., SONTHEIMER, H., KÖHLER, M., SCHOFIELD, P.R. and SEEBURG, P.H. (1989). Two novel GABA$_A$ receptor subunits exist in distinct neuronal subpopulations. *Neuron*, **3**, 327–337.

SIEGHART, W. (1989). Multiplicity of GABA$_A$-benzodiazepine receptors. *TiPs*, **10**, 407–411.

SIGEL, E., BAUR, R., TRUBE, G., MÖHLER, H. and MALHERBE, P. (1990). The effect of subunit composition of rat brain GABA$_A$ receptors on channel function. *Neuron*, **5**, 703–711.

TANAKA, K. and MATSUMURA, F. (1985). Action of avermectin B$_{1a}$ on the leg muscles and the nervous system of the American cockroach. *Pestic. Biochem. Physiol.*, **24**, 124–135.

VERDOORN, T.A., DRAGUHN, A., YMER, S., SEEBURG, P.H. and SAKMAN, B. (1990). Functional properties of recombinant rat GABA$_A$ receptors depend upon subunit composition. *Neuron*, **4**, 919–928.

von BLANKENFELD, G., YMER, S., PRITCHETT, D.B., SONTHEIMER, H., EWERT, M., SEEBURG, P.H. and KETTENMAN, H. (1990). Differential benzodiazepine pharmacology of mammalian recombinant GABA$_A$ receptors. *Neurosci. Lett.*, **115**, 269–273.

WAFFORD, K.A., SATTELLE, D.B., GANT, D.B., ELDEFRAWI, A.T. and ELDEFRAWI, M.E. (1989). Noncompetitive inhibition of GABA receptors in insect and vertebrate CNS by endrin and lindane. *Pestic. Biochem. Physiol.*, **33(3)**, 213–219.

WATSON, A.H.D. and BURROWS, M. (1987). Immunocytochemical and pharmacological evidence for GABAergic spiking local interneurons in the locust, *J. Neurosci.*, **7 (6)**, 1741–1751.

21
Progress in Structure-Activity on Cage Convulsants and Related GABA Receptor-Chloride Ionophore Antagonists

G. T. BROOKS

Department of Biochemistry and Physiology, University of Reading, PO Box 228, Whiteknights, Reading RG6 2AJ, UK

There is substantial evidence (Obata *et al.*, 1988; Ozoe *et al.*, 1990) that polychloroalkane insecticides (PCCA, i.e. cyclodienes, toxaphene, lindane) exert convulsant actions by interfering with the gamma-aminobutyric acid (GABA)-receptor associated chloride ionophore, although non-GABA-dependent chloride ion channels may also be affected. Using the displacement of binding $[^{35}S]$-t-butylbicyclophosphorothionate (TBPS) from putative GABA-receptors by these toxicants, differences between vertebrates and insect GABA-receptors have become evident, so that selective toxicity involving this nerve target site is possible. Cyclodiene resistance, a special form of selective toxicity arising through selection by insecticide pressure, appears to involve an insensitive site of action that is not the TBPS-binding site in a dieldrin resistant strain of housefly, *Musca domestica*, examined by Anthony *et al.* (1991). However, the cyclodiene resistance gene from a highly dieldrin resistant strain of *Drosophila melanogaster* has been cloned and shown to code for a GABA$_A$-receptor (ffrench-Constant and Roush, 1991). With this background, new insights regarding the nature of the chloride ionophore and the mechanisms of resistance to cage convulsants may be forthcoming and a brief discussion of their structure-activity relationships is relevant.

Early reports (see references in Brooks and Mace, 1987) showed

Insecticides: Mechanism of Action and Resistance
© 1992   Intercept Ltd, P.O. Box 716, Andover, Hants SP10 1YG, UK

that certain chlorine atoms in dieldrin were not essential for toxicity, that an aldrin analogue containing only two chlorine atoms is highly toxic to the German cockroach, *Blatella germanica*, and that trichloro-trimethyl analogues of endosulfan acted more rapidly than the parent endosulfan. Moreover, if one of the dichloromethane bridge chlorine atoms in cyclodiene insecticides is replaced by hydrogen, the isomer with chlorine *anti-* to the chlorinated double bond (Y = Cl, Z = H in Figure 1) is almost invariably more toxic than the *syn*-chloro-isomer (Z = Cl, Y = H), with very large differences for some structural types.

If we suppose that the dieldrin surface (Figure 1) delineates a binding site receptor in or near to the chloride ionophore, with a pocket (P) that accommodates the essential bulky *anti-* chlorine atom (Y), then the *syn*-chlorine atom (Z) evidently makes a less significant contribution to interaction with the receptor. When other cyclodiene molecules are fitted in the same orientation (e.g. endrin, endosulfan isomers, isobenzan, heptachlor epoxide enantiomers, chlordene epoxide enantiomers), a three dimensional zone of electronegative moieties is seen adjacent to area E of the receptor, with which interactions (e.g. hydrogen bonding) may occur. There appears to be a second region of electronegative interactions in the region of the chlorinated double bond. In contrast, if the *syn*-Cl atom (Z) is fitted into pocket (P), with the spatial position of the chlorinated norbornene nucleus otherwise maintained as for dieldrin in Figure 1, then the epoxide ring of dieldrin adopts position B and can no longer interact closely with the receptor zone E. For endrin and beta-endosulfan (the 'endrin-like' isomer), however, the compact, cage-like structure places the epoxide ring, or the S = O moiety respectively, near to position C, whether chlorine atom Y or Z is presented to the pocket P. For these molecules, but not for the extended (dieldrin-like) alpha-endosulfan, there appears to be little difference in insect toxicity between the respective pairs of bridge mono-dechloro-isomers (Brooks and Mace, 1987).

Considering Figure 1, if selection for dieldrin resistance modifies the receptor so as to preclude interaction of the dieldrin epoxide ring with the adjacent (upper) region of zone E, then more compact molecules such as endrin, isobenzan and beta-endosulfan may still be able to interact with the lower region of zone E in proximity to C. Lindane resembles the chlorinated region of the cyclodienes structurally and has a pentagonal arrangement of chlorine atoms in which Z = H. However, its compact form may permit interaction with the receptor in more than one orientation. It may be noted that in the classical pattern of cross-resistance to cyclodienes, these four

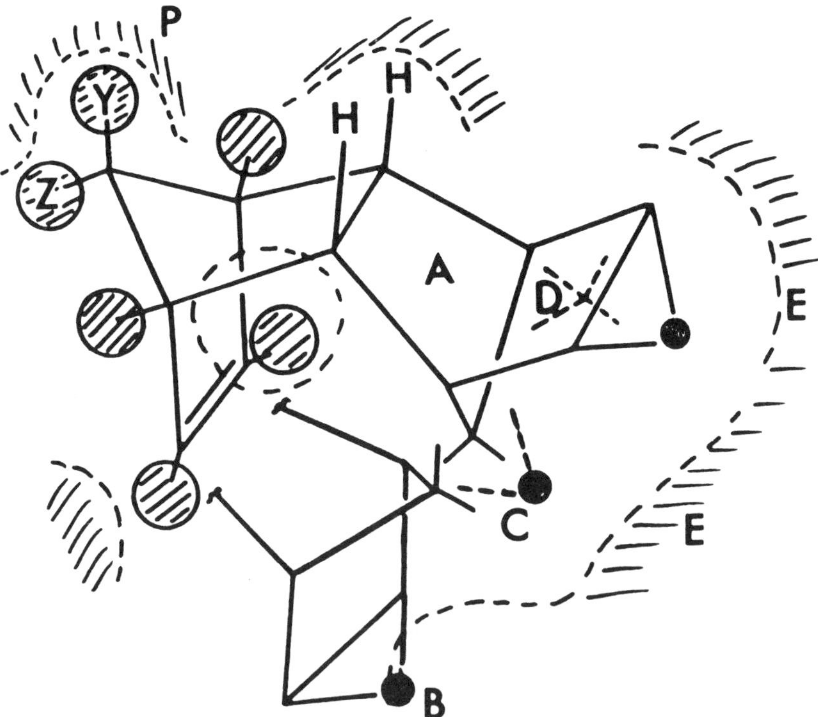

Figure 1 Dieldrin (A) aligned in a hypothetical binding site (called receptor in the text for convenience) with essential chlorine Y (*anti-*) presented to a pocket (P) which accommodates a bulky substituent. The epoxide ring is then close to a shaded area E which accommodates the electronegative groups of other active cyclodienes in a three-dimensional space, when they are similarly orientated. For orientation B of dieldrin, see text. C indicates the approximate position of the endrin epoxide ring (and beta-endosulfan sulphur) when *either* Y or Z is presented to P. D shows the position of the endrin methano-bridge when chlorine Y is presented to the pocket P, corresponding to the dieldrin orientation A. Figure is about 10Å across.

structural types retain significant toxicity when resistance to dieldrin is very high (Busvine, 1964).

Figure 2 shows the convergence of structural development between partly (Brooks and Mace, 1987) and totally (Ozoe *et al.*, 1990) dechlorinated alpha-endosulfan (I), which retains insect

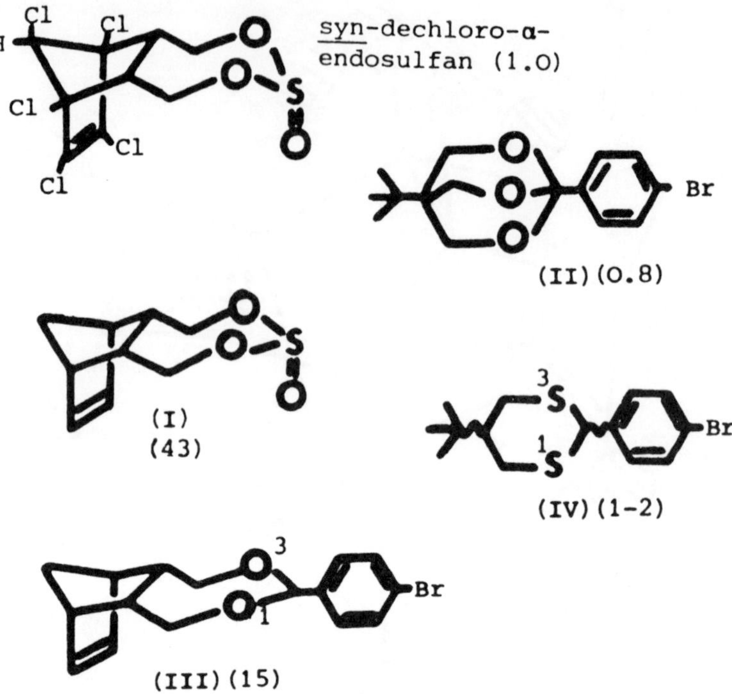

Figure 2 Structures of compounds discussed in the text with housefly topical LD50s (μgg^{-1}), measured with PBO except for (IV). Partial dechlorination of α-endosulfan retains toxicity (Brooks and Mace, 1987), which is not entirely lost in dechloro-α-endosulfan (I). Ozoe *et al.* (1990) devised the hybrid molecule (III), based on (I) and the bicycloorthobenzoate (II) of Palmer and Casida (1985). Dithiane (IV) resulted from attempts to simplify the ring system of compound (II) (Elliot *et al*, 1990).

toxicity, the toxic *t*-butyltrioxabicyclo-octanes (TBO) such as (II) (Palmer and Casida, 1985) and the recently described dithianes related to compound (IV) (Elliott *et al.*, 1990). An examination of the published information indicates that the conformationally flexible 1,3-dioxan analogue (IVa: not illustrated) of IV, is weakly toxic to houseflies and is not potentiated by piperonyl butoxide (PBO), whereas (IV) is as toxic as alpha-endosulfan but is antagonized by PBO (Elliott *et al.*, 1990). This suggests that oxidation of sulphur occurs as a bioactivation and may *inter alia*, confer a beneficial conformational stability on the molecule. Notably, the 1,3-dioxan derivative (III) (Ozoe *et al.*, 1990), in which the bulky *t*-butyl group in (IVa) is replaced by fusion of a conformationally rigid norbornene nucleus, is apparently considerably more toxic than (IVa) and about sevenfold less toxic than (IV). Changes in the direction of conformational stability appear to improve toxicity in dioxanes and dithianes but full chlorination of the norbornene nucleus of the type III structure, as in cyclodienes, virtually eliminates toxicity (Ozoe *et al.*, 1990). Accordingly, it seems that these molecules cannot be extended in the P-direction (Figure 1) by chlorination, whereas the fully chlorinated cyclodienes can only be extended in the E-direction in a very limited way, without loss of insect toxicity. This discussion supposes that the same receptor is involved but it remains to be determined whether the PCCAs and the TBO derived compounds act at the same or at closely associated, interacting sites. Future progress in this area presents an exciting interdisciplinary challenge.

References

ANTHONY, N.M., BENNER, E.A., SATTELLE, D.B. and RAUH, J.J. (1991). GABA receptors of insects susceptible and resistant to cyclodienes. Paper presented at the SCI International Symposium 'Neurotox 91', University of Southampton, April 7–11, 1991. *Pesticide Science*, **33**, 223–230.

BROOKS, G.T. and MACE, D.W. (1987). Toxicity and mode of action of reductively dechlorinated cyclodiene insecticide analogues on houseflies (*Musca domestica* L.) and other diptera. *Pesticide Science*, **21**, 129–142.

BUSVINE, J.R. (1964). The insecticidal potency of gamma–BHC and the chlorinated cyclodiene compounds and the significance of resistance to them. *Bulletin of Entomological Research*, **55**, 271–288.

ELLIOTT, M., PULMAN, D.A. and CASIDA, J.E. (1990). 2,5-Disubstituted, 1,3-dithianes, a new group of insecticides. Seventh International IUPAC Congress of Pesticide Chemistry, Hamburg, 5–10 August, 1990. Abstract Vol. 1, No 01A-17, p. 29.

ffRENCH-CONSTANT, R.H. and ROUSH, R.T. (1991). The cloning and transformation of cyclodiene resistance in *Drosophila melanogaster*: an invertebrate GABA receptor? Abstracts of the SCI International Symposium 'Neurotox 91', University of Southampton, 7–11 April, 1991, p. 27

OBATA, T., YAMAMURA, H.I., MALATYNSKA, E., IKEDA, M., LAIRD, H., PALMER, C.J. and CASIDA, J.E. (1988). Modulation of gamma-aminobutyric acid-stimulated chloride influx by bicycloorthocarboxylates, bicyclophosphorus esters, polychlorocycloalkanes and other cage convulsants. *Journal of Pharmacology and Experimental Therapeutics*, **244**, 802–806.

OZOE, Y., SAWADA, Y., MOCHIDA, K., NAKAMURA, T. and MATSUMURA, F. (1990). Structure-activity relationships in a new series of insecticidally active dioxatricycloalkenes derived by structural comparison of the GABA antagonists bicycloorthocarboxylates and endosulfan. *Journal of Agricultural and Food Chemistry*, **38**, 1264–1268.

PALMER, C.J. and CASIDA, J.E. (1985). 1,4-Disubstituted 2,6,7-trioxabicyclo[2.2.2]octanes: a new class of insecticides. *Journal of Agricultural and Food Chemistry*, **33**, 967–980.

22
The Detection of GABA-like and Proctolin-like Immunoreactivity in the Fifth Abdominal Ganglion of *Agrotis segetum* Larvae

U. STARK, D. OTTO AND G. CASPERSON

*Biological Research Centre Berlin, Stahnsdorfer Damm 81,
0-1532 Kleinmachnow, Germany*

The mechanisms of action and structure–activity relationships of insecticidal compounds can be investigated not only via biological screening but also with the help of biochemical and physiological test systems.

In electrophysiological tests to determine the effect of insecticidal compounds on neurone activity, the simple-structured fifth abdominal ganglion of larvae of *Agrotis segetum* Schiff. (Lepidoptera Noctuidae) was used. This required the analysis of its structural and functional organization.

The structure of the ganglion was elucidated using histological methods on paraffin sections. In total, 400 neurones were found, the perikarya of which are morphologically subdivided into three groups. The majority of the perikarya are situated in ventral and lateral positions, with only a few in the dorsal position. They occur more frequently in the posterior than in the anterior part of the ganglion. A certain number of immunolabelled perikarya are present in characteristic groups.

GABA and proctolin, two transmitter substances occurring in the insect nervous system, were detected through immunohistochemistry, i.e. with the PAP-procedure (Sternberger, 1970) on a series of paraffin sections (8 μm). GABA-anti-rabbit-serum and proctolin-anti-rabbit-serum were used as primary antibodies and goat-anti-rabbit-IgG-serum was used as the source of secondary antibodies. Immunocytochemical GABA detection was performed electronmicroscopically on ultra-thin sections and according to the above-mentioned immunohistochemical method.

The immunohistochemical test showed a clear difference between immunolabelled and non-labelled elements. Of the 400 neurones, 150 were identified as GABA-like and only one as a proctolin-like immunoreactive neurone. The number of immunolabelled perikarya remained constant in the ganglia investigated.

The GABA-like immunoreactive perikarya are generally laterally- and ventrally-situated and belong to all three morphological cell groups. Only a few labelled perikarya belong to the group of giant nerve cells (diam. 30–40 μm), but most of them are medium (diam. 15–25 μm) or small (diam. 5–10 μm).

In the anterior part of the ganglion there are no perikarya with GABA-like immunoreactivity. In the posterior end, however, at the base of the second lateral nerve pair, immunolabelled perikarya occasionally occur individually and in groups (Figures 1–4). They are located in all ganglia in relatively constant numbers.

Electron microscopy revealed GABA-like immunoreactivity in vesicles of the perikarya and axons (Figure 6).

The neuropile and the connectives of the ganglion contain numerous GABA-like immunoreactive axons and fibrous regions displaying high GABA-like immunoreactivity (Figure 2).

Labelled fibres could not be found in the lateral nerves, and no GABA-like immunoreactive axons were found in the median nerve.

Similar results are described in the literature for other insects but not for noctuids.

The fact that 40% of the neurones of the ganglion show GABA-like immunoreactivity emphasizes the great importance of this transmitter in the central nervous system. The absence of GABA-like axons in the lateral nerves supports the assumption that GABA-like processes are important for interneurone connections, but have hardly any peripheral function.

The proctolin content of the ganglion is far lower than the GABA content. Only one labelled perikaryon with proctolin-like immunoreactivity was found in the posterior part of the ganglion, in the dorso-median position (Figure 5), but it did not occur in all ganglia investigated.

The neuropile, ganglion connections and lateral nerves did not exhibit proctolin-like immunoreactivity. The small number of

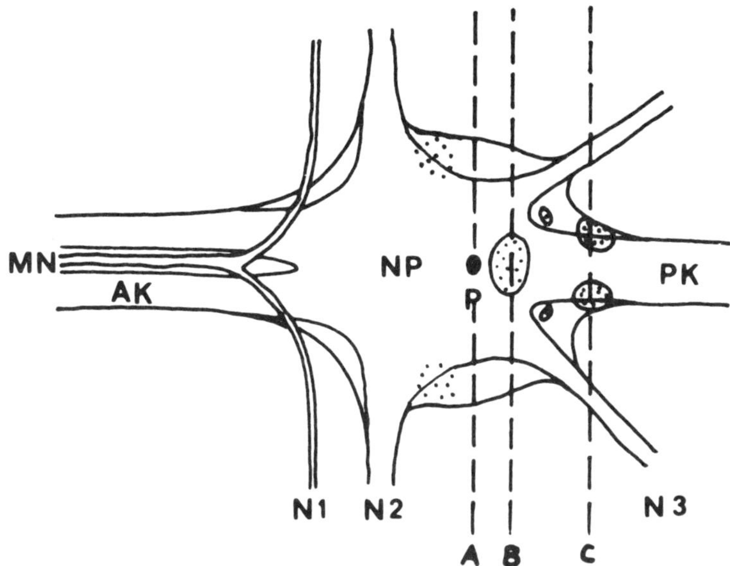

Figure 1 Distribution (pattern) of perikarya groups with GABA-like and proctolin-like immunoreactivity in the 5th abdominal ganglion (longitudinal section). ○ = GABA-like immunoreactivity; ● = proctolin-like immunoreactivity; AK = anterior connectives; PK = posterior connective; N 1, 2, 3 = lateral nerves; MN = median nerve; NP = neuropile; A, B, C = section plane.

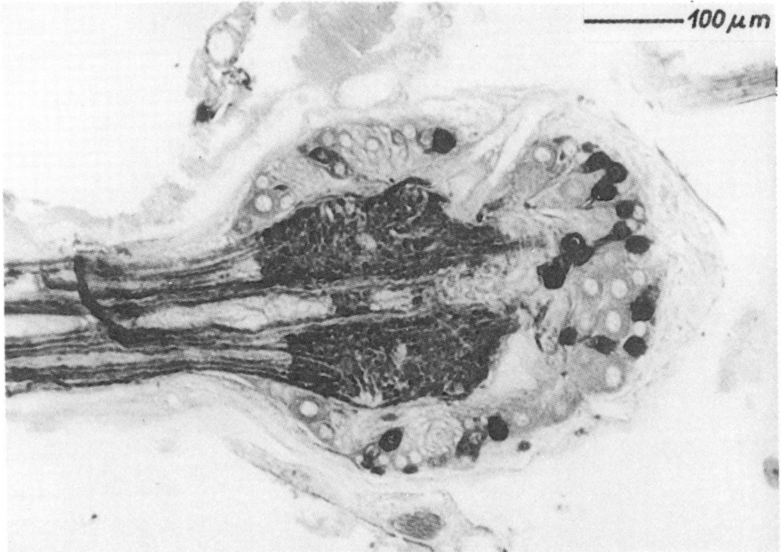

Figure 2 GABA-like immunoreactivity in perikarya and axons of the ganglion (longitudinal section).

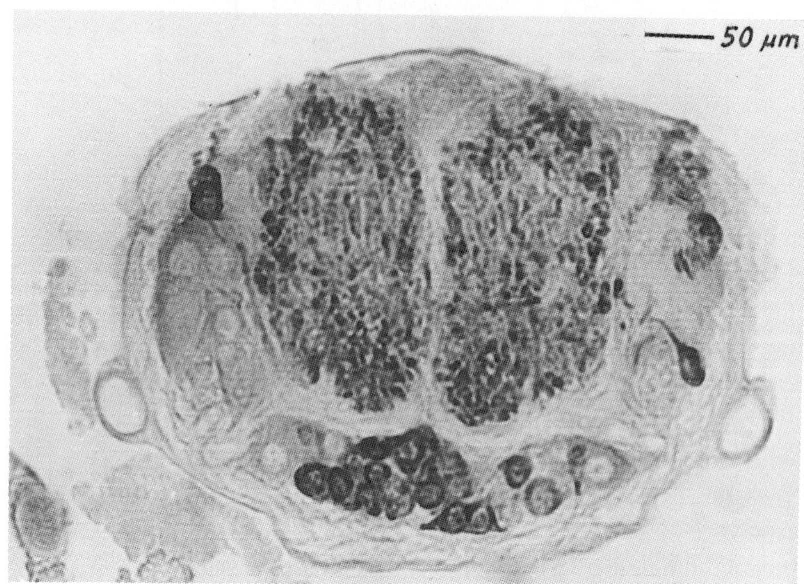

Figure 3 Ventro-median GABA-like immunoreactive group of perikarya in the posterior region of the ganglion (sagittal section).

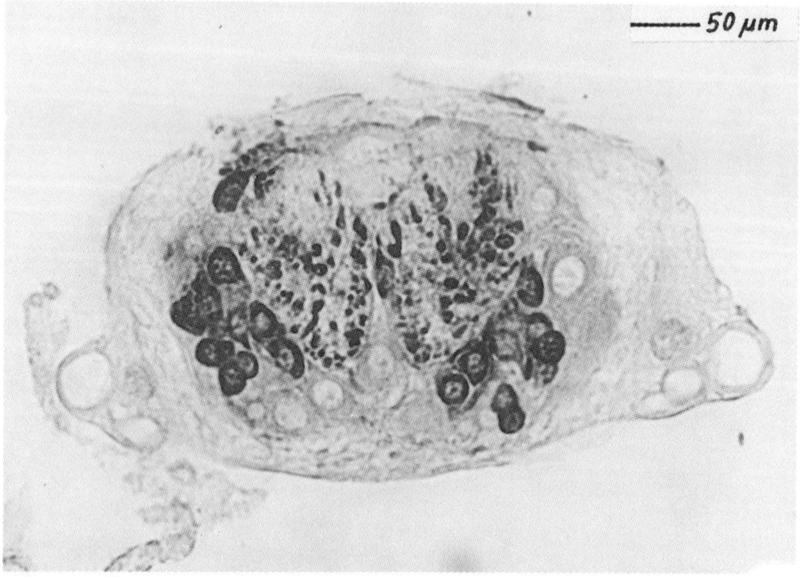

Figure 4 Ventro-lateral groups of GABAergic prikarya at the posterior end of the ganglion (sagittal section).

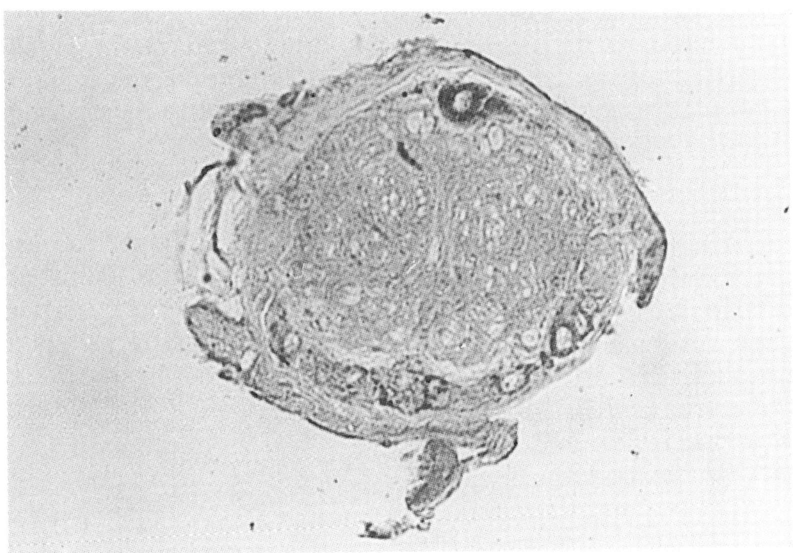

Figure 5 Proctolin-like immunoreactivity in the dorso-median prikaryon in in the posterior region (sagittal section).

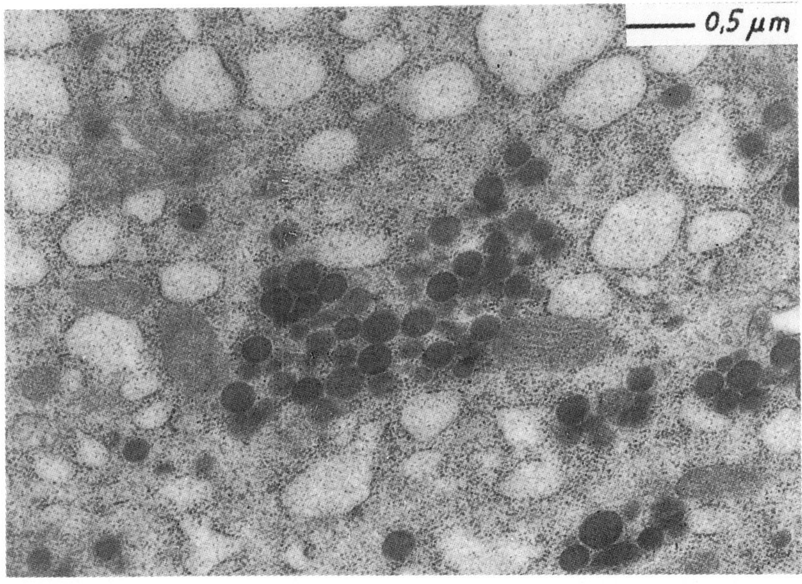

Figure 6   GABA-like immunoreactive vesicles in the perikaryon.

neurones with proctolin-like immunoreactivity in the abdominal ganglion has also been described for other lepidoptera. Some of these species also possessed only one proctolin-like neurone. This indicates that proctolin has little or no importance in the abdominal nerve processes.

Note

For a detailed contribution on this subject and references to literature see STARK, U., OTTO, D. and CASPERSON, G. (1991). Investigations of the structure of the 5th abdominal ganglion of *Agrotis segetum* Schiff. larvae (Lepiloptera, Vocturidae) and immunohistochemical detection of the neurotransmitters γ-aminobutyric acid (GABA) and proctolin. *Zoologisches Jahrbuch, Anatomie*, **121**, 381–390.

23
The Influence of Nereistoxin Derivatives on the Ultra-structure of Nervous Cells in Insects

G. CASPERSON, D. OTTO, B. WEBER AND
L. BANASIAK

*Biological Research Centre Berlin, Stahnsdorfer Damm 81,
0-1532 Kleinmachnow, Germany*

Nereistoxin (Figure 1) was isolated from the marine annelid *Lumbriconereis heteropoda* Marenz (Sakai, 1969) and synthesized by Hashimoto and Okaichi (1960), Hagiwara *et al.* (1965) and Numata Hagiwara (1968). Nereistoxin and its derivatives are known to block the nicotinic acetylcholine receptors of insects (Sakai, 1966, 1969; Nagawa and Saji, 1971; Satelle, 1985). A feature of this blocking is the knockdown effect, which occurs shortly after application of the toxin. High doses are lethal to the insect, but sublethal doses cause a short-lived knockdown effect followed by a recovery of the insect (Otto *et al.*, 1989; Richter *et al.*, 1989). The ultrastructural changes in *Periplaneta americana* and *Leptinotarsa decemlineata* were studied in relation to the dose of the toxin and the time of exposure (Figures 2–9). Initially, the mitochondriae are more strongly attacked in the perikarya than in the neuropile. This is visible as a swelling and destruction of the cristae. These effects can be repaired in the recovery period. Higher doses lead to a greater destruction of mitochondriae and cytoplasm, marking the beginning of the lethal process. Dicytosomes, neurovesicles and glial cells are not involved in the destruction process. Nereistoxin derivatives do not attack muscle cells.

Insecticides: Mechanism of Action and Resistance
© 1992   Intercept Ltd, P.O. Box 716, Andover, Hants SP10 1YG, UK

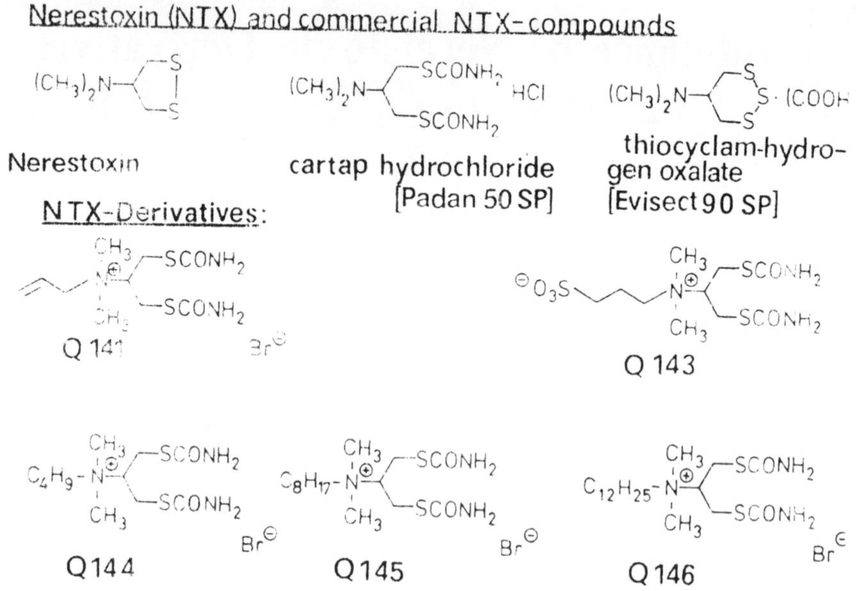

Nerestoxin (NTX) and commercial NTX-compounds

Nerestoxin

cartap hydrochloride
[Padan 50 SP]

thiocyclam-hydro-
gen oxalate
[Evisect 90 SP]

NTX-Derivatives:

Q 141

Q 143

Q 144

Q 145

Q 146

Figure 1   Chemical structures of the compounds investigated.

The mode of action of commercial products, Evisect and Padan
(Figure 1) is quite similar. The modification of the Cartap structure
by the quaternization of the dimethylamino moiety with selected
alkyl halogenides resulted in various quaternary ammonium
compounds with different hydrophobicity. One of the derivatives
represents a sulphur betaine structure (Figure 1, Q143). The
insecticidal efficiency of these compounds was screened to prove
whether the essential principle of action of Cartap is determined by
the free amine structure or the derived hydrochloride with a
positively charged nitrogen atom.

First results from various bioassays show that, in general, the
quaternization of the nitrogen atom of the Cartap molecule reduces
insecticidal action. Experiments on potato beetle larvae revealed a
correlation in which an enhanced hydrophobicity of alkyl substituents of the derivatives leads to an increased insect toxicity. The
highly hydrophilic N-methyl and N-3-sulfopropyl derivatives were in
each case ineffective, but the insecticidal activity of the more
hydrophobic N-butyl-, N-octyl- and N-dodecyl derivatives was
comparable to Cartap hydrochloride. The existence of a non-
quaternized nitrogen atom in the Cartap molecule seems to be
essential for binding to the receptor site because an irreversible
blocking of the free amine structure by quaternization, resulting in
compounds with strong salt character, causes a total loss of

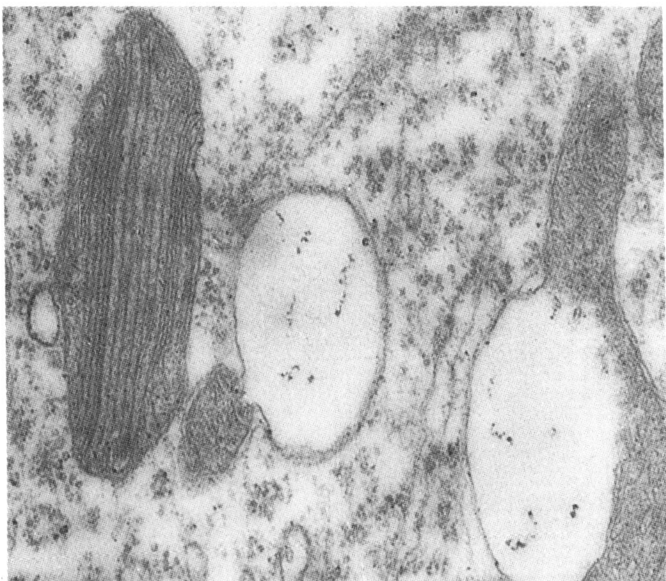

Figure 2 *Leptinotarsa decemlineata* (L4). Perikaryon of the 3rd thorakalganglion. Larvae (L4) placed for 1 hour on leaves which were treated with a solution of 3 g l^{-1} Evisect. The mitochondriae are attacked, swollen and the cristae particularly destroyed.

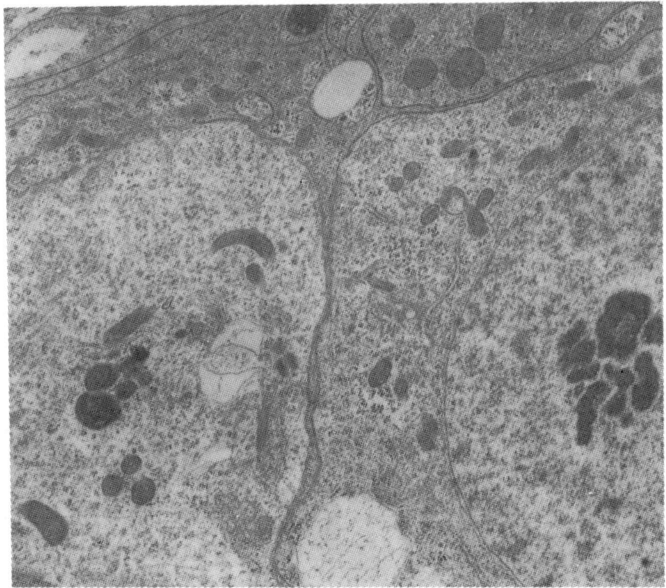

Figure 3 *Leptinotarsa decemlineata* (L 4). Parikaryon. Untreated control shows no effects.

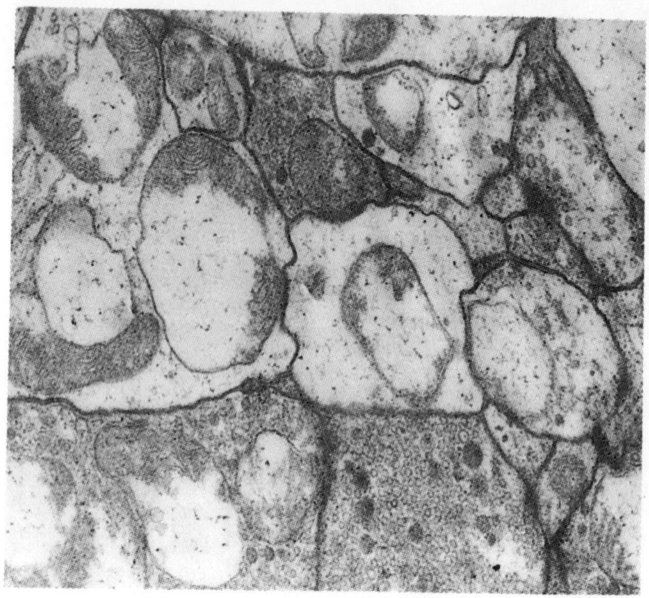

Figure 4   *Periplaneta americana*. Neuropile of the sixth abdominal ganglion. 3 hours after injection of 48 μg Padan the mitochondriae are swollen and the cristae are particularly destroyed.

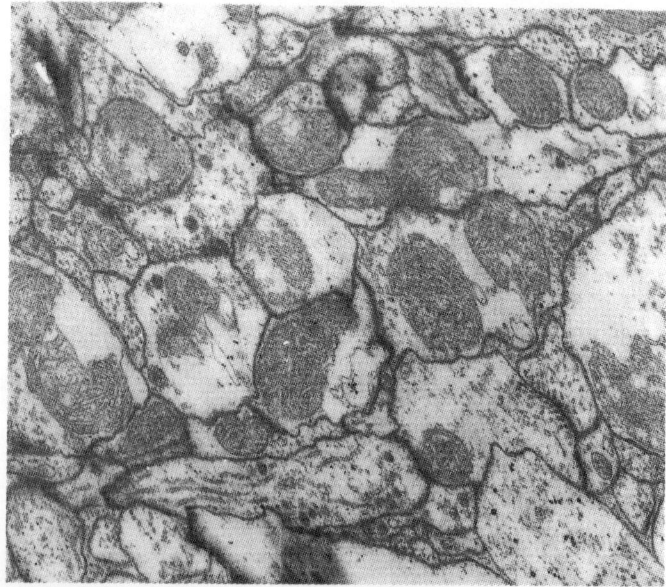

Figure 5   *Periplaneta americana*. Neuropile. 14 hours after injection of 48 μg Padan the attacked mitochondriae can be repaired in the recovery period.

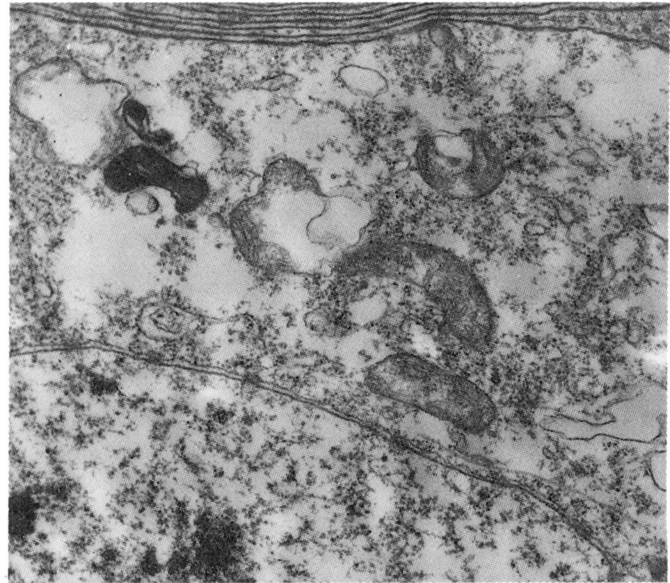

Figure 6 *Periplaneta americana*. Perikaryon. 3 hours after injection of 478 µg Padan the mitochondriae are swollen and the cristae destroyed, vacuolization and destruction of the cytoplasm.

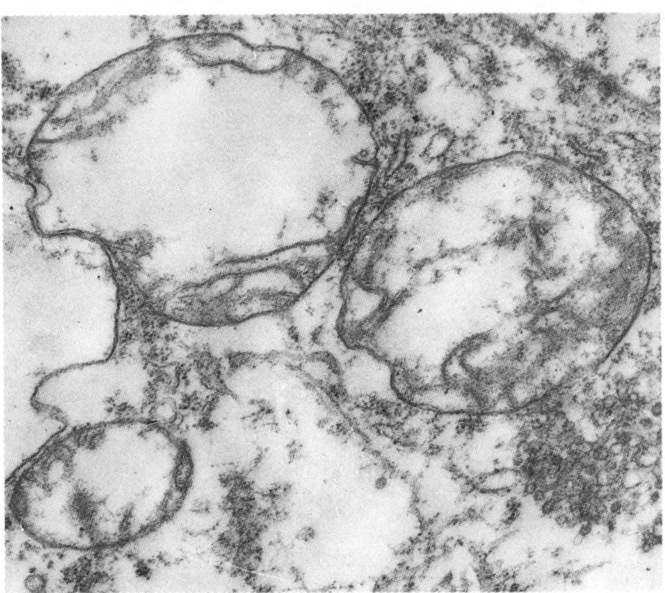

Figure 7 *Periplaneta americana*. Perikaryon. 24 hours after injection of 478 µg Padan. The beginning of lethal processes leads to a greater destruction of the cell, the cristae in the swollen mitochondriae are destroyed.

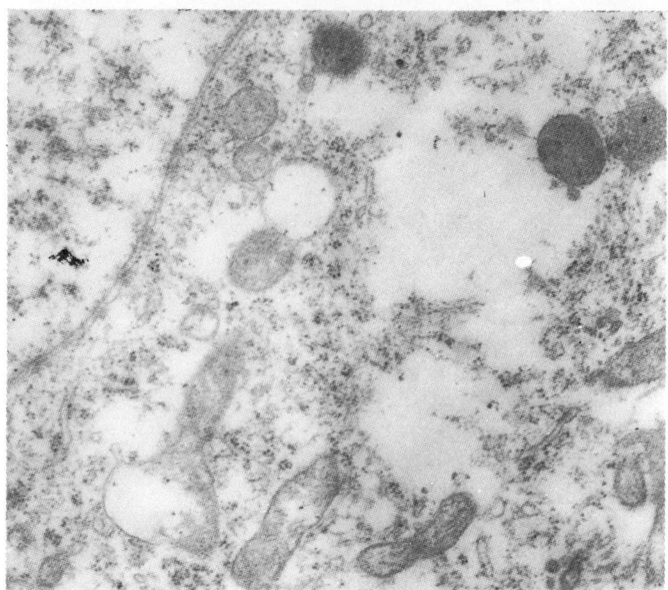

Figure 8   *Periplaneta americana*. Perikaryon. 3 hours after injection of 500 µg nereistoxin derivative with C_4-alkyl group the destruction is smaller than in Figure 5.

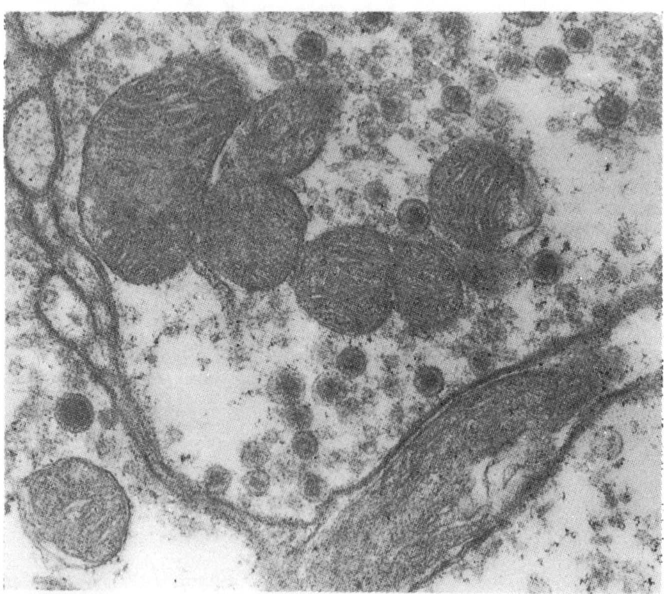

Figure 9   *Periplaneta americana*. Neuropile. The untreated control shows no effects.

insecticidal activity. Cartap derivatives with long alkyl chain substituents correspond in their activity to the common conception of the mode of action of quaternary ammonium compounds.

References

HAGIWARA, H., NUMATA, M., KONISHI, K. and OKA, Y. (1965). Synthesis of nereistoxin and related compounds. *I. Chemical and Pharmaceutical Bulletin*, **13**, 253–260.

HASHIMOTO, Y. and OKAICHI, T. (1960). Some chemical properties of nereistoxin. *Annals of the New York Academy of Sciences*, **90**, 667–673.

NAGAWA, Y. and SAJI, Y. (1971). Blocking of the cholinergic receptor by nereistoxin. *Japan Journal of Pharmacology*, **21**, 185–197.

NUMATA, H. and HAGIWARA, H. (1968). Synthesis of nereistoxin and related compounds. *II. Chemical and Pharmaceutical Bulletin*, **16**, 311–317.

OTTO, D., RICHTER, R. and KARABENSCH, K.-H. (1989). The acetylcholine receptor as a target of insecticides. In *Proceedings* of the First International Symposium 'Insecticides – Mechanisms of Action and Resistance', Reinhardsbrunn, 1988. Tagungsbericht der Akademie der Landwirtschaftswissenschaften der DDR, **274**, 47–58.

RICHTER, R., OTTO, D. and MENGS, H.-J. (1989). Insecticide compounds, acting on the acetylcholine receptor of the insect nervous system. In *Chemistry of Plant Protection* Volume **2**, pp. 157–195. Springer Verlag, Berlin.

SAKAI, M. (1966). Role of the anti-cholinesterase activity in the insecticidal action to housefly, *Musca domestica* L. *Applied Entomology and Zoology*, **1**, 73–82.

SAKAI, M. (1969). Nereistoxin and Cartap. Their mode of action as insecticides. *Review of Plant Protection Research*, **2**, 17–29.

SATELLE, D.B. (1985). Nereistoxin actions on the central nervous system acetylcholine receptor ion-channel in the cockroach *Periplaneta americana* L. *Journal of Experimental Biology*, **118**, 37–52.

24
Inhibitory Effects of Insecticides on the Oxygen Uptake of Isolated Cockroach Ganglia

M GUNDEL,[1] H. LEISNER[2] AND H. PENZLIN[1]

[1]*Friedrich Schiller University Jena, Institute of General Zoology and Animal Physiology, Ebertstrasse 1, 0-6900 Jena, Germany*
[2]*Medical Academy Erfurt, Institute of Pathophysiology, Nordhäuser Strasse 74, 0-5010 Erfurt, Germany*

Introduction

The influence of insecticides on respiration in insects has been investigated mainly with whole animals (reviewed by Keister and Buck, 1974; Gerolt, 1983). In comparison, there is relatively little information concerning the oxygen uptake of isolated insect tissues, especially of nervous tissue. In some insect nervous preparations a high rate of oxygen consumption has been found, e.g. in isolated ventral ganglia of locusts (Clement and Strang, 1978), cockroaches (Steele and Chan, 1980) and in brains of various holometabolous insects (Wegener, 1983). However, these authors did not consider the effects of insecticides on the oxygen uptake of the isolated insect ganglia. Therefore, this was investigated in the present study.

Materials and Methods

Isolated terminal ganglia from adult male cockroaches, *Periplaneta americana* L., were used. The oxygen uptake was determined amperometrically by means of a Clark O_2 electrode attached to a micro-chamber at 37°C according to Leisner and Tiedt (1987).

Insecticides: Mechanism of Action and Resistance
© 1992   Intercept Ltd, P.O. Box 716, Andover, Hants SP10 1YG, UK

Ganglia were incubated in insect saline (pH 7.3), which contained trehalose (20mM) and the tested insecticides. Measurements of the oxygen uptake were performed successively over an experimental period of one to six hours.

Tested chemicals were potassium cyanide (Merck), rotenone (Serva) and conventional insecticides obtained from the Chemie AG Bitterfeld in a purity of at least 98%, with the exception of Bancol (containing as active 50% bensultap) (for structure see inset in Figure 3). Substances were dissolved in distilled water (trichlorfon), acetone (Bancol and rotenone) and the other insecticides in dimethyl sulfoxide (DMSO) and then resuspended into saline immediately before each experiment. The tested concentrations of water-insoluble insecticides represent the added amounts. The final concentration of solvents were not more than 1% (v/v) and without any significant effect on the oxygen uptake of ganglia. The statistical significance was determined using the Mann-Whitney U test.

In most cases (except when potassium cyanide and rotenone were tested) desheathed ganglia were used in the experiments, because in these cases a faster inhibition of the oxygen uptake was observed than in ganglia with intact sheath.

Results and Discussion

Terminal ganglia of *Periplaneta americana* show a high rate of oxygen uptake in air-saturated insect saline (initial value $V_{0_2} = 9.7 \pm 0.4\ \mu l\ O_2 \cdot cm^{-2} \cdot h^{-1}$, n = 32). Over the experimental period no significant differences were found between intact and desheathed ganglia.

The oxygen uptake measured was highly sensitive to cyanide (Figure 1).

Rotenone, an insecticidally active constituent of certain plants, blocked the oxygen uptake of ganglia extremely rapidly (Figure 2). All other insecticides tested revealed a delayed and more gradual reduction of oxygen uptake. The most striking inhibitory effect among these insecticides was caused by the nereistoxin derivative Bancol and by the pyrethroid deltamethrin (Figure 3). The carbamate carbaryl and the chlorinated hydrocarbons dieldrin, DDT and lindane blocked the oxygen uptake of ganglia only weakly (Figure 4). No effect could be found with the organophosphate trichlorfon, tested in concentrations up to 10^{-2}M.

It is apparent that insecticides cause a complex set of physiological responses. This is also reflected by inhibitory effects on the oxygen

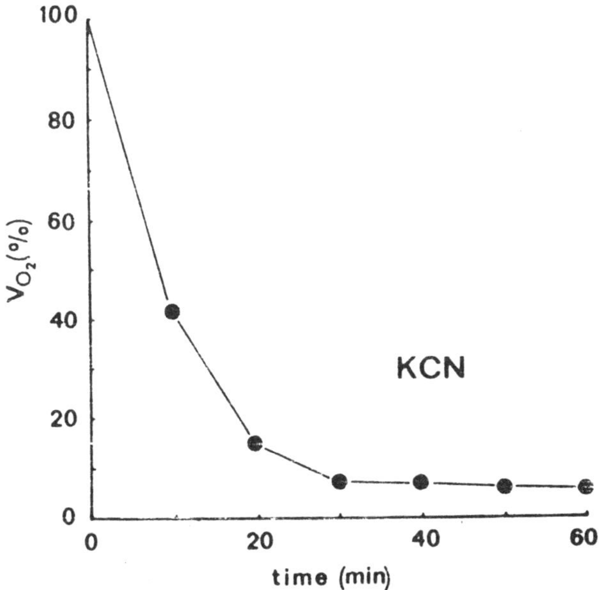

Figure 1   Inhibition of the oxygen uptake (V_{O_2}) of isolated intact terminal ganglia from *Periplaneta americana* in the presence of 10^{-2}M potassium cyanide (KCN). The values are means of 5 ganglia.

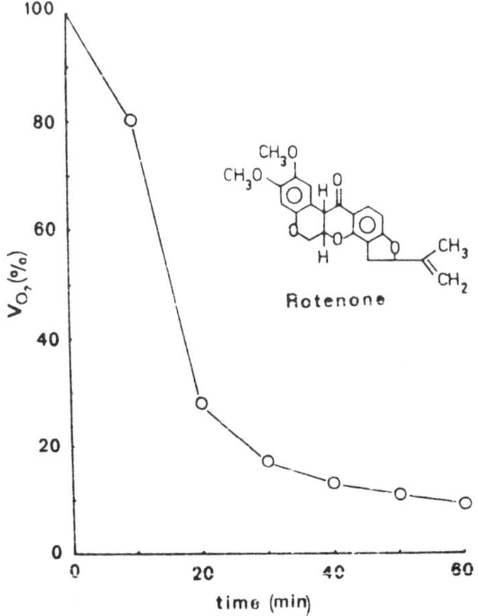

Figure 2   Reduction of the oxygen uptake (V_{O_2}) of the intact terminal ganglia by 10^{-2}M synthetic rotenone. See inset for structure formula. The values are means of 7 ganglia.

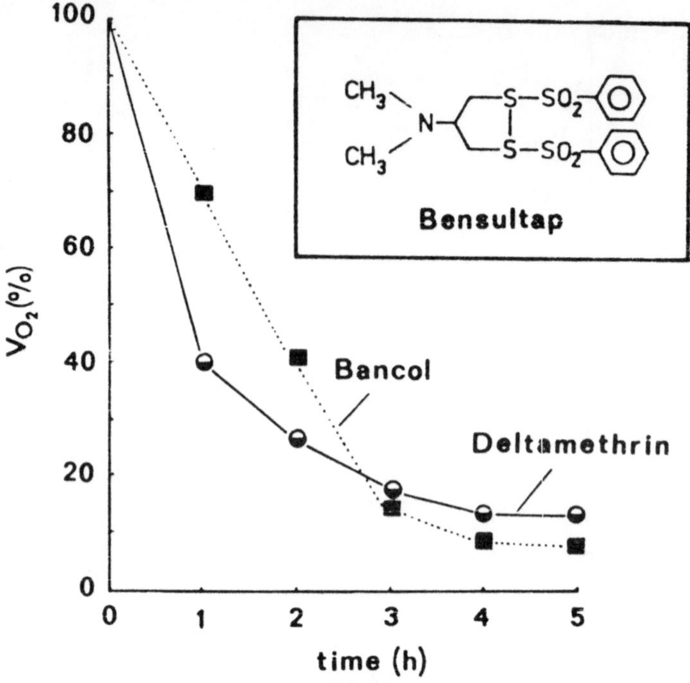

Figure 3 Inhibitory effects of deltamethrin and Bancol on the oxygen uptake (V_{O_2}) of desheathed terminal ganglia *in vitro*. The insecticide concentration in the bathing solution was 5×10^{-3}M. Each value is the mean of 9 experiments. Inset shows the chemical structure of synthetic nereistoxin derivative bensultap, the active ingredient of the commercial insecticide Bancol.

uptake of isolated terminal ganglia. For most insecticides the direct effect on cell respiration is unknown. But it was reported that rotenone acts as a specific inhibitor of a pyridine nucleotide linked oxidase of the respiratory chain (reviewed by Fukami, 1985). We also demonstrated a drastic reduction of oxygen uptake by this naturally occurring substance. The inhibitory effect of synthetic rotenone was only slightly lower than that observed after exposure of the ganglia to the classical respiration poison potassium cyanide.

We propose that for most of the insecticides tested, indirect effects on cell respiration are occurring, for instance a destruction of the mitochondrial structure and/or blocking of mitochondrial ATPases. This would explain the delayed and often only partial inhibition of oxygen uptake. Some of these secondary effects have already been observed in poisoned nervous tissue from insects (Lane, 1973; Desaiah *et al.*, 1975; Cutkomp *et al.*, 1976; Singh and Singh, 1984;

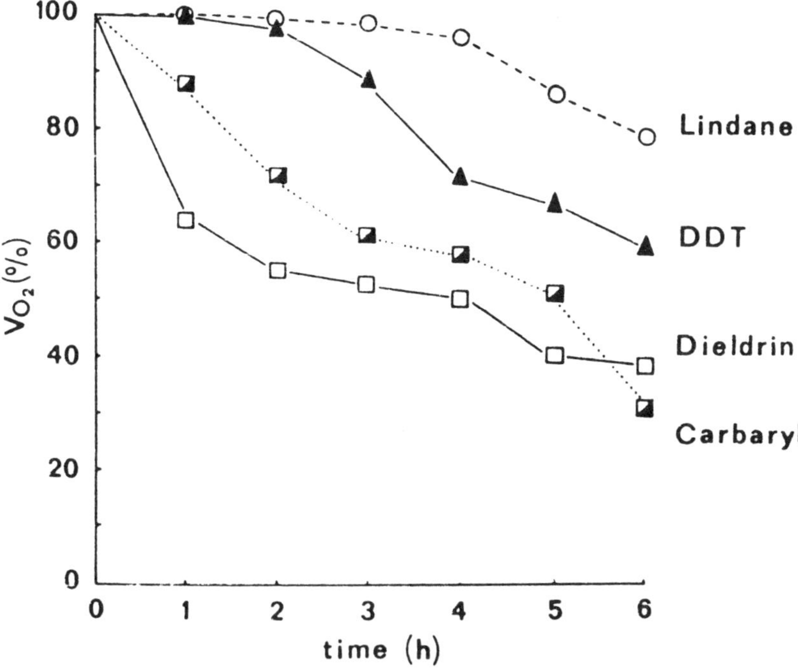

Figure 4 Blocking effects of some insecticides on the oxygen uptake (V_{O_2}) of desheathed terminal ganglia *in vitro*. Each value represents the mean of 9 (lindane), 8 (DDT), 6 (dieldrin) and 6 (carbaryl) experiments. Note a delayed onset of inhibition after exposure of ganglia to lindane and DDT. Insecticide concentration is the same as in Figure 3.

Casperson *et al.*, 1989). Moreover, the slower time course in the blocking effect of intact ganglia indicates an important role played by the ganglionic sheath in the process of intoxication. This problem, however, must be investigated in further detail.

References

CASPERSON, G., KARABENSCH, K.-H. and OTTO, D. (1989). Influence of nereistoxin on the ultrastructure of ganglia of the American cockroach (*Periplaneta americana* L.) *Tagungsberichte der Akademie der Landwirtschaftswissenschaften der DDR*, **274**, 97–102.
CLEMENT, E.M. and STRANG, R.H.C. (1978). A comparison of some aspects of the physiology and metabolism of the nervous system of the locust *Schistocerca gregaria in vitro* with those *in*

vivo. Journal of Neurochemistry, **31**, 135–145.

CUTKOMP, L.K., DESAIAH, D., CHENG, E.Y., VEA, E.V. and KOCH, R.B. (1976). The *in vivo* sensitivity of ATPases to DDT, DDE, and physical immobilization in American cockroaches. *Pesticide Biochemistry and Physiology*, **6**, 203–208.

DESAIAH, D., CUTKOMP, L.K. VEA, E.V. and KOCH, R.B. (1975). The effect of three pyrethroids on ATPases of insects and fish. *General Pharmacology*, **6**, 31–34.

FUKAMI, J.-I. (1985). Rotenone and rotenoids. In *Comprehensive Insect Physiology Biochemistry and Pharmacology* (G.A. Kerkut and L.I. Gilbert, eds), Volume **12**, pp. 291–311. Pergamon Press, Oxford.

GEROLT, P. (1983). Insecticides: their route of entry, mechanism of transport and mode of action. *Biological Reviews*, **58**, 233–274.

KEISTER, M. and BUCK, J. (1974). Respiration: some exogenous and endogenous effects on rate of respiration. In *The Physiology of Insecta* (M. Rockstein, ed), second edition, Volume **6**, pp. 469–509. Academic Press, New York.

LANE, N.J. (1973). Effects of insecticides on the fine structure of the nervous system of the common housefly. *Journal of Cell Biology*, **59**, 158a.

LEISNER, H. and TIEDT, N. (1987). Sauerstoffaufnahmeverhalten der Gefäßwand isolierter Koronararterien. *Ergebnisse der experimentallen Medizin*, **48**, 152–160.

SINGH, G.J.P. and SINGH, B. (1984). Action of dieldrin and trans-al-drindiol upon the ultrastructure of the sixth abdominal ganglion of *Periplaneta americana* in relation to their electro-physiological effects. *Pesticide Biochemistry and Physiology*, **21**, 102–126.

STEELE, J.E. and CHAN, F. (1980). Na^+-dependent respiration in the insect nerve cord and its control by octopamine. In *Insect Neurobiology and Pesticide Action* (M. Sherwood, ed), pp. 347–350. Society of Chemical Industry, London.

WEGENER, G. (1983). Brains burning fat: Different forms of energy metabolism in the CNS of insects. *Naturwissenschaften*, **70**, 43–45.

25
The Mode of Action of the Insecticide/ Acaricide Diafenthiuron

FRANZ J. RUDER, JACK A. BENSON AND HARTMUT KAYSER

R + D Insect Control, Agricultural Division, CIBA-GEIGY Ltd, CH-4002 Basel, Switzerland

Introduction

The thiourea diafenthiuron has a novel mode of action – its carbodiimide product inhibits mitochondrial ATPase *in vitro*.

Thioureas are a relatively new class of insecticides of hitherto unknown mode of action. As one of these, diafenthiuron (CGA 106630; 3-(2,6-diisopropyl-4-phenoxyphenyl)-1-t-butyl-thiourea; Figure 1), a product especially effective against sucking pests (Streibert *et al.*, 1988), has recently been introduced into the market.

Observations of field biologists with diafenthiuron told an interesting story:

(1) The efficacy in field trials exceeded that of greenhouse experiments.
(2) An excellent vapour phase activity was observed although diafenthiuron has only a low vapour pressure.
(3) The pesticidal activity of diafenthiuron increased with time.

Possible Targets

Therefore, the idea emerged that a highly active photoproduct of diafenthiuron is responsible for its pesticidal activity.

Insecticides: Mechanism of Action and Resistance
© 1992   Intercept Ltd, P.O. Box 716, Andover, Hants SP10 1YG, UK

CGA 106630 - a Fingerprint

Chemical name:	3-(2,6-Diisopropyl-4-phenoxyphenyl)-1-tert-butyl urea
Common name:	Diafenthiuron
Trade name:	POLO (cotton)
	PEGASUS (edible crops)
Targets:	Mites, aphids, whiteflies, jassids, some Lepidoptera
Crops:	Cotton, vegetables, ornamentals fruits/citrus, tea, hops

Figure 1

Indeed, chemodynamic studies revealed diafenthiuron is processed to the carbodiimide CGA 140408 by singlet oxygen which can be generated by sunlight (Figure 2) (Steinemann *et al.*, 1989). CGA 140408 is known as a compound with strong acaricidal/insecticidal activity even exceeding that of diafenthiuron itself.

Derived from these results we focused our research about the mode of action of diafenthiuron on the following issues:

(1) Has diafenthiuron itself any insecticidal activity?
(2) Is there also a biological desulphuration of diafenthiuron *in vivo* to yield the carbodiimide? This question is important because of the observation that diafenthiuron also acts in the absence of sunlight (although with minor efficacy).
(3) What is the mode of action of the carbodiimide CGA 140408?

To get some ideas about possible targets we applied test systems established in our laboratory which include the targets of the major commercial insecticides (Figure 3).

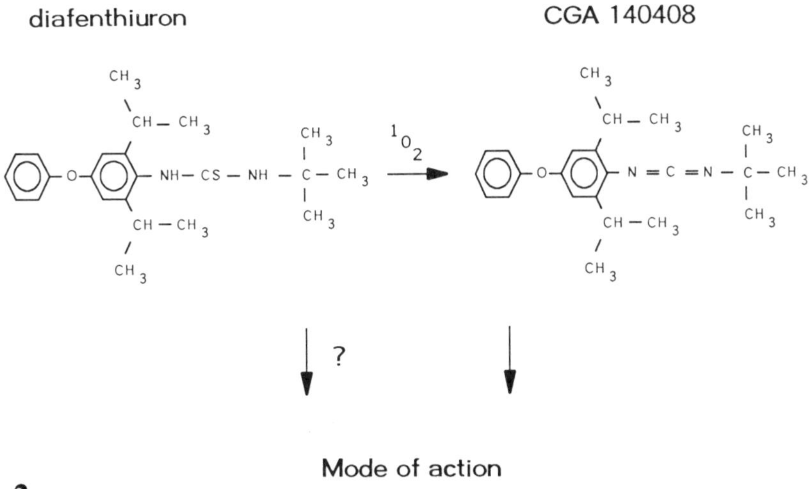

diafenthiuron   CGA 140408

Figure 2

Mode of action

target	assay system	diafenthiuron	CGA 140 408
mitochondria	*Calliphora*	NE	<u>block of the coupling site</u>
	rat liver	NE	<u>block of the coupling site</u>
P-450 spectroscopy	rat liver	<u>Type 1 spectrum</u>	NE
cuticle formation	*Calliphora*	NE	NE
chitin synthesis	*Spodoptera*	NE	NE
acetylcholinesterase	bovine erythrocytes	NE	NE
axonal Na^+ channel	*Periplaneta*	NE	NE
nicotinic cholinergic receptor	*Locusta*	NE	NE
muscarinic cholinergic receptor	*Locusta*	-	NE
GABA receptor	*Locusta*	-	NE
	Limulus	-	NE
dopamine receptor	*Locusta*	-	NE
	Limulus	-	NE
serotonin receptor	*Locusta*	NE	NE
	Limulus	-	NE
octopamine receptor	*Locusta*	-	NE
	Limulus	-	NE
glutamate receptor	*Musca*	NE	NE

NE = no effect   - = not tested

Note: the neurobiological transmitter receptor preparations were tested electrophysiologically so that both the ligand recognition site and the associated ionophore were assayed.

Figure 3

- Neither compound affects any of the targets of commercial insecticides.
- CGA 140408 has potent activity against mitochondrial ATP-synthesis.
- Diafenthiuron interferes with none of the potential targets but binds to cytochrome P-450 like a substrate (indicated by a type I spectrum).

In vivo Metabolism of Diafenthiuron

(1) Feeding tests reveal a decreased toxicity of diafenthiuron in the presence of piperonyl-butoxide (PBO) which is known as an inhibitor of cytochrome P-450, and, consequently, an insecticide synergist (Figure 4). This indicates an important contribution of P-450 to metabolize diafenthiuron to a toxic product (presumably to CGA 140408).

(2) HPLC analysis of insects treated with diafenthiuron shows that CGA 140408 is indeed a metabolic product of diafenthiuron.

(3) In the cytochrome P-450 *in vitro* test, under normal conditions, surprisingly no desulphuration but aromatic p-hydroxylation is observed (Figure 5).

The fact that singlet oxygen is able to desulphurate the thiourea encouraged us to investigate whether other active oxygen species which are of more biological significance could manage the desulphuration. Indeed, we found that to a less extent hydrogen peroxide and especially well hydroxy radicals (Fenton cycle: hydroxyradicals generated from hydrogen peroxide in the presence of iron ions) could mediate the desulphuration with a good yield.

It is known that cytochrome P-450 generates hydrogen peroxide as a by-product during its normal enzymatic cycle. In the presence of iron, hydrogen peroxide is transformed to hydroxyradicals. Therefore, we investigated the reaction of diafenthiuron with cytochrome P-450 in the presence of iron ions *in vitro*. In the presence of either free iron ions or iron complexes (micromolar range), there is no longer hydroxylation but desulphuration of diafenthiuron to its carbodiimide (Figure 5). Therefore, there is a transformation of the cytochrome P-450 reaction pathway from hydroxylation to desulphuration.

So far, we have no clear idea yet about the role of iron in the cytochrome P-450 enzyme cycle in detail (hydroxyradicals, hydrogen peroxide, change of enzymatic properties of P-450). There is some reason to assume that hydroxyradicals participate in one or the other way.

Growth

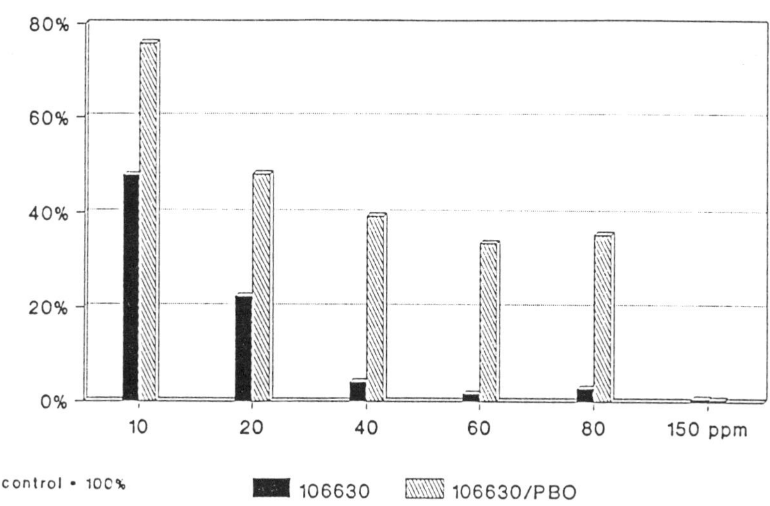

Survival rate

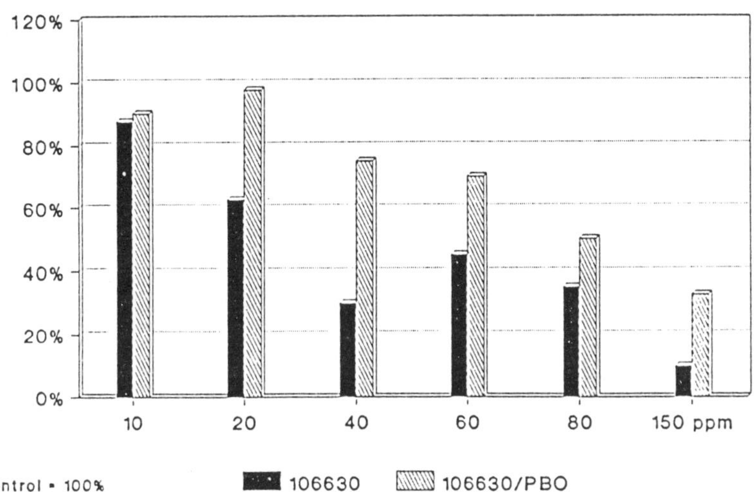

Figure 4 *Feeding test with diafenthiuron (CGA 106630)*
The growth and survival rate was measured with the indicated amount of diafenthiuron (in ppm) in the diet. The toxicity of diafenthiuron was compared without and with 100 ppm piperonyl-butoxid (PBO) in the diet. The data indicate that PBO partially protects against toxic effects of diafenthiuron.

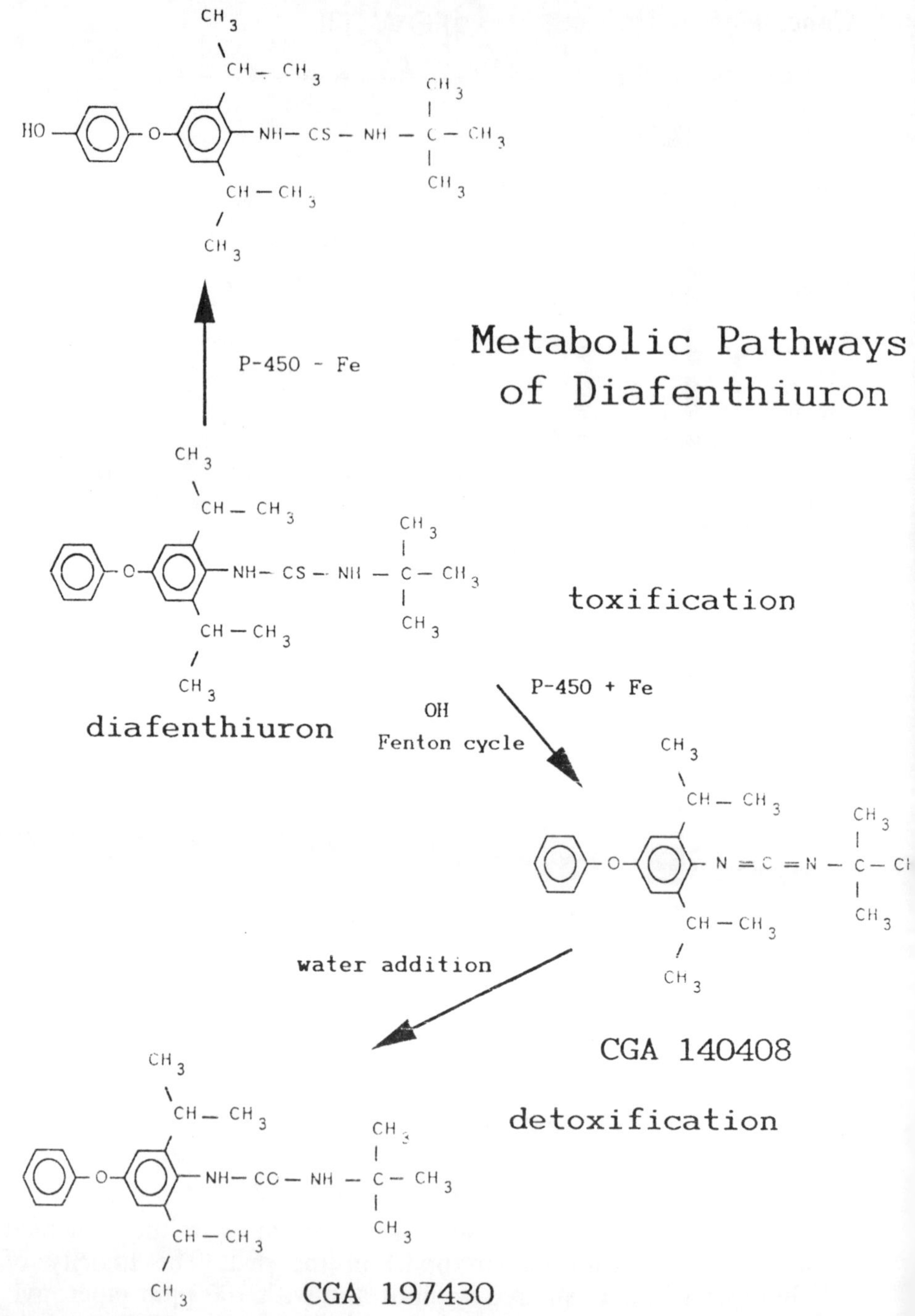
Metabolic Pathways of Diafenthiuron
P-450 - Fe
toxification
P-450 + Fe
OH
Fenton cycle
diafenthiuron
CGA 140408
water addition
detoxification
CGA 197430

Figure 5

Conclusion of Previous Results

- No evidence for a toxicity of diafenthiuron itself.
- diafenthiuron is desulphurated to CGA 140408, and CGA 140408 is the toxic compound.

Therefore, if we know the mode of action to CGA 140408 then we shall know the mode of action of diafenthiuron.

In vitro Targets of CGA 140408

As already mentioned, mitochondria are a major target *in vitro*. The site affected by CGA 140408 is respiring mitochondria reminds to the well-known reactivity of the carbodiimide DCCD (dicyclohex-ylcarbodiimide): like CGA 140408, DCCD is a hydrophobic carbodiimide and, at low concentrations, covalently reacts with two proteins:

(1) With the proteolipid of the FO-part of the ATPase (a Glu side COOH in hydrophobic environment) which is an integral part of the inner mitochondrial membrane. The labelling of the proteolipid is paralleled by a block of mitochondrial ATP-synthesis and ATPase respectively.
(2) With porin, a voltage-dependent channel forming protein of the outer mitochondrial membrane. The biochemical function of porin is not completely understood but besides its channel-forming properties, porin can bind cytosolic hexokinase, a key regulatory enzyme of the glycolytic pathway. The binding of DCCD to porin is paralleled by a block of hexokinase binding to porin.

ATPase test with CGA 140408
- CGA 140408 is indeed a potent inhibitor of mitochondrial ATPase at concentrations of 0.5 nmole/mg protein. CGA 140408 is even slightly more potent than DCCD.
- CGA 140408 reacts irreversibly with the ATPase with a pseudo-first order kinetics similar to DCCD.

Therefore we followed the reaction of radiolabelled CGA 140408 with mitochondrial proteins (Figures 6 and 7). Rat liver mitochon-dria (Figure 6) were incubated with (14C)-DCCD or (14C)-CGA 140408 under various conditions, and labelled proteins separated by SDS gel electrophoresis, and visualized by fluorography. We observed the well-known reactivity of DCCD with two rat liver

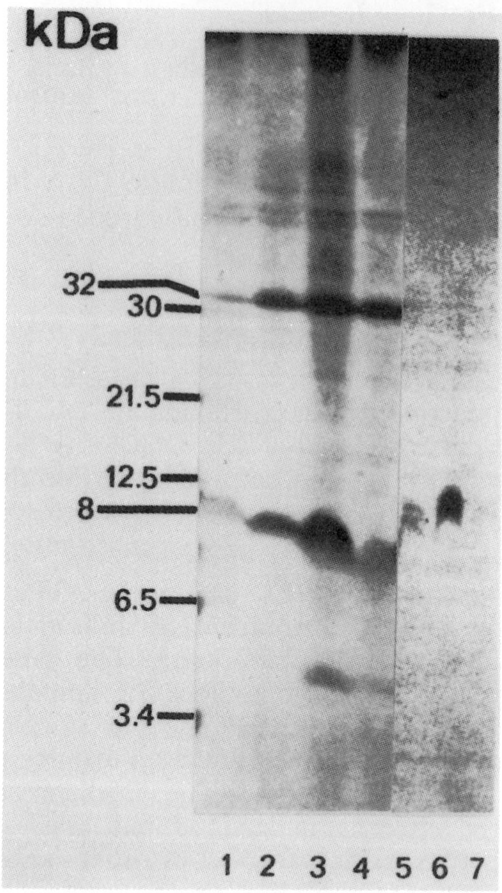

Figure 6 *Fluorograms of carbodiimide-labelled rat liver mitochondrial proteins separated by SDS gel electrophoresis*

Rat liver mitochondria were incubated with either (^{14}C)-DCCD or (^{14}C)-CGA 140408, separated by SDS gel electrophoresis, and label was visualized by fluorography. The positions of marker proteins are indicated. Heavily labelled protein bands are indicated by their molecular weights, as derived from their migration behaviour. In all lanes, identical amounts of protein from the same mitochondrial preparation were used. Therefore all lanes can be compared with one another directly.

Lanes 1–4: labelling by (^{14}C)-DCCD (lane 3) was inhibited by unlabelled DCCD (lane 1), unlabelled CGA 140408 (lane 2), and venturicidine (lane 4). *Lanes 5–7*: labelling by (^{14}C)-CGA 140408 (lane 6) was inhibited by unlabelled DCCD (lane 5) and venturicidine (lane 7).

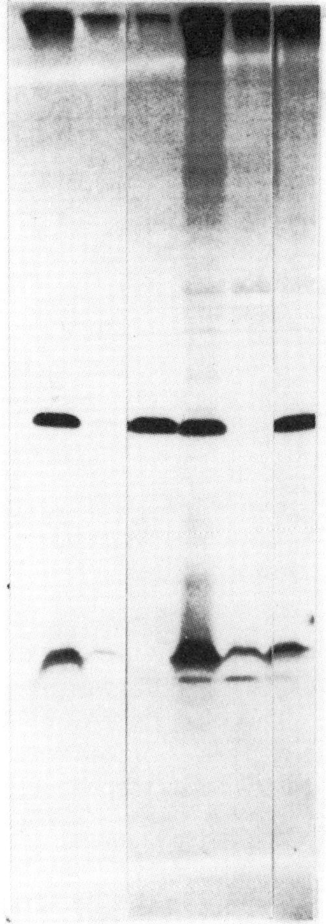

Figure 7 *Fluorograms of carbodiimide labelled Calliphora proteins separated by SDS gel electrophoresis*
Calliphora flight muscle mitochondria were incubated with either (^{14}C)-DCCD or (^{14}C)-CGA 140408, separated by gel electrophoresis, and label was visualized by fluorography. The positions of marker proteins are indicated. Identical amounts of protein from the same mitochondrial perparation were used. Therefore, all lanes can be compared directly with each other.

Lanes 9–14: proteins labelled by (^{14}C)-DCCD (lanes 9–11) or (^{14}C)-CGA 140408 (lanes 12–14). Labelling (lanes 9 and 12) was inhibited by unlabelled DCCD (lanes 10 and 13) and venturicidine (lanes 11 and 14).

mitochondria proteins (discussed by Ruder and Kayser, 1992 and Ruder *et al.*, 1992):

(1) the 8 kDa FO-proteolipid;
(2) porin (32 kDa).

CGA 140408 only reacts with an 8 kDa protein but *not* with the 32 kDa protein in rat liver mitochondria.

Several features indicate that CGA 140408 labels the mitochondrial proteolipid in rat liver:

(1) Labelling of 8 kDa protein by both, DCCD and CGA 140408, is sensitive to venturicidin (venturicidine binds reversibly to the same site of the proteolipid as DCCD thereby inhibiting access of DCCD to the proteolipid).
(2) Labelling by (14C)-CGA 140408 can be completed by excess of unlabelled DCCD and vice versa.
(3) Labelling of the 8 kDa protein by both CGA 140408 and DCCD is paralleled by the disappearance of mitochondrial ATPase activity.

When *Calliphora* mitochondria (Figure 7) were incubated with radiolabelled DCCD or CGA 140408, two proteins always become labelled (30 kDa and an 8 kDa). Several features give strong evidence that the labelled proteins are homologous to the rat liver mitochondrial proteolipid and porin respectively.

(1) Labelling of both proteins by each carbodiimide can be completed by excess of unlabelled DCCD or CGA 140408.
(2) In the presence of venturicidin, only the 30 kDa protein but not the 8 kDa protein becomes labelled by each carbodiimide.
(3) The 8 kDa protein can be extracted by chloroform/methanol in accordance with published procedures for isolation of this proteolipid.
(4) The 8 kDa protein is enriched in the inner mitochondrial membrane fraction.
(5) The 30 kDa protein is enriched in the outer mitochondrial membrane fraction.

A comparison of labelled proteins in rat liver and *Calliphora* mitochondria shows that CGA 140408 has a higher chemical specificity than DCCD, because CGA 140408 only labels porin in *Calliphora*.

From our data, the following picture emerges (Figure 8): diafenthiuron is converted abiotically (singly oxygen) and/or biotically (cytochrome P-450) to CGA 140408. Consequently, CGA 140408 covalently reacts with mitochondrial porin and the ATPase.

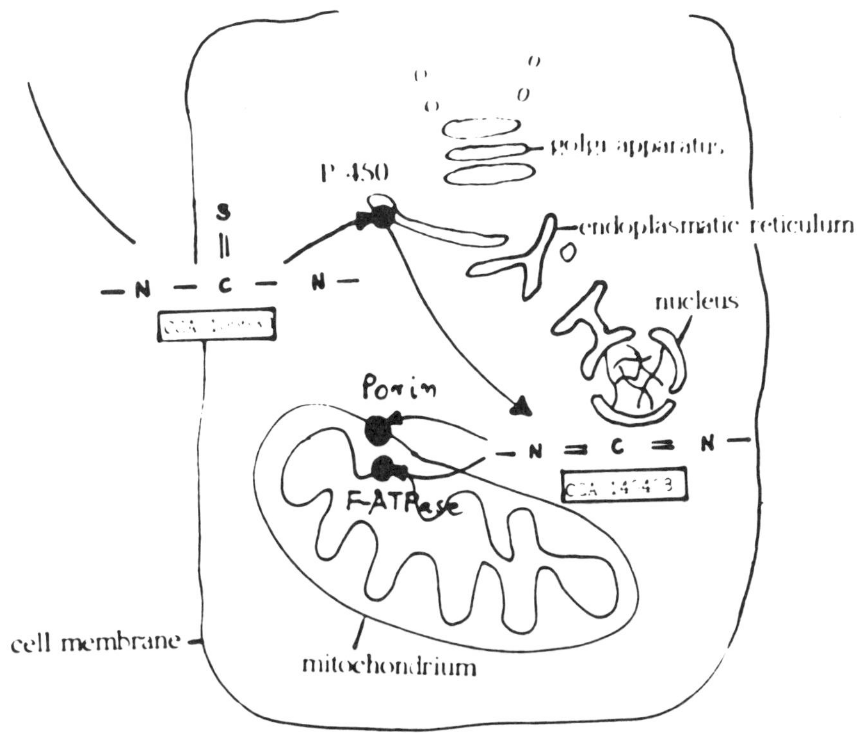

Porin: A channel forming protein of the outer mitochondrial membrane

F-ATPase: Mitochondrial ATP forming enzyme (coupling site)

Figure 8

In vivo Studies

Are the data obtained *in vitro* studies really significant for the *in vivo* pesticidal activity of diafenthiuron and CGA 140408? As indicated in Figure 9, intoxication with both diafenthiuron and CGA 140408 leads to covalent labelling of the FO-proteolipid and porin *in vivo*, thus showing that the same proteins are labelled *in vitro* and *in vivo*.

But very recent results indicate:

(1) Not more than 10% of total *Calliphora* thorax mitochondria are labelled and not more than 10% of total thorax mitochondrial-ATPase activity is blocked after application of a lethal dose of diafenthiuron or CGA 140408.

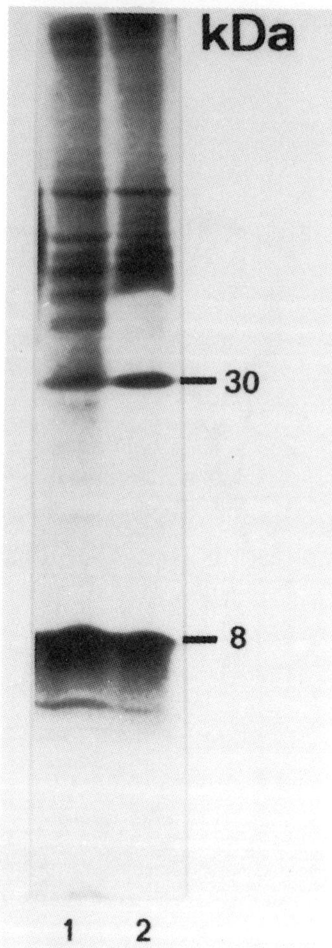

Figure 9   Fluorograms of *in vivo* and *in vitro* labelled *Calliphora* proteins separated by SDS gel electrophoresis (*in vivo* labelled proteins were analysed after treatment of *Calliphora* with (^{14}C)-CGA 140408)

Lane 1: *in vivo* labelled proteins. *Lane 2*: *in vitro* labelled proteins (see Figure 7). Labelled proteins in the molecular weight region above 30 kDa are due to incomplete denaturation of the probes. After complete denaturation these protein bands disappear with a associated stronger labelling of the 8 kDa band. Comparison of *in vivo* and *in vitro* labelled proteins reveals no obvious difference.

(2) We find no significant reduction of the concentration of ATP or arginine phosphate in the thorax. We even find a significant increase in the concentration of the arginine phosphate, as indicated by 31P-NMR studies.
(3) There is a 50% reduction of the level of glucose-6-phosphate which is the product of the hexokinase reaction.

The data show that treatment by diafenthiuron or CGA 140408 does not significantly inhibit total energy metabolism, although glucose-6-phosphate level is reduced. We do not know whether the observed decrease of glucose-6-phosphate formation is correlated with a reduced hexokinase activity and with a possible inability of CGA 140408-modified porin to bind hexokinase.

Concluding Remarks

1 The presented data show that diafenthiuron is a pro-pesticide which can be converted to the carbodiimide CGA 140408 both abiotically (singlet oxygen) and biotically (cytochrome P-450). CGA 140408 is the final active product of diafenthiuron.

2 CGA 140408 is a potent inhibitor of mitochondrial ATP-synthesis *in vitro* and reacts covalently with the proteolipid subunit of the ATPase and porin, the outer membrane channel forming protein in mitochondria.

3 *In vivo* only a few of the total mitochondria are inhibited after a lethal dose of CGA 140408. It may be that mitochondria in essential tissues (e.g. nervous system) are preferentially blocked.

4 Besides mitochondrial ATPase, other targets, such as porin, have to be discussed.

5 Diafenthiuron and CGA 140408 have a novel mode of action,different from all commercial insecticides.

References

RUDER, F.J. GUYER, W., BENSON, J.A. and KAYSER, H. (1992).The thiourea insecticide/acaricide diafenthiuron has a novel mode of action – inhibition of mitochondrial respiration by its carbodiimide product. Submitted.
RUDER, F.J. and KAYSER, H. (1992). The carbodiimide product of diafenthiuron reacts covalently with two mitochondrial proteins, porin and the FO-proteolipid, and inhibits mitochon-

drial ATPases *in vitro*. *Pesticide Biochemistry*, submitted.
STEINEMANN, A., STAMM, E. and FREI, B. (1989). The effect of chemodynamic parameters on pesticide performance. *Aspects Appl. Biol.*, **21**, 203.
STREIBERT, H.P., DRABEK, J. and RINDLISBACHER, A. (1988). Diafenthiuron – a new type of acaricide/insecticide for the control of the sucking pest complex in cotton and other crops. *Proceedings* of the 1988 Brighton Crop Protection Conference – Pest and Diseases, 25.

26
Development and Application of an Enzyme-linked Immunosorbent Assay (ELISA) for the Determination of the Insecticidal Substance Dimethoate

B. DOROBEK

Biological Research Centre Berlin, Stahnsdorfer Damm 81, 0-1532 Kleinmachnow, Germany

Introduction

The analysis of chemicals in environmental samples is always a costly task. A time-consuming clean-up and often a derivatization are necessary. In addition, expensive equipment (GC-MS, HPLC) is required which must be operated by specially trained staff. An attractive alternative to conventional analysis is the enzyme immunoassay. This sensitive immunological method is based on the highly specific binding of a pesticide to antibodies which are raised against a pesticide. The amount of binding is determined by means of the reaction of an enzyme coupled to either the pesticide or the antibody. The reaction is simple to monitor by generation of a coloured product, and unknown concentrations of samples are evaluated by means of a standard curve.

Until the present the insecticide Dimethoate, a main product of the Chemie AG Bitterfeld in Germany, has had a large field of application in Europe and Asia. This necessitates an efficient residual monitoring for both the producer and the user.

The classical analytical methods (thin-layer chomatography or GC) are highly sensitive procedures to determine Dimethoate

Insecticides: Mechanism of Action and Resistance

residues in the statutory detection range, but are inadequate for extensive sample screening (Ferreira *et al.*, 1987).

The production of polyclonal antibodies against a Dimethoate-protein conjugate and the development of a quantitative enzyme-linked immunosorbent assay for the determination of Dimethoate residues in water samples and in several homogenates of plant material allow use of this analytical technique for an easy sample screening.

Preparation of Antigens

Low-molecular-weight substances, like pesticides, cannot evoke an immunoresponse *per se* in an animal. They are so-called haptens. Therefore, it is necessary to prepare conjugates by coupling these small molecules to macromolecular carriers, like proteins (BSA, thyroglobuline, hemocyanin), which are immunogenic and able to induce an antibody immunoresponse.

For immunogen synthesis the carboxylic acid derivative of Dimethoate was used. The derivative was synthesized by a method described by Bertin and Geralt (1963) and used for the preparation of the immunogenic conjugates. For the covalent coupling to the amino groups of the carrier proteins, BSA or thyroglobuline, the terminal carboxyl group of the modified Dimethoate molecule was coupled via an active ester by carbodiimide reaction as described by Goodfriend *et al.*, (1964) and is shown in Figure 1.

Various coupling rates and procedures were tested. The following synthesis turned out to be the best one:

Solution A
50 mg carboxylic acid derivative was dissolved in a 5 ml borate-HCL-buffer (0.05M, pH 9.0) 150 mg 1-ethyl-3-(3-dimethylamino-propyl)carbodiimide - EDC (Sigma) and 170 mg n-hydroxysuccini-mide (Serva) were then added.

Solution B
150 mg BSA or ovalbumin or thyroglobuline was dissolved in 2 ml borate-HCL buffer (0.05M, pH 9.0).

Reaction
Solution A was dropwisely added under stirring to solution B and incubated for four hours at 4°C. The incubates were dialysed for three days in PBS (phosphate-buffered solution, pH 7.4) with four changes of buffer. The conjugates were then frozen at -20°C.

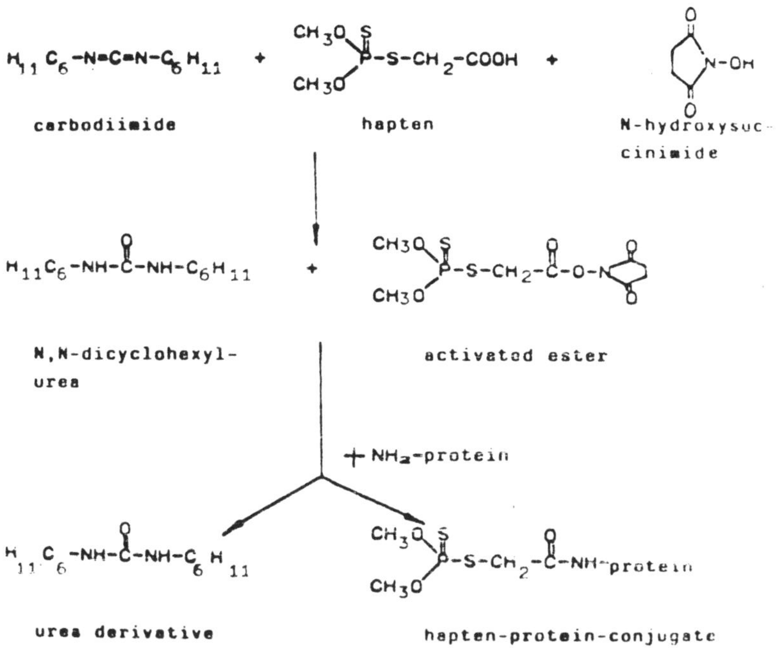

Figure 1 Scheme of the carbodiimide reaction.

The epitome density of the immunoconjugates was measured according to trinitrobenzensulphonic acid method (TNBS) of Habeeb (1966) and was calculated as 25 molecules hapten residues per molecule BSA and as 14 for ovalbumin and 64 for thyroglobuline respectively.

The coating antigens were prepared using the same hapten and proteins but coupled via an acid-anhydride reaction as described by Erlanger (1959).

In the immunoassay a total difference between the immunization antigen and the coating antigen referring to the coupling procedure as well as the binding protein was necessary.

Immunization

Rabbits were first injected with 1 mg of Dimethoate-thyroglobuline conjugate (prepared by the carbodiimide reaction) by an immunization scheme as described by Vaitukaitis (1980) suspended in PBS and emulsified in Freund's Complete Adjuvant (Figure 2). After five months the rabbits were boosted with 1 mg conjugate in Freund's Incomplete Adjuvant. The boosts were repeated at a three-day interval. The rabbits were bled ten days after the last boost injection.

Figure 2 Scheme of the rabbit imunization. Balck dots in the shaved area represent intradermal injection sites of 30-50 µl.

Enzyme-linked Immunosorbent Assay (ELISA)

ELISAs were performed in microtitre plates (NUNC or MLW) with 96 wells at room temperature. The assays were run with 96 replications. The pH of the used samples (standard or unknown samples) were adjusted to 7.2–7.4.

The following steps were carried out for the optimized assays (Figure 3):

1 Coating. 200 µl of an appropriate Dimethoate-protein conjugate dilution in sodium carbonate buffer (pH 9.6) was incubated at 4°C overnight.

2 Washing. Five washes were performed with 300 µl PBS washing buffer (pH 7.2).

3 Immunoreaction I. The unknown samples or standards were incubated with a constant amount of antibodies in a separate tube. 200 µl standards or samples are then added for three hours to each well.

4 Washing. See step 2.

5 Immunoreaction II. Goat-anti-rabbit antibodies (GaR-AP) labelled with alkaline phosphates of an appropriate dilution in

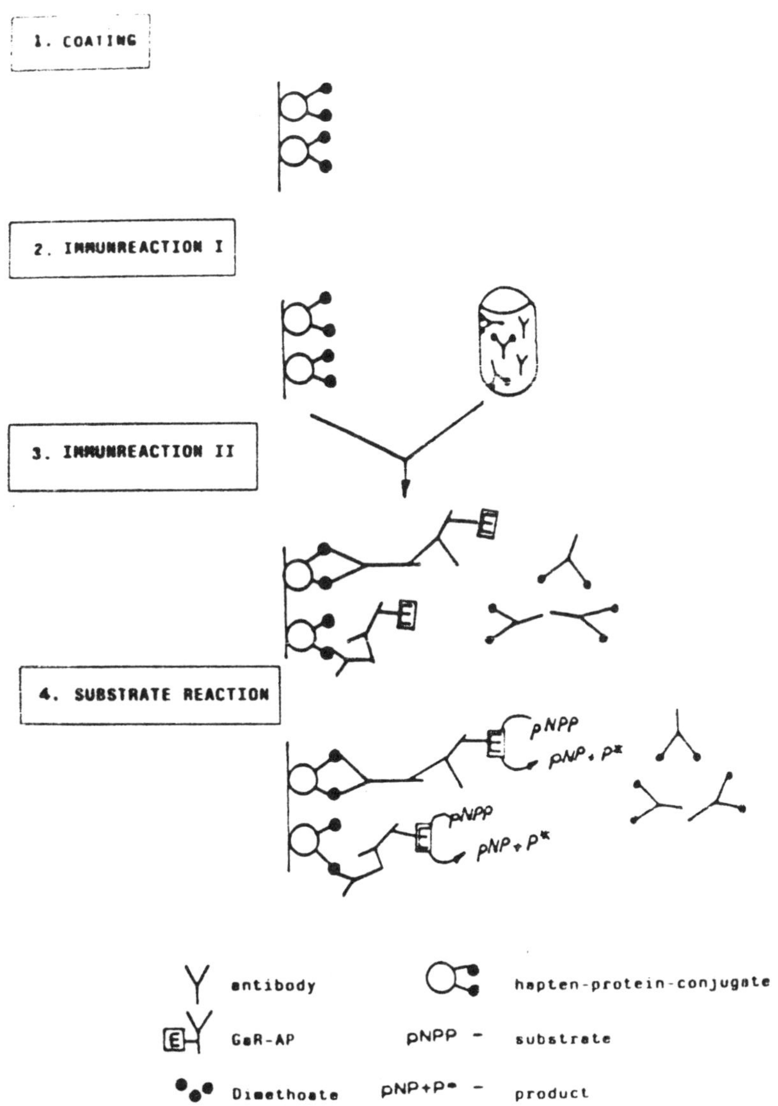

Figure 3   Scheme of the indirect competitive ELISA.

Tris-HCl buffer (pH 3.0) were added for an incubation time of one hour.

6 Washing. See step 2.

7 Substrate reaction. 200 µl substrate (p-nitrophenylphosphate) in diethanolamine solution (pH 9.8) was added and the absorption was read at 405 nm after 30–60 minutes.

Optimized Dimethoate ELISA

The antiserum was diluted 1:10,000 and the microtitre plates were coated with 3 µl hapten-protein conjugate per ml. The absorption was measured 40 minutes after the addition of the substrate. The standard curve with the described parameters is shown in Figure 4.

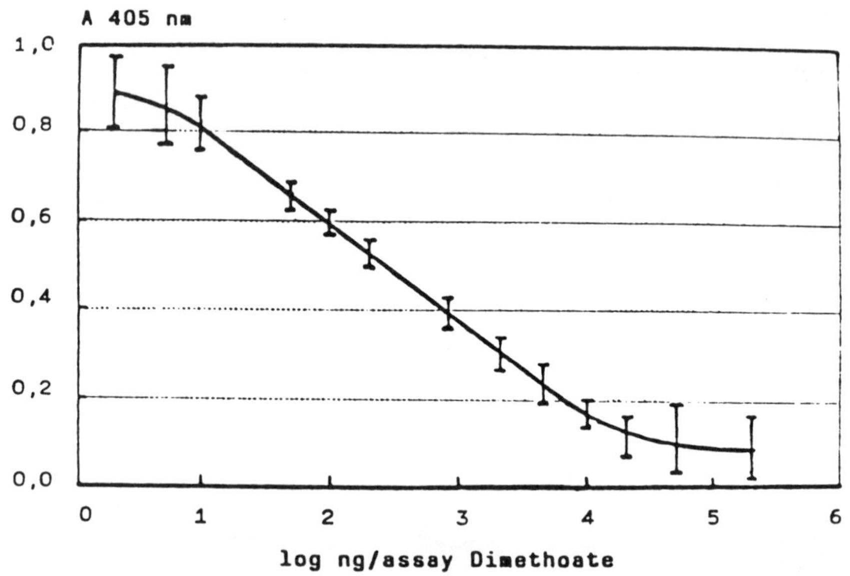

Figure 4   Calibration curve of Dimethoate.

Results of important criteria of the assay:

50% B/Bo: 1.5 µg ml^{-1} for Dimethoate
Detection range: between about 5 ng and 200 µg per assay
Linear range: between about 5 ng and 10 µg per assay

Precision
The linear detection ranges from 10 to 200 ng. Dimethoate in the within assay has a variation coefficient of less than 10% (Figure 5), in the between assay, however, of less than 13% (Figure 6).

Cross reactivities
The cross reactivities of antisera were related to Dimethoate (= 100%). The data were obtained from standard curves with different

Dimethoate (ng)	n	R̄	± s	s%
5	6	0,85	0,15	11,7
10	6	0,82	0,03	3,7
20	6	0,73	0,06	8,2
50	6	0,69	0,03	4,4
200	6	0,55	0,04	7,3
2000	6	0,31	0,03	9,7
10000	6	0,20	0,05	25,0

Figure 5   Within assay of Dimethoate.

Dimethoate (ng)	n	R̄	± SD	s%
2	7	10	5	50,0
10	7	17	3	17,6
20	7	25	4	16,0
200	7	48	6	12,5
2000	7	68	8	11,8
20000	7	92	10	10,9

Figure 6   Between assay of Dimethoate.

organophosphates and two metabolites of the Dimethoate and calculated according to the formula:

$$\text{Cross reactivity} = \frac{\text{hapten concentration at 50\% B/Bo}}{\text{concentration of the cross-reacting hapten at 50\% B/Bo}}$$

The results of the investigation into specificity are shown in Figure 7.

Analysis of samples (Application of the test)
With the dimethoate-ELISA water samples, plant extracts and homogenates of poisoned honeybees were measured.

1 Recovery. The results of the determination of the recovery from Dimethoate in PBS are shown in Figure 8.

2 Analysis of plant extracts. Aqueous extracts of tomato, grape, cucumber and cabbage with a constant amount of 25 ng, 100 ng,

Substance	Cross reactivity (%)
Dimethoate	100
Carboxylic acid derivative	0,6
o-Dimethoate	4,7
Methamidophos	-
Trichlorfon	-
Naled	-
Parathionmethyl	-
Standardformulation Bi 58	-

Figure 7   Specificity.

Dimethoate (ng/ml)	n	x (ng/ml)	Recovery SD	%
2	6	1,5	± 2,6	75,0
20	6	18,7	± 10,0	99,5
200	6	189,0	± 15,5	94,5
2000	6	2031,0	± 24,0	101,6
20000	6	21300,0	± 56,0	106,5
200000	6	235000,0	± 135,0	117,5

Figure 8   Analysis of Dimethoate in PBS.

1 µg, 10 µg or 50 µg Dimethoate were analysed. In comparison with the Dimethoate standard curve, small matrix effects were observed (Figure 9).

3 Analysis of honeybees. Homogenates of poisoned honeybees were analysed for Dimethoate residues. Because of the strong interactions between bee homogenates and antibodies, a correlation between the results of the immunoassay and the GC analysis was not found. The comparison of ELISA and GC results is shown in Figure 10.

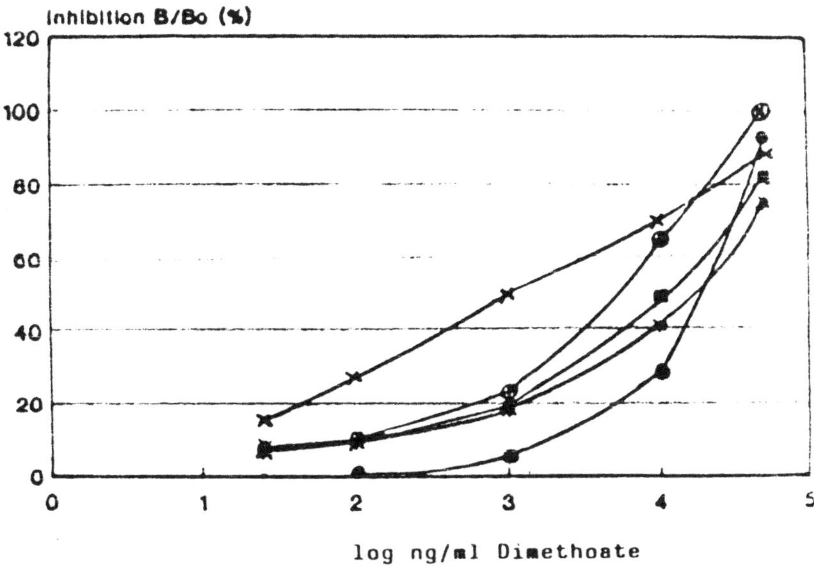

Figure 9 Analysis of plant extracts.

Sample No.	ELISA ng/bee	GC	f factor of overrate
1	480	41	11,7
2	150	10	15,0
3	70	5	14,0
4	370	35	10,6
5	300	26	11,5

Figure 10 Analysis of honeybees.

Conclusion

The presented ELISA for the analysis of Dimethoate is sensitive and very specific. Water samples and plant extracts could be measured directly by the ELISA. There is no clean-up procedure required. A monitoring of water samples with this test according to the upper limit of the EC guidelines and the drinking water regulations of Germany is not yet possible without any enrichment steps. But the determination of Dimethoate residues in several vegetable samples in the statutory detection range can be guaranteed.

For the screening of Dimethoate-poisoned honeybees, a correlation factor between the immunological investigations and the GC should be found. The easy handling, the small sample volumes needed, the omission of clean-up and concentration steps in most cases, the relatively fast measurement, the high sample turnover, the lowdetection limit and the acceptable costs are important criteria. Therefore, this test can be used as a screening method for the analysis of Dimethoate.

Acknowledgment

I thank Dr H. Poleschner from Biologische Zentralanstalt Berlin, Germany, for kindly providing the carboxylic acid derivative of the Dimethoate molecule.

References

BERTIN, D. and GERALT, P. (1963). *Neue Carbamoyldithiophosphate und Verfahren zu ihrer Herstellung*. DE-PS (Deutsche Patentschrift) 19554235.

ERLANGER, B.F., BREK, F., BEISSER, S.M. and LIEBERMAN, S. (1959). Steroid-protein conjugates. 3. Preparation and characterization of conjugates of bovine serum albumin with progesterone, deoxycorticosterone, and estrone. *Journal of Biological Chemistry*, **234**, 1090–1094.

FERREIRA, J.R., FALCAO, M.M. and TAINHA, A. (1987). Residue of dimethoate and omethoate in peaches and apples following repeated applications of dimethoate. *Journal of Agricultural Food Chemistry*, **35**, 506–508.

GOODFRIEND, T.L., LEVINE, L. and FASMAN, G.D. (1964). Antibodies in bradykinin and angiotensin. A use of carbodiimides in immunology. *Science* **144**, 1344–1348.

HABEEB, A.F.S.A. (1966). Determination of free amino groups in proteins by trinitrobenzensulphonic acid. *Analytical Biochemistry*, **14**, 328–336.

VAITUKAITIS, J. (1980). Production of antisera with small doses of immunogen: Multiple intradermal injections. In *Methods of*

Enzymology (H. van Nunakis, J.J. Langone, eds), Volume **70**, pp. 46–52. Academic Press, New York.

Part 4
Biochemical and Genetic Bases of the Development, Detection and Management of Insecticide Resistance Relationship Studies

27
Status of IRAC – the Insecticides Resistance Action Committee from Industry

W. KNAUF
(Chairman of IRAC)

Höchst AG, Agricultural Division, Biological Research, Frankfurt a. Main, Germany

After more than seven years of IRAC's existence (Voss, 1989) this group of specialists from industry is well established and active and forms an instrument to avoid or delay the resistance problems of insecticides and acaricides.

Today the different subgroups of IRAC are working at different activity levels depending on the actual status of resistance problems in the specific crop or chemical group envisaged. It is interesting to note that an additional working group dealing with possible resistance towards *Bacillus thuringiensis* joined IRAC recently and broadened IRAC's responsibilities towards biologicals.

In the meantime IRAC is engaged in a network of groups with industrial input in countries around the world, and with its expertise is respected as a valuable discussion partner.

During recent years some of the groups inside IRAC invited experts to investigate special questions, e.g. to test series of products against a variety of strains with different sensibility resistance levels. This was done with spider mites in the Fruit Working Group.

If there are research programmes that fit the strategy of IRAC working groups, such proposals can be made to the group chairman, who will discuss them inside IRAC. Interesting research proposals

---

Insecticides: Mechanism of Action and Resistance
© 1992   Intercept Ltd, P.O. Box 716, Andover, Hants SP10 1YG, UK

can also be granted from the IRAC central fund.

One of the main functions of IRAC was to publish a whole package of testing methods designed to detect changes of sensibility or resistance in insects and spider mites. This was published recently by the International Group of National Associations of Agrochemical Manufacturers (GIFAP) (reprints in GIFAP, Brussels, available). It is planned that after a period of two years the methods will be updated with the input of experts and published again in the final version. By these mans the researcher can make tests, the results of which can be compared with results of other published tests.

Because of the growing interest in the phenomenon of resistance, IRAC was also invited to send experts from the working groups to international conferences. In any case IRAC offers help and expertise to explain, to combat and to avoid resistance towards a real resistance management.

References

VOSS, G. (1989). Insecticide and acaricide resistance – Industry's efforts to cope. In *Tag.-Ber., Akad. Landwirtsch.-Wiss. DDR, Berlin,* **274** (s), 203–210.

BUHOLZER, H. and VOSS, G. (1991). Insektizied-Resistenz: Ursachen – Bedeutung –Lösungen, Wien, Deutscher Entomologenkongress.

IRAC (1990). Proposed Insecticide/Acaricide Susceptibility Tests, developed by Insecticide Resistance Action Committee. *EPPO Bulletin,* **20**, 389–404.

28
Use of Biochemical Markers to Study the Interaction of Insecticide Resistance Genes

I. DENHOLM, A.L. DEVONSHIRE, K.J. GORMAN and G.D. MOORES

AFRC Institute of Arable Crops Research, Rothamsted Experimental Station, Harpenden, Herts AL5 2JQ, UK

Introduction

The increasing incidence and severity of insecticide resistance poses a serious threat to the continued success of chemicals for controlling many insect and mite pests. The potential economic and social consequences of resistance are now widely accepted, but the exact nature of factors that influence, both positively and negatively, the rate of resistance build-up in a particular species is generally poorly understood. As a result, much of the existing conceptual framework for combating resistance is based on theoretical insights into how various options for deploying pesticides are likely to affect selection rates. These modelling studies have been valuable for identifying the major determinants of resistance dynamics, but have proved notoriously difficult to relate to the idiosyncrasies of specific pest problems.

Despite this shortage of genetic and ecological data, the principles of population genetics predict that resistance will be far easier to circumvent when genes conferring the trait are still rare and/or very localized in their distribution (Roush and Daly, 1990). Yet resistance development is insidious and often goes unnoticed at this critical early stage, only becoming apparent once the proportion of resistant individuals is sufficiently high to impair the efficacy of pesticide

Insecticides: Mechanism of Action and Resistance
© 1992   Intercept Ltd, P.O. Box 716, Andover, Hants SP10 1YG, UK

applications. By then the options for modifying control regimes to combat the threat are extremely limited and often ineffective. In addition, scope for switching to new chemical groups once resistance becomes established is steadily diminishing as novel chemicals become increasingly difficult to discover and register, and cross-resistance conferred by broad-spectrum detoxication enzymes and insensitive target proteins becomes increasingly pervasive.

Monitoring programmes to detect resistance phenotypes as early as possible and to document their distribution should therefore be a key component of any resistance management strategy. Whole-organism bioassays, involving topical application or exposure to pesticide residues, have long been the cornerstone of such programmes, but are limited in their application and resolution (Roush and Miller, 1986). Comparisons of LD_{50} or LD_{90} values of samples from populations – the most widely adopted approach – may be useful for detecting a high frequency of resistant insects but are far too insensitive for detecting incipient resistance. Use of a discriminating dose corresponding to the LD_{99} or higher of 'susceptible' populations, although better, is still subject to severe statistical constraints. Firstly, the estimation of these doses incorporates a wide margin of error because the fitting of probit or logit models is very imprecise at the extreme ends of response distributions. Secondly, unless doses are perfectly diagnostic (which is rarely the case) sample sizes required for the reliable detection of even 1% resistance may be prohibitively large.

Biochemical Assays for Insecticide Resistance

In principle, many of these shortcomings can be overcome by developing more incisive tests that can be applied rapidly on a large scale and which distinguish unambiguously between susceptible and resistant individuals. With recent progress in resolving the biochemical and molecular basis of resistance mechanisms, *in vitro* assays that diagnose the genes or enzymes implicated in resistance are being widely advocated for this purpose (Brown and Brogdon, 1987). Such assays have the considerable advantage over bioassays of providing direct information on the genetic constitution of a population, rather than just a phenotypic measure of changes in response to pesticides. In particular, assays capable of detecting resistance heterozygotes, and distinguishing these from homozygotes, offer exciting prospects for combating resistance in the early stages of selection when virtually all resistance alleles are present in heterozygous condition.

In vitro assays are none the less also subject to important constraints on their use with field populations. They are obviously entirely dependent on a thorough understanding of the biochemical changes responsible for resistance in a particular species, usually acquired through detailed studies on populations from areas where resistance is particularly well advanced. In addition, biochemical techniques usually depend on the use of a model substrate for the enzyme under consideration, in order to achieve the required sensitivity. Such substrates are often affected by many similar enzymes, and it is essential that the differences in activity measured have a proven bearing on resistance, or at least correlate fully with differences in tolerance of insecticides. Assays requiring least validation in this respect, and which can be transferred with greatest confidence from one species to another, are those based on simple measurement of the insecticide inhibition kinetics of acetylcholinesterase (Moores *et al.*, 1988b), the target enzyme of organophosphorus (OP) and carbamate compounds.

In contrast to bioassays, which measure differences in overall response to a toxicant and thereby recognize any type or combination of resistance factors, biochemical and DNA diagnostics are, by definition, applied to identify changes in specific metabolic enzymes, target proteins or genes. Hence, when testing field populations, the absence of an *in vitro* marker shows only that the insects lack that particular resistance mechanisms; it does not necessarily signify the absence of resistance. Indeed, reliance on one specific assay could be dangerously misleading if the population contains other, unsuspected, mechanisms of equal or greater practical significance. This danger gets more acute as assays become more finely tuned by moving from robust 'spot tests', through immunoassays to DNA diagnostics. The ultimate specificity to be achieved by diagnosing a point mutation would be of very restricted application in a field monitoring programme. In most cases, robust biochemical assays will be better equipped than DNA diagnostics to provide the information required.

Therefore, *in vitro* assays cannot be regarded as a panacea for problems encountered with conventional bioassay procedures. However, this reservation about their large-scale utility should be tempered with a full appreciation of their potential role in studying the build-up of known resistance genes in defined populations under controlled laboratory or field conditions. In such cases, assays allowing rapid and precise diagnosis of single mechanisms may be invaluable research tools, not only for monitoring individual resistance genes but also for investigating the co-selection of mechanisms conferring resistance to the same selecting agent. Work done at Rothamsted to investigate interactions between two biochemical markers of OP resistance in houseflies (*Musca domestica* L.) under simulated field conditions in the laboratory provides an example of this approach.

Experimental Procedure

Resistance markers and their detection

In houseflies, as in many other pests, resistance to OP insecticides is conferred by an intricate multifactorial system that includes a decrease in cuticular penetration, enhanced detoxication involving hydrolases, glutathione-*S*-transferases and mixed function oxidases, and reduced sensitivity of the target acetylcholinesterase (AChE) enzyme. The study reported here involved two OP-resistance markers that commonly coexist in field populations, esterase $E_{0.39}$ and one of the forms of insensitive AChE (AChE-R) – both of which have been isolated in homozygous strains and can be monitored biochemically in single insects, with different degrees of resolution.

$E_{0.39}$ describes an esterase variant implicated in resistance to the OPs malathion and trichlorfon, and in slight (*c.* three-fold) resistance to several pyrethroids (Sawicki *et al.*, 1984). Since this enzyme occurs as a distinct bank on electrophoresis gels stained with 1-naphthyl acetate, there is no difficulty in distinguishing flies with $E_{0.39}$ from susceptible ones homozygous for the corresponding null allele. The main drawback with this qualitative assay is that it is generally not feasible to separate $E_{0.39}$ heterozygotes and homozygotes, both of which exhibit the band. The differences in staining intensity expected of genotypes possessing either one or two copies of the $E_{0.39}$ allele have proved too small for categorizing individuals from mixed populations. Hence, without the major effort needed to purify the enzyme and develop a specific immunoassay, it is only possible to monitor changes in the frequency of the $E_{0.39}$ phenotype over time.

In contrast, the quantitative kinetic assay developed to diagnose the AChE-R mechanism not only allows a faster throughput of test insects than electrophoresis but also clearly resolves different alleles, whether homo- or heterozygous, at the AChE locus (Moores *et al.* 1988b). This technique entails homogenizing insects in separate wells of a 96-well microtitre plate, and then assaying portions of each homogenate with acetylthiocholine in both the presence and the absence of an appropriate insecticide inhibitor. An on-line microcomputer takes absorbance readings throughout the incuba-tion period and fits linear regressions to the reaction curves. The average slope of each curve in the presence of inhibitor is then expressed as a percentage of the corresponding uninhibited rate, and distribution histograms are generated to distinguish heterozygotes and homozygotes.

Since several mutant forms of AChE with contrasting patterns of insensitivity occur in houseflies, the type and concentration of test inhibitor must be tailored to the particular enzyme variant being investigated. The CH_2 enzyme employed in this study was best characterized by its insensitivity to dichlorvos (Moores *et al.*, 1988a), and a 10 μM concentration of this chemical gave the clearest discrimination between susceptible and resistant genotypes (Figure 1). Unlike $E_{0.39}$, data for all three genotypes could then be used to calculate actual gene frequencies of the sensitive (S) and CH_2 alleles in selected populations.

Since the $E_{0.39}$ and AChE-R assays require only 10% and 5% of a single fly homogenate respectively, both markers can easily be diagnosed in the same individual, so optimizing the use of resources

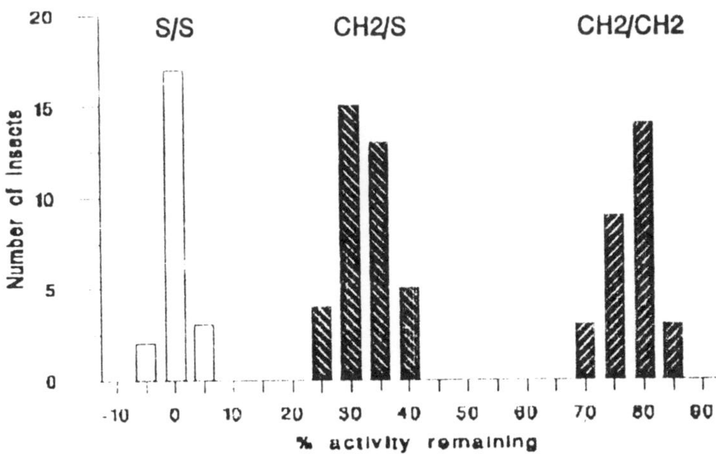

Figure 1   Distribution of mean percentage activity remaining during inhibition of AChE by 10μM dichlorvos for three housefly genotypes derived from a sensitive (S) and an OP-insensitive (CH_2) allele.

Choice of selecting agent

In conventional topical application bioassays, the cross-resistance pattern of a strain homozygous for the CH_2 enzyme alone to a range of OP insecticides (Table 1) correspond with its AChE insensitivity profile as disclosed by the microplate assay. Marked insensitivity (fifteen-fold) of AChE dichlorvos was expressed as only low (*c.* five-fold) resistance to its more stable and widely used precursor, trichlorfon. Similar or even lower levels of resistance to parathion, malathion and fenitrothion were correlated with insensitivity to their respective -oxon metabolites (Devonshire, 1987; Denholm *et al.*, 1990). Resistance to trichlorfon and malathion (twelve- and six-fold respectively) was greater in a strain possessing $E_{0.39}$ alone, but increased very dramatically in flies homozygous for both markers. Interaction between the loci was therefore multiplicative rather than additive, highlighting its potential importance in determining the overall resistance phenotype of field populations. On the basis of these considerations, trichlorfon was chosen as the selecting agent for studying these effects further under more realistic exposure conditions.

Table 1 Resistance factors in topical bioassays for housefly strains homozygous for $E_{0.39}$ and the CH_2 AChE-R variant, alone and in combination

Insecticide	AChE-R only	$E_{0.39}$ only	AChE-R $+ E_{0.39}$
Trichlorfon	5.4	12	92
Parathion	5.0	1.7	7.2
Malathion	2.5	5.8	53
Dimethoate	1.2	1.5	2.5
Fenitrothion	5.3	2.6	10
Azamethiphos	1.1	1.3	2.7

Selection procedure

Populations, initially containing a low frequency of one or both resistance markers, were established in large cages (80 × 40 × 40 cm) constructed from aluminium sheeting, and cultured continuously by collecting a prescribed volume of eggs three times each week, rearing larvae on a standard bran-based medium, and replacing pupae inside the cage (Denholm *et al.*, 1986). The populations were allowed to equilibrate at densities of 2000–3000 adults/cage before being exposed to residues of formulated trichlorfon ('Dipterex 80'; Bayer UK) sprayed on to 15 cm square aluminium panels that were then attached to the cage walls to cover *c.* 4% of the total surface area of the cage. Population numbers were assessed at weekly intervals throughout the experiment by filming and counting adults settled on a grid etched into the untreated rear wall, yielding a 'population size index' that could be calibrated with the total number of adults present. For example, indices of 5, 10 and 20 corresponded approximately to 1000, 2000 and 4000 insects/cage respectively. Frequencies of the two resistance markers were monitored every two weeks in samples of 96 adult females emerging from eggs laid by the treated populations. Testing of F_1 insects was necessary to avoid flies whose AChE might already be partly inhibited through direct exposure to the insecticide.

Details of the experimental regime are given in Figure 2. Freshly sprayed panels introduced on four occasions were left in the cages for two weeks before being replaced. Those added on days 0 and 14 were sprayed at 0.5 g a.i./m^2, corresponding to the lower end of the range of recommended field application rates. On days 28 and 42 this was increased to 2.0 g/m^2 to impose a stronger selection pressure and to compare responses at the two concentrations.

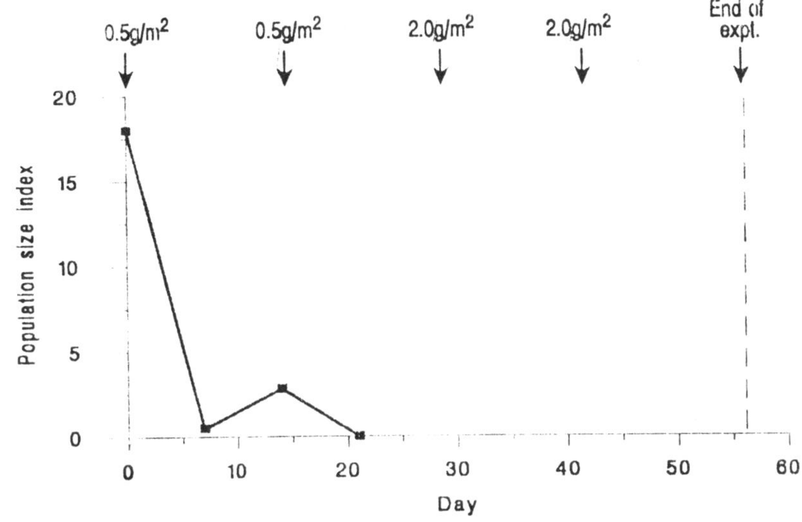

Figure 2 Changes in adult numbers in a fully susceptible housefly population under treatment with trichlorfon. Arrows identify when treated panels were introduced, and the concentration of trichlorfon residues.

Results

No resistance markers present

Figure 2 shows that a population derived wholly from a standard susceptible strain was unable to withstand even the lower application rate. Numbers decreased dramatically as soon as trichlorfon treatment commenced, and despite a slight recovery by day 14 (reflecting development and emergence from eggs laid prior to the start of treatment), the population died out by day 20.

Single resistance markers present

In the populations containing AChE-R or $E_{0.39}$ alone, trichlorfon selected very effectively for each resistance allele. However, the resulting changes in adult numbers depended on the type of resistance present (Figure 3). After the initial, rapid elimination of susceptible homozygotes, numbers in the AChE-R population remained very low despite the increasing predominance and eventual fixation of the AChE-R allele (Figure 3(a)). Hence, although resistance genotypes were selectively favoured in the population, the protection conferred was clearly very slight, and insufficient to permit the population to build up, even at the lower application rate.

In agreement with the higher resistance exhibited by flies with $E_{0.39}$ in topical bioassays (Table 1), selection of this gene did allow a build-up in numbers at the lower concentrations (Figure 3(b)). Increasing the application rate four-fold on day 28 none the less overcame much of this advantage, severely reducing the size of the population for the remainder of the experiment.

Both resistance markers present

With both markers present, the population showed a repetitive pattern of decline and recovery at the two concentrations (Figure 4). As a consequence of being able to monitor each marker separately, these changes could be attributed to the sequential build-up of resistance mechanisms allowing the population to withstand increasingly greater exposure to the insecticide. As it is the more potent factor, $E_{0.39}$ was initially selected in preference to AChE-R and mostly accounted for the recovery phase at the lower concentration. As before, however, this mechanism was incapable of protecting insects against the higher application rate. Recovery in this case required additional selection of the AChE-R gene to generate multiple resistance genotypes exploiting the powerful

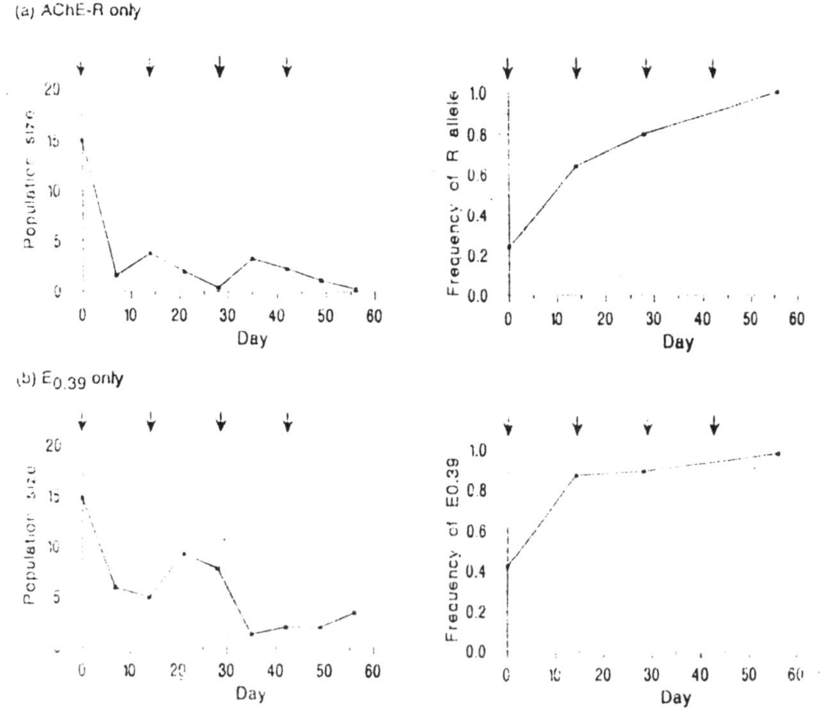

Figure 3 Changes in adult numbers and the frequency of the resistance marker populations containing (a) AChE-R alone, and (b) $E_{0.39}$ alone. Arrows identify the timing of trichlorfon treatments.

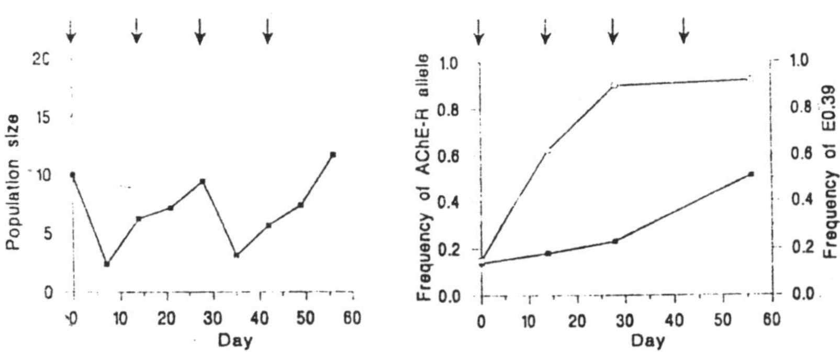

Figure 4 Changes in adult numbers and the frequency of resistance markers in a population containing both AChE-R (■) and $E_{0.39}$ ((□). Arrows identify the timing of trichlorfon treatments.

interaction between the two loci. By the end of the period, 71% of insects were homozygous or heterozygous for the AChE-R allele and it appeared that both these genotypes, in combination with $E_{0.39}$, were capable of surviving exposure to simulate a control failure at the higher application rate.

Discussion

These results illustrate well the disparity that can arise between differing perceptions of resistance (Sawicki, 1987). All three genetically heterogeneous populations responded rapidly to selection with a build-up of resistance genes, but this was not necessarily matched by a corresponding increase in adult numbers. Thus, although selection of each gene alone could be detected by biochemical techniques (or if necessary, sensitive bioassays), this was unlikely to be of practical significance, even for the more effective marker $E_{0.39}$. Whether or not such populations are deemed to be 'resistant' depends on whether the term is defined using genetic or economic criteria (Sawicki, 1987).

Under stong, simulated field selection with trichlorfon, AChE-R and $E_{0.39}$ acted much more effectively as components of a bifactorial system than as major resistance mechanisms *per se*. This is probably far from atypical, since there is now an increasing number of reports that insects such a bollworms (Sawicki and Denholm, 1989; Gunning *et al.*, 1992), diamondback moth (Cheng, 1988) and mosquitoes (Raymond *et al.*, 1986), when subjected to intensive chemical control, accumulate an array of genetically independent resistance mechanisms. Interactions between these loci therefore become increasingly significant, both in raising resistance levels to individual compounds and in broadening the cross-resistance spectrum. While these interactions can be studied indirectly using pharmacokinetic models (Raymond *et al.*, 1989) or by bioassays against multiresistant strains reconstituted from isolated genotypes (Plapp and Hoyer, 1968, Sawicki, 1970), biochemical or molecular tests that diagnose individual genes provide the only incisive means of investigating the acquisition and dynamics of multifactorial resistance. This approach, so far applied to houseflies and mosquitoes (Bonning and Hemingway, 1991), should become more widely applicable as knowledge of resistance mechanisms in different species improves. Exploited in this way, *in vitro* techniques will not only complement bioassays but also cast light on a phenomenon that is virtually intractable using conventional toxicological methods.

References

BONNING, B.C. and HEMINGWAY, J. (1991). Identification of reduced fitness associated with an insecticide resistance gene in *Culex pipiens* by microtitre plate tests. *Medical and Veterinary Entomology*, **5**, 377–379.

BROWN, T.M. and BROGDON, W.G. (1987). Improved detection of insecticide resistance through conventional and molecular techniques. *Annual Review of Entomology*, **32**, 145–162.

CHENG, E.Y. (1988). Problems of control of insecticide-resistant *Plutella xylostella*. *Pesticide Science*, **23**, 177–188.

DENHOLM, I., SAWICKI, R.M., FARNHAM, A.W. and WHITE, J.C. (1986). Evaluation of a method for maintaining age-structured housefly populations to study the evolution of insecticide resistance. *Bulletin of Entomological Research*, **76**, 297–302.

DENHOLM, I., GORMAN, K.J., MOORES, G.D. and DEVONSHIRE, A.L. (1990). Biochemical assays for insecticide resistance: strengths and limitations. *Proceedings Brighton Crop Protection Conference*, **3**, 1175–1180.

DEVONSHIRE, A.L. (1987). Biochemical studies of organophosphorus and carbamate resistance in houseflies and aphids. In *Combating Resistance to Xenobiotics: Biological and Chemical Approaches* (M.G. Ford, D.W. Holloman, B.P.S. Khambay and R.M. Sawicki, eds), pp. 239–255. Ellis Horwood, Chichester.

GUNNING, R.V., EASTON, C.S., BALFE, M.E. and FERRIS, I.G. (1992). Pyrethroid resistance mechanisms in Australian *Helicoverpa armigera*. *Pesticide Science*, in press.

MOORES, G.D., DENHOLM, I., BYRNE, F.J., KENNEDY, A.L. and DEVONSHIRE, A.L. (1988a). Characterizing acetylcholinesterase genotypes in resistant insect populations. *Proceedings Brighton Crop Protection Conference*, **1**, 451–456.

MOORES, G.D., DEVONSHIRE, A.L. and DENHOLM, I. (1988b). A microtitre plate assay for characterizing insensitive acetylcholinesterase genotypes of insecticide resistant insects. *Bulletin of Entomological Research*, **78**, 537–544.

PLAPP, F.W. and HOYER, R.F. (1968). Insecticide resistance in the housefly: decreased rate of absorbtion as the mechanism of action of a gene that acts as an intensifier of resistance. *Journal of Economic Entomology*, **61**, 1298–1303.

RAYMOND, M., FOURNIER, D., BRIDE, J.M., CUANY, A., BERGE, J., MAGNIN, M. and PASTEUR, N. (1986). Identification of resistance mechanisms in *Culex pipiens*

(Diptera: Culicidae) from southern France: insensitive acetylcholinesterase and detoxifying oxidases. *Journal of Economic Entomology*, **79**, 1452–1458.

RAYMOND, M., HECKEL, D.G. and SCOTT, J.G. (1989). Interactions between insecticide genes: model and experiment. *Genetics*, **123**, 543–551.

ROUSH, R.T. and DALY, J.C. (1990). The role of population genetics in resistance research and management. In *Pesticide Resistance in Arthropods* (R.T. Roush and B.E. Tabashnik, eds), pp. 97–152. Chapman and Hall, New York.

ROUSH, R.T. and MILLER, G.L. (1986). Considerations for the design of insecticide resistance monitoring programs. *Journal of Economic Entomology*, **79**, 293–298.

SAWICKI, R.M. (1970). Interaction between the factor delaying penetration of insecticides and the desethylation mechanism of resistance in organophosphate-resistant houseflies. *Pesticide Science*, **1**, 84–87.

SAWICKI, R.M. (1987). Definition, detection and documentation of insecticide resistance. In *Combating Resistance to Xenobiotics: Biological and Chemical Approaches* (M.G. Ford, D.W. Holloman, B.P.S. Khambay and R.M. Sawicki, eds), pp. 105–117. Ellis Horwood, Chichester.

SAWICKI, R.M. and DENHOLM I. (1989). Insecticide resistance management revisited. In *Progress and Prospects in Insect Control* (N.F. McFarlane, ed.), pp. 193–203. British Crop Protection Council, Farnham, Surrey.

SAWICKI, R.M., DEVONSHIRE, A.L., FARNHAM, A.W., O'DELL, K., MOORES, G.D. and DENHOLM I. (1984). Factors affecting resistance to insecticide in houseflies (*Musca domestica* L) (Diptera: Muscidae). 2. Close linkage on autosome 2 between an esterase and resistance to trichlorphon and the pyrethroids. *Bulletin of Entomological Research*, **74**, 197–206.

29
Detection of Insecticide Resistance in Peach Potato Aphids on Sugar Beet in East Germany by Biological, Biochemical and Immunological Investigations

B. DOROBEK, A. MUELLER and D. OTTO

Biological Research Centre Berlin, Stahnsdorfer Damm 81, 0-1532 Kleinmachnow, Germany

Introduction

The peach potato aphid (*Myzus persicae* Sulz.) is the main vector of virus diseases in East Germany. In the last two years, infection pressure has increased. Half of the sugar beet plants have been infected by aphid-transmitted viruses. This equals a yield loss of 20% or 200 000 tonnes sugar in this region.

For vector control several organophosphates, like dimethoate, parathion-methyl and methamidophos, were used, and pyrethroids and carbamates only in small quantities.

Since the early 1980s there has been information that chemical control has been less effective than desired and needed (Fritzsche *et al.*, 1985, 1987; Müller *et al.*, 1987). Therefore, the existence of insecticide resistance in practice has been investigated.

Over the last three years, aphids have been collected from beets, spinach, rape and other crops and the samples were reared on china cabbage. Biological, biochemical and immunological methods were used for resistance assay.

Insecticides: Mechanism of Action and Resistance
© 1992   Intercept Ltd, P.O. Box 716, Andover, Hants SP10 1YG, UK

Biological Assay with Dimethoate

For each dimethoate concentration sixty adult aphids were assayed on treated leaf discs. Three days after treatment the proportion of dead animals was measured and calculated by probit analysis. The resistance index of field populations varied between 3 and 117. A resistance index of 83 is shown in Figure 1. All field populations had a reduced susceptibility to dimethoate.

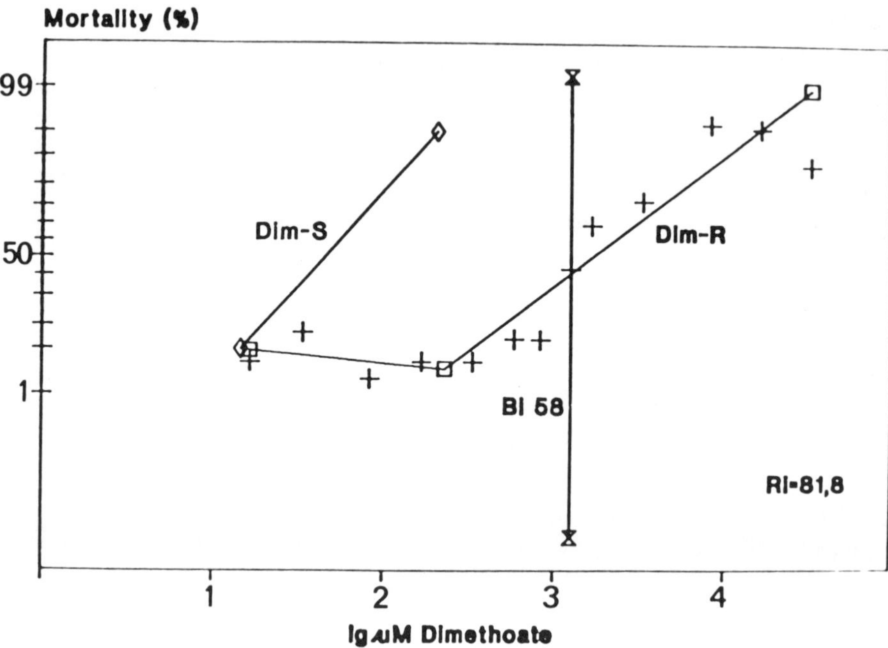

Figure 1   Susceptibility to dimethoate in field populations of *Myzus persicae* Sulz.

Biochemical Assay

Since the publications of Needham and Sawicki (1971) and Devonshire (1973, 1975) twenty years ago, it has been known that insecticide resistance in *Myzus persicae* is conferred by increased production of naphthylacetate-splitting carboxylesterase in aphids. The carboxylesterase E4 as a source of resistance was analysed and characterized by Devonshire (1977). It predominately splits α-naphthylacetate. Its activity was used to estimate the resistance level of aphid populations.

In our experiments we used microtitre plates and an equivalent of 0.1 aphid. Different esterase activities were found in biochemical assays for field populations from a region near Magdeburg and from a region near Aschersleben in the south-west of East Germany. For comparison, in this experiment the enzyme levels of aphid standard clones (Rothamsted Experimental Station, Harpenden/England) were investigated (Figure 2). In another example for the biochemical determination of insecticide resistance in a wild strain of *M. persicae*, a pronounced distinction in esterase levels between the susceptible strain and the field population as measured by esterase activity was found (Figure 3).

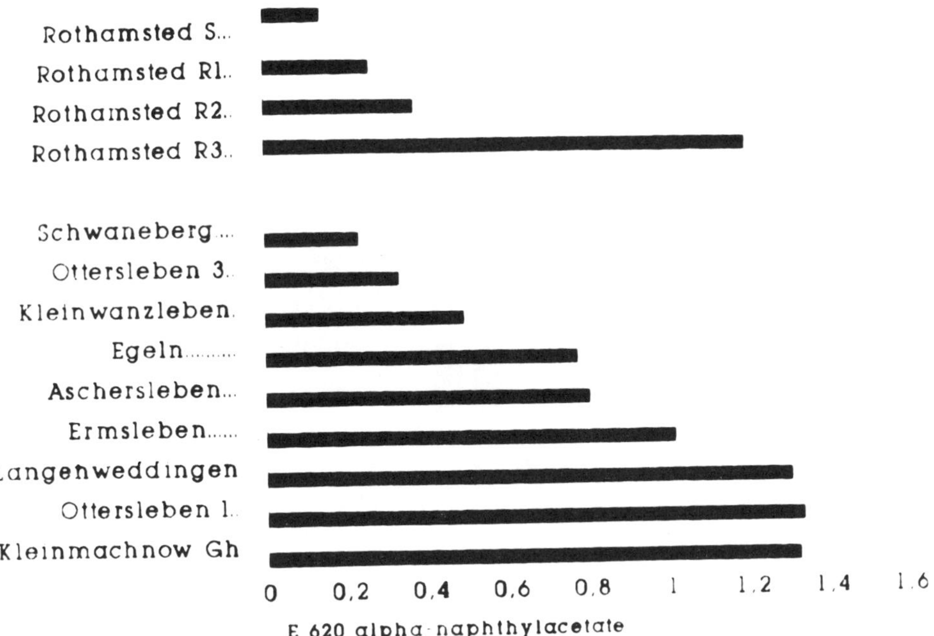

Figure 2 Esterase activity in field populations of *Myzus persicae* Sulz.

The esterase activities of several individuals of an aphid clone vary in a wide range. The assayed animals possessed different weights, numbers of embryos and ages and the leaves of the host plants were at different physiological stages. In comparison to the susceptible clone all field populations showed higher esterase activities.

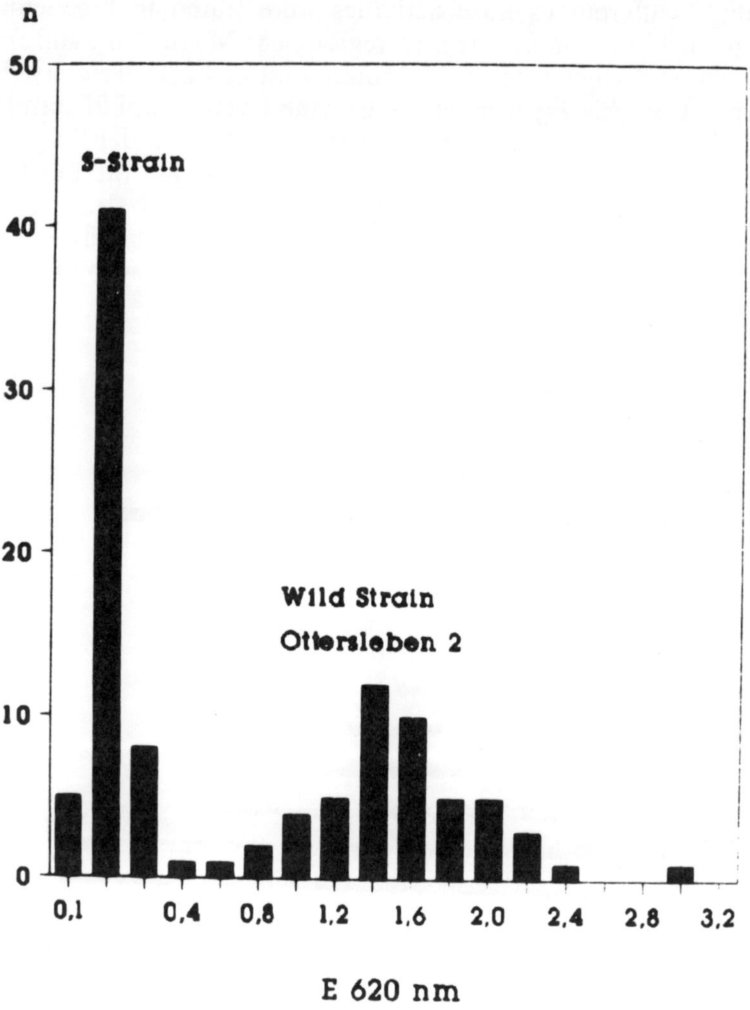

Figure 3   Esterase activity in field populations of *Myzus persicae* Sulz.

Electrophoresis

Homogenates of aphids were separated by PAGE according to the method of William and Reisfeld (1964) without sample and spacer gel. For visualization, phosphate buffer at pH 6 was incubated with 1 mM α- and β-naphthylacetate or with the butyrate analogue together with fast Blue RR.

On the gel the E4 is visible as a black–blue band, E1 and E2 are represented as red-coloured bands due to β-naphthylacetate (Figure 4).

Figure 4 Electrophoresis results obtained from homogenates of *Myzus persicae* Sulz.

We found that in resistant aphids the hydrolytic activity for α-naphthylacetate was several times higher than in susceptible ones, whereas the esterase-splitting β-naphthylacetate remained unchanged in both (Figure 5).

An electrophoretic technique which estimates the amount of E4 in a single aphid and thereby its resistance level was used for detecting the E4 in field populations from Germany in comparison to susceptible, moderately (R1) and very resistant (R2) standards from Rothamsted Experimental Station. Aphids from Ermsleben were mainly of moderately resistant level R1 with some susceptible animals, and aphids from Ottersleben were predominately R1.

However, electrophoresis suffers from relatively low throughput and depends on a subjective assessment of the band intensity. It provides a good discrimination between susceptible and moderately resistant aphids, but the large amount of activity in more resistant individuals (very resistant and extremely resistant) can cause difficulties in the subjective interpretation of band intensity. These limitations have recently been overcome by the development of an immunoassay by Devonshire and Moores (1984). Besides the high throughput of analysed aphids per day, the immunoassay has the advantage of giving a quantitative measure of the amount of E4 by which the range of esterase activities in susceptible and moderately resistant aphids is clearly resolved, whereas total esterase assay would show considerable overlapping between both forms.

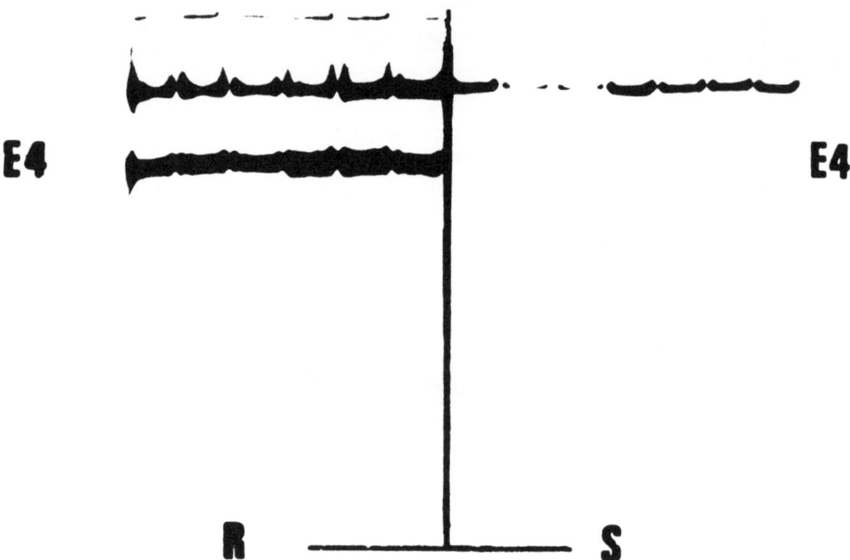

Figure 5 Electrophoresis results obtained from homogenates of *Myzus persicae* Sulz.

Immunoassay

To monitor the distribution of resistance in field populations of aphids in Germany, an immunoassay was used. The immunoassay protocol was similar to that of Devonshire and Moores (1984). The carboxylesterase E4 (C-E4) was extracted from most resistant aphids of a greenhouse aphid population of Kleinmachnow. The in rabbits produced C-E4 antibodies were used in the immunoassay and the

C-E4 binding capacity of the purified immunoglobulin fraction was compared first with the absorbance of standard laboratory clones, classified according to E4 quantity as reference markers from Rothamsted Experimental Station, and second with the E4 binding capacity of the anti-E4-IgG from Rothamsted used for the discrimination and classification of resistant variants (Figure 6) (Dorobek, 1991).

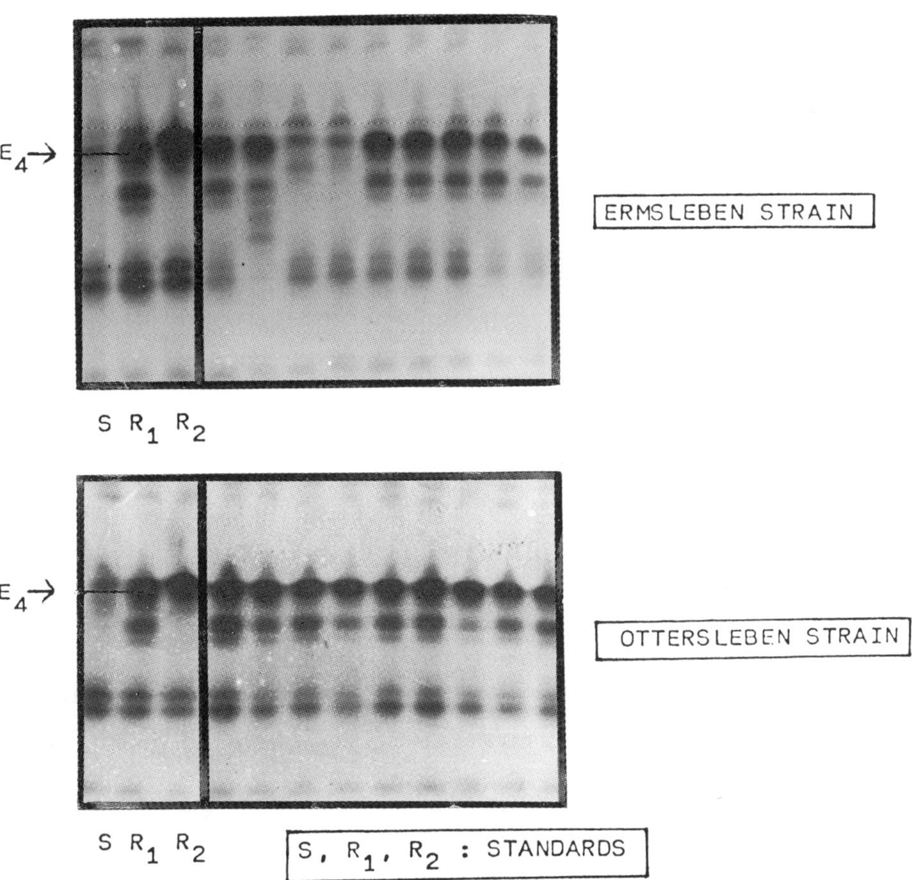

Figure 6   Immunoassay results.

Because of the large difference in esterase activity between susceptible and most resistant aphids it was necessary to assay two portions of each aphid in order to obtain accurate estimates over the whole range of resistance. The 0.08 portion distinguished susceptible from resistant aphids, whereas the 0.02 portion discriminated between the resistant variants.

The interrupted curve in Figure 7 shows the distribution of standard clones from Rothamsted Experimental Station, the unbroken line characterizes the IgG-binding capacity of prepared anti-E4-IgG from Kleinmachnow/Germany. E4 activity distributions for susceptible aphids show no significant difference between both IgGs. The characterization of moderately resistant aphids with an absorbance between 0.2 and 0.6 at the 0.02 dilution of one aphid in the immunoassay is comparable with Rothamsted IgG. Both IgGs are comparable at the very resistant level (R2) with an absorbance between 0.6 and 2.0 in the immunoassay. At the R3 level, with an absorbance of more than 2.0, a differentiation between R2 and R3 aphids is not possible with the Kleinmachnow IgG. The prepared antibody can be used to discriminate between S, R1 R2 but not between R2 and R3. Maximum binding capacity had been reached at the R2 level.

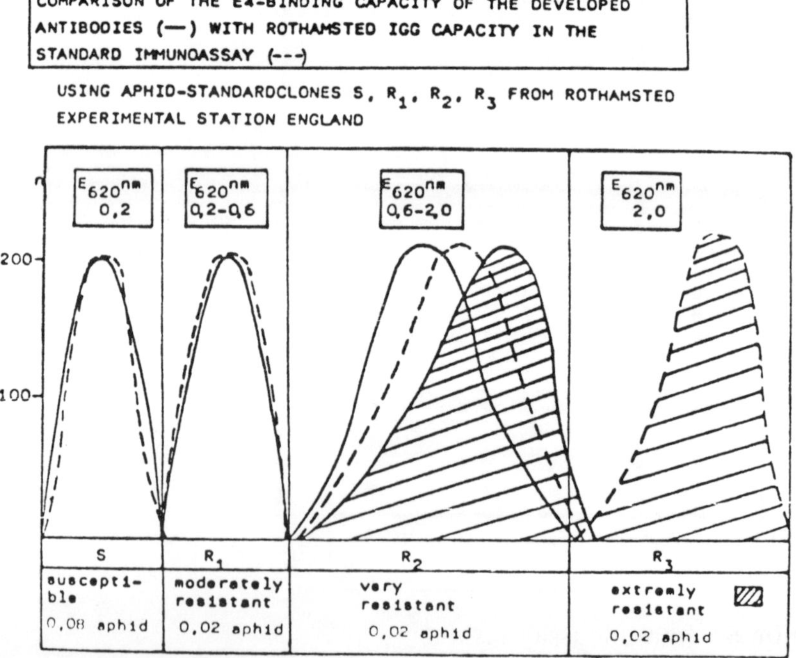

Figure 7 Comparison of resistance situation in different field populations.

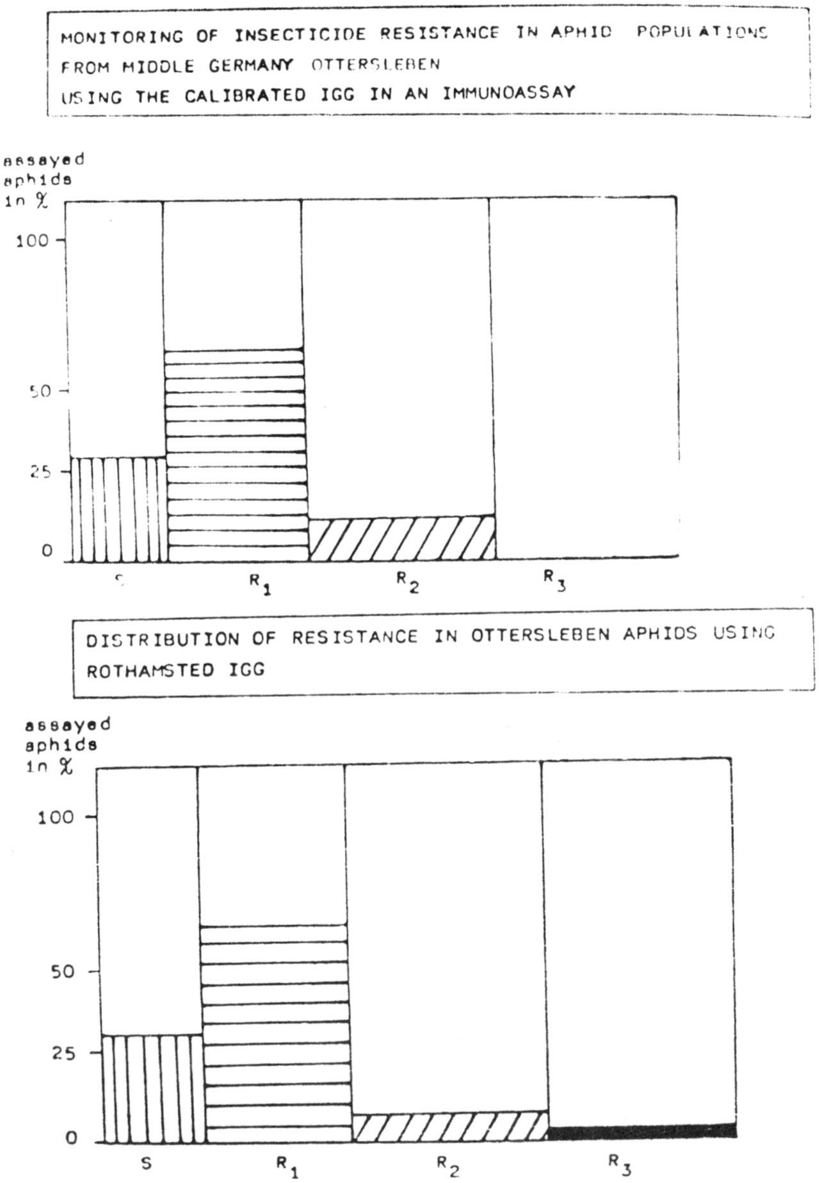

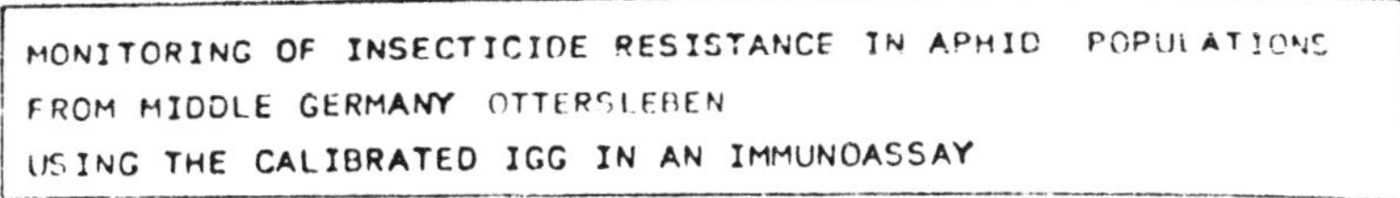

Figure 8 Comparison of resistance situation in different field populations.

For the accurate determination of R3 variants in the field a more sensitive antiserum must be prepared. But for treatment strategies with insecticides in the field the results of the immunoassay may be sufficient information.

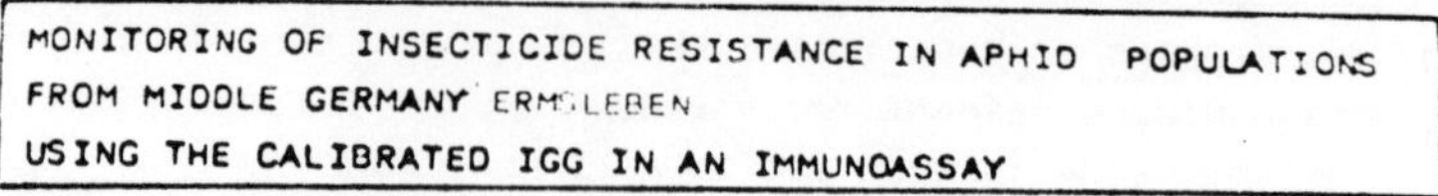

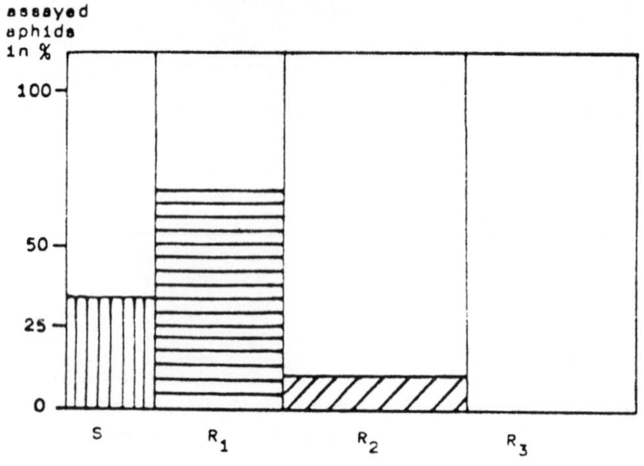

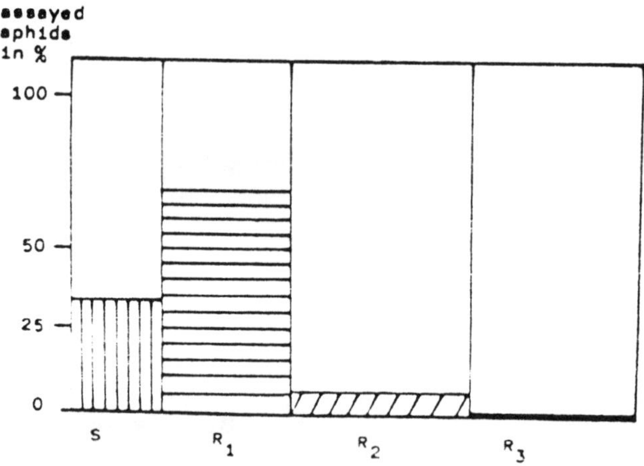

Figure 9 Comparison of resistance situation in different field populations.

Application of the immunoassay

The immunoassay has been used to monitor the distribution of resistance in two field populations of aphids in Germany. Therefore, cultures of the Ottersleben and the Ermsleben strains were reared and 100 individuals from each one were tested.

The E4 activities of the Ottersleben strain corresponded to the susceptible and the R1 level with 30% and 66% of the population, respectively. These results confirmed those from electrophoresis.

Only 7% of the population are very resistant R2 variants (Figure 7). Because of the missing R3 binding capacity of the prepared IgG the R3 level of resistance was detectable with Rothamsted IgGs only. But in this case a small amount, only 2% of the aphids with the highest level of resistance, were detected (Figure 8).

Almost the same situation of resistance was found in the Ermsleben population (Figure 9). The immunoassay results correspond also to that from electrophoresis.

The results of the immunoassay confirm, firstly, a successful preparation of the C-E4 antiserum for a useful determination of resistance in *M. persicae* in Germany and, secondly, the existence of the same factor of resistance in *M. persicae* in Germany as in Great Britain. It can be summarized that the Ottersleben and Ermsleben results, as representative samples, show that the field populations comprise mainly R1s and that very resistant R2s do not form a major proportion.

Conclusion

With the use of biological, biochemical and immunological methods detection, characterization and differentiation of resistance are possible also for aphid populations in Germany. The occurrence of resistance in Germany is also associated with the increasing level of the carboxylesterase – the known factor of resistance in aphids – although there was a different treatment for vector control, using different types of insecticides.

Acknowledgement

For leaving the aphid standard clones and the support in the development and application of the immunoassay as well as for providing the possibility to work at Rothamsted Experimental Station/England we wish to thank Alan Devonshire and his group.

References

DEVONSHIRE, A.L. (1973). The biochemical mechanism of resistance to insecticides with especial reference to the housefly, *Musca domestica*, and aphid, *Myzus persicae*. *Pesticide Science*, **4**, 521–529.

DEVONSHIRE, A.L. (1975). Studies of the carboxylesterases of *Myzus persicae* resistant and susceptible to organophosphorus insecticides. In *Proceedings of the Eight British Crop Protection Conference – Pests and Diseases, Brighton, 17–20 November 1975*, Volume 1, 63–67.

DEVONSHIRE, A.L. (1977). The properties of a carboxylesterase from the peach potato aphid, *Myzus persicae* (Sulz.), and its role in conferring insecticide resistance. *Biochemical Journal*, **167**, 675–683.

DEVONSHIRE, A.L. and MOORES, G.D. (1984). Immunoassay of carboxylesterase activity for identifying insecticide resistant *Myzus persicae*. In *Proceedings of the British Crop Protection Conference – Pests and Diseases, Brighton, 19–22 November 1984*, Volume 2, 6A–13, 515–520.

DOROBEK, B. (1991). Entwicklung immologischer Diagnoseverfahren am Beispiel des Pflanzenschutzmittels Dimethoat (Bi 58). *Ph.D. Thesis University Leipzig*, 103pp.

FRITZSCHE, R., GIERSEMEHL, I. and THIELE, S. (1985). Wirkung von Insektiziden auf anholozyklisch überwinterte *Myzus persicae* Sulz. als Vektor der Vergilbungsviren der Zuckerrüben. *Nachrichtenblatt für den Pflanzenschutz der DDR*, **39**, 43.

FRITZSCHE, R., KARL, E., GIERSEMEHL, I. and THIELE, S. (1987). Probleme des effektiven Einsatzes von Insektiziden gegen Virusvektoren in Beta-Rüben. *Nachrichtenblatt für den Pflanzenschutz der DDR*, **41**, 38–39.

MÜLLER, G. ECKERT, H. and KARL, E. (1987). Auftreten dimethoatresistenter *Myzus persicae* (Sulz.) inZuckerrübenvermehrungsbeständen im Jahre 1985. *Nachrichtenblatt für den Pflanzenschutz der DDR*, **41**, 69–71.

NEEDHAM, P.H. and SAWICKI, R.M. (1971). Diagnosis of resistance to organophosphorus insecticides in *Myzus persicae* (Sulz.). *Nature*, **230**, 125–126.

WILLIAM, D.E. and REISFELD, R.A. (1964). Disc electrophoresis in polyacrylamide gels: extension to new conditions of pH and buffer. *Annals of the New York Academy of Sciences*, **121**, 373–381.

30
The Role of Cytochrome P 450 in the Metabolism of Substituted *S*-ethynyl Thio- and Dithiophosphates

S.V. NEDELKINA,[1] L.A. VICHREVA,[2] T.A. PUDOVA,[2] N.N. GODOVIKOV[2] and R.I. SALGANIK[1]

[1]*Institute of Cytology and Genetics, Siberian Branch of the USSR Academy of Sciences, Lavrenteva prosp. 10, 630090 Novosibirsk*
[2]*A.N. Nesmeyanov, Institute of Organo-element Compounds, ul. Vavilova 28, 117334 Moscow, USSR*

Many organophosphorus compounds (OPCs), highly effective as insecticides and acaricides, are unsuitable to be used because of their high toxicity to warm-blooded animals. Substitution of the thiophosphoryl group for the phosphoryl group in the OPC molecule reduces the toxicity of the compound as the thioanalogue has no anti-acetylcholinesterase (AChE) activity (the thionic effect):

$$\frac{LD_{50} \text{ or } LC_{50} \; (\mathord{>}\!\overset{\displaystyle S}{\overset{\|}{P}}\!-)}{LD_{50} \text{ or } LC_{50} \; (\mathord{>}\!\overset{\displaystyle O}{\overset{\|}{P}}\!-)} = \text{THIONIC EFFECTS}$$

Its toxicity is manifested only upon oxidative desulphuration yielding the oxo analogue (Agosin, 1985). As oxidative desulphuration takes place at a much slower rate in warm-blooded animals than

Insecticides: Mechanism of Action and Resistance
© 1992   Intercept Ltd, P.O. Box 716, Andover, Hants SP10 1YG, UK

in arthropods, this reduces the toxicity of the thioanalogues (by 2–20 times) and indeed, these compounds make up the vast majority of OPCs currently in use.

In the Moscow Institute of Organo-element Compounds OPC were synthesized with acetylenic bond-substituted *S*-ethynyl thio- and dithiophosphates of the common formula:

$$(RO)^2P(Y)S\text{-}C \equiv C\text{-}X$$

where $R = CH_3, C_2H_5$
$Y = O, S$
$X = C_4H_9, CH_2SC_2H_5, CH_2SC_4H_9, CH_2OCOCH_3, C_6H_5$

These compounds were found to have a greater AChE-inhibiting activity to arthropods than their saturated analogues (Balashova *et al.*, 1983) and in some cases they were as toxic as standard insecticides or even more toxic (Vichreva *et al.*, 1988) (Table 1).

Table 1 Values of LC_{50} and LD_{50} for the compounds $(C_2H_5O)_2P(Y)S-R$

R	LD_{50}(mg kg^{-1}) Mice		LC_{50} (%) *Aphis fabae* S.		LC_{50} (%) *Calandra oryzae* L.	
	$Y=O$	$Y=S$	$Y=O$	$Y=S$	$Y=O$	$Y=S$
$-C \equiv CC_4H_9$	0.45	235	0.0045	0.0030	0.020	0.009
$-C \equiv CCH_2OCOCH_3$	0.50	279	0.0012	0.0150	0.070	0.700
$-(CH_2)_2C_4H_9$	160	400	very low toxicity			
$-(CH_2)_2CH_2OCOH_3$	210	400	very low toxicity			

The introduction of a triple bond in the alpha position of the thioester radical of oxo analogues increases the toxicity to warm-blooded animals, e.g. from 160 to 0.45 mg kg^{-1} and from 210 to 0.5 mg kg^{-1}. The introduction of a triple bond into the thioanalogues also increases the toxicity of these compounds, e.g. from 400 to 279 mg kg^{-1} and from 400 to 235 mg kg^{-1}. They thus have a high thionic effect in warm-blooded animals (reaching values of several hundred in some cases) and a low thionic effect in arthropods (Table 2).

Table 2 Thionic effect (see equation 1) for compounds $(C_2H_5O)_2P(Y)-S-R$

R	Mice	*Aphis fabae*	*Calandra oryzae*
	$\dfrac{LD_{50}(Y=S)}{LD_{50}(Y=O)}$	$\dfrac{LC_{50}(Y=S)}{LC_{50}(Y=O)}$	
$-C\equiv CC_4H_9$	470	0.7	0.5
$-C\equiv CCH_2OCOCH_3$	540	12	10
$-(CH_2)_2C_4H_9$	2.5		
$-(CH_2)_2CH_2OCOH_3$	1.9		

The toxic effect of thionic esters of phosphoric acids largely depends on their activation and inactivation in organisms. Esterases and glutathione *S*-transferases (GST) catalyse reactions which lead to a reduction of toxicity. Cytochrome P 450 catalyses dealkylation and the cleavage of P–S bonds as well, but also oxidative desulphuration, which leads to the formation of AChE inhibitors and, consequently, to an increase in toxicity (Figure 1).

DIMETHOATE

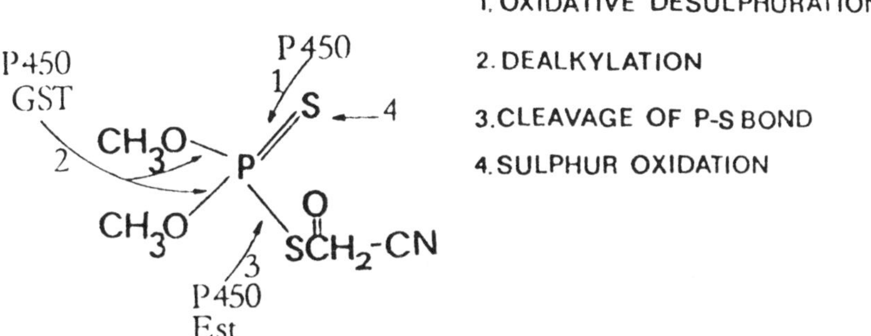

Figure 1 Enzyme-catalysed reactions with the organophosphorus compound dimethoate.

We supposed that the high selectivity is connected with cytochrome P 450 and studied the interaction of substituted *S*-

ethynyl thio- and dithiophosphates with P 450 from arthropods (abdomen of houseflies, *Musca domestica*) and from warm-blooded animals (rat liver).

Hepatic microsomal preparations were obtained according to Testai *et al.* (1982). Microsomes from fly abdomens were isolated in 0.1 potassium phosphate buffer pH 7.5 with 0.1 mM EDTA and 0.1 mM dithiothreitol containing 20% glycerol. Cytochrome P 450 concentrations were determined according to Omura and Sato (1964) with a spectrophotometer (Specord M 40). The protein concentration was measured according to Lowry *et al.* (1951) with bovine serum albumin as standard.

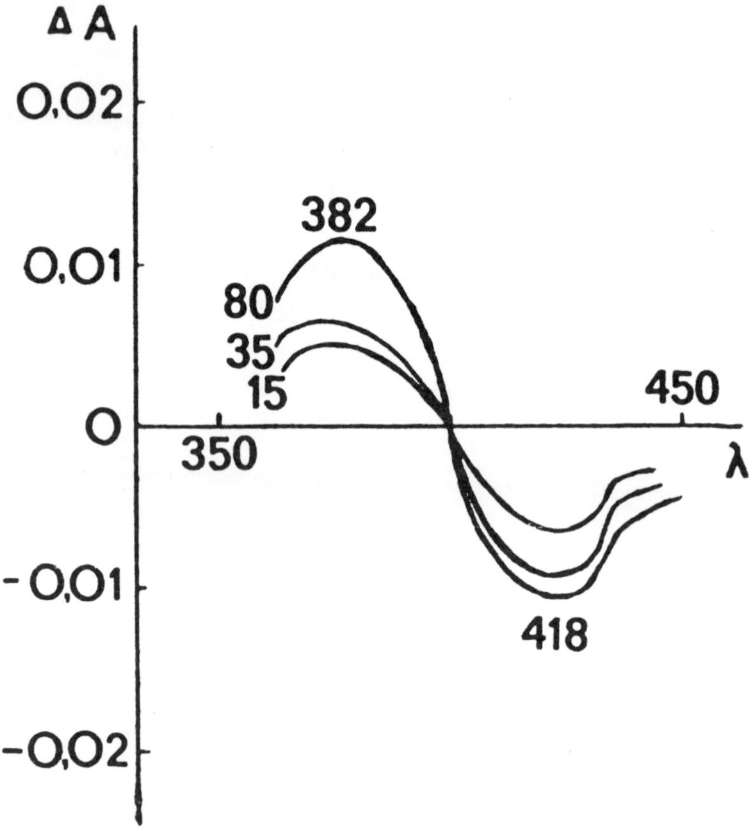

Figure 2   Differential spectrum of cytochrome P 450 binding in rat liver with $(C_2H_5O)_2P(O)SC=C_4H_9$. The lines correspond to the addition of 1 mM solution of the compound to the microsomal suspensions of 15, 35, 80 µl. The content of cytochrome P 450 is 1–2 nmol ml^{-1}.

The evaluation of destructive influences of compounds on P 450 was carried out under conditions providing maximum destruction. The results revealed an affinity of the compounds to the active site of cytochrome P 450 and to type I binding spectra (λmin 418, λmax 382), i.e. binding at the protein part of the site. Figure 2 shows the spectrum of cytochrome binding with one of the compounds of this set. K_S is 3.3 × 10^{-5} mM, which corresponds to a moderate affinity to the cytochrome P 450 (λmax 382, λmin 418).

Furthermore, P 450 was irreversibly damaged (Figure 3). The destruction takes place only in the presence of NADPH, which is absolutely indispensable for the function of cytochrome P 450 as an electron donor. That indicates that destructive abilities do not belong to an initial molecule but to the metabolite.

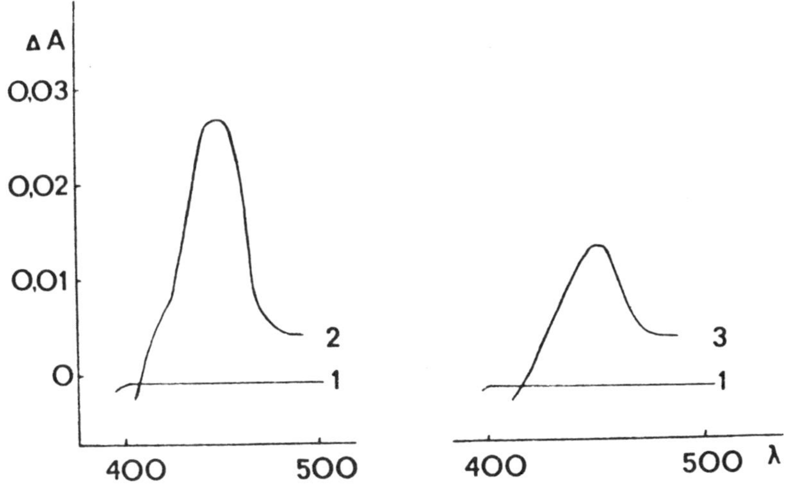

Figure 3   Influence of $(C_2H_5O)_2P(S)SC=C_4H_9$ on the content of cytochrome P 450 in housefly microsomes. Microsomes were incubated for 20 min with 1 mM NADPH at a temperature of 30°C in the absence of (2) or presence (3) of 1 mM of the compound.

The destruction of P 450 was additionally examined in the presence or absence of the acetylenic bond and thiolic sulphur in a few substituted *S*-ethynyl esters of thio- and dithiophosphorous acids differing by the presence or lack of the acetylenic bond and thiolic sulphur in the molecule (Table 3). The destruction of P 450 in warm-blooded animals and arthropods is characteristic only of compounds which have both P–S and acetylenic bond in the chipped-off fragment (compounds **1** and **2** in Table 3). Additionally, thioanalogues without an acetylenic bond cause destruction in arthropods (compound **3**). Thus, compounds **5** and **6** do not destruct P 450. From this arises the question of why destruction occurs.

Table 3 Interaction of compounds $(C_2H_5O)_2$ $P(Y)-R$ with cytochrome P 450

Compound	R	Y	Decrease in the content of cytochrome P 450 in microsomes (%)	
			Rat liver	Housefly abdomen
1	$SC{\equiv}C-C_4H_9$	S	22.2 + 2.5	60.0 + 6.5
2	$SC{\equiv}C-C_4H_9$	O	22.2 + 1.7	27.5 + 1.5
3	$SCH_2CH_2C_4H_9$	S	0	10.5 + 3.7
4	$SCH_2CH_2C_4H_9$	O	0	0
5	$C{\equiv}C-C_4H_9$	O	0	0
6	$CH_2CH_2C_4H_9$	O	0	0

Compounds with a terminal acetylenic bond, which alkylates protoporphyrine rings to haem, are known to possess such an ability (Wilkinson and Murray, 1984). In the compounds investigated the acetylenic bond becomes a terminal bond after $P-S$ cleavage with the formation of alkinyl mercaptan.

Alkinyl mercaptans are highly unstable compounds and even at the moment of their formation they are converted to thioketenes.

$$>P(S)SC{\equiv}C-C_4H_9 \; ----\rightarrow \; >P(O)SC{\equiv}C-C_4H_9$$

$$PSOH + HSC{\equiv}C-C_4H_9 \qquad\qquad HSC{\equiv}C-C_4H_9 + >P(O)OH$$

$$C_4H_9CH{=}C{=}S$$

Thioketenes are highly reactive compounds, which are quickly involved in binding reactions.

We supposed that the 'killer' particles destroying P 450 are thioketenes. A few more compounds of the common formula were studied, in which the triple bond is situated in the alpha or beta position of the phosphoryl mercapto group (Table 4).

Table 4 Interaction of compounds $(C_2H_5O)_2$ $P(O)-S-R$ with cytochrome P 450

| R | K_S | Decrease in the content of cytochrome P 450 in microsomes of rat liver |
		With NADPH
A $C\equiv CC_4H_9$	3.4×10^{-5} M	22.2
B $CH_2C\equiv CH$	4.7×10^{-3} M	11.7
C $CH_2C\equiv C-C_3H_7$ $\quad\mid$ $\quad CH_3$	2.5×10^{-5} M	0
D $-C-C=C-C_3H_7$ $\quad\mid$ $\quad CH_3$	2.5×10^{-4} M	0

In the case of compound **A** in Table 4 cleavage of the $P-S$ bond forms alkinyl mercaptan, which is transformed into thioketene. Compound **B** has the ability of prototropic isomerization, forming ethynyl mercaptan and then thioketene. Compounds **C** and **D** cannot form alkynyl mercaptan and, thus, also no thioketenes.

Destruction, however, is caused only by compounds which form alkynyl mercaptans (compounds **A** and **B**).

Compounds **5** and **6** (Table 3) do not cause the destruction of cytochrome P 450, because their metabolites obviously cannot be converted into thioketenes.

To study the direct interaction of P 450 with 'killer' particles, lithium hexynyl mercaptide, which does not require a metabolic reaction for its conversion to hexynyl mercaptan and which is quickly hydrolysed in an aqueous medium, was used.

$$C_4H_9C\equiv CSLi + HOH \longrightarrow C_4H_9C\equiv CSH \longrightarrow C_4H_9CH=C=S$$

In fact this compound was found to considerably destroy cytochrome P 450 in the absence of NADPH (i.e. there is no need for metabolic conversion to manifest 'killer' abilities; Figure 4). Thus, thioketene finally proved to be the 'killer' particle.

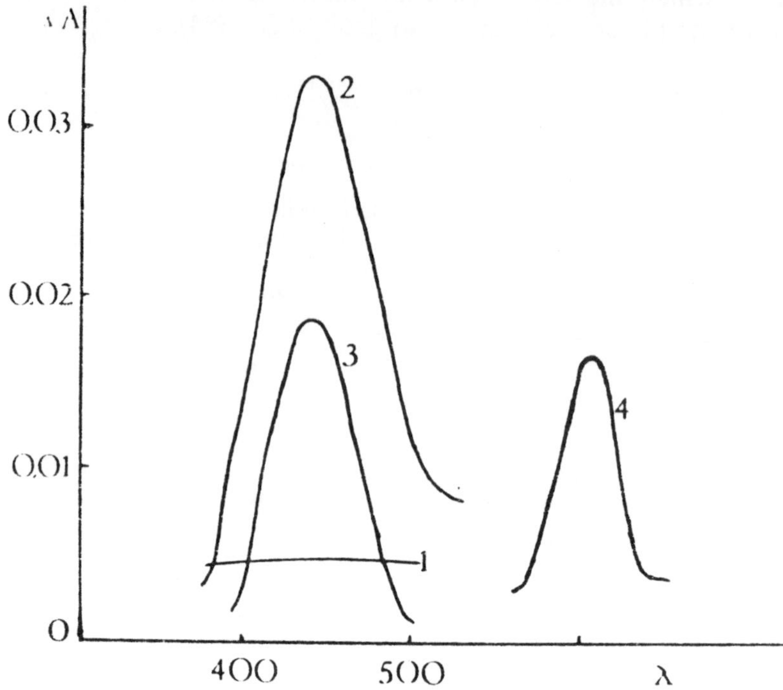

Figure 4 Influence of $C_4H_9CH = C = S$ on the content of cyto-chrome P 450 in rat liver microsomes. Microsomes were incubated for 20 min with 1 mM NADPH in the absence of compound (2), in the presence of 1 mM of compound (3) and without NADPH in the presence of compound (4). (1) is the base line.

The reason for the destruction of cytochrome P 450 by thioanalogues is the release of the sulphur atom during oxidative desulphuration, which can bind with the active site of cytochrome P 450 causing the degradation of the latter (Furakawa *et al.*, 1986). Thus, the destruction of cytochrome P 450 from arthropods by compound **3** may be explained by the covalent binding with the atomic sulphur.

This indicates that oxidative desulphuration is faster in arthropods than in warm-blooded animals, whereas in the latter the content of isoforms causing this reaction is small. This is proved by the similar level of destruction of cytochrome P 450 by compounds **1** and **2** in rat liver (oxo- and thioanalogues), i.e. the sulphur atom seems to have no influence in warm-blooded animals, that means that the velocity of the reaction is low.

Table 3 also shows that the destructive action of these compounds in arthropods differs considerably.

Conclusions

1. The oxidation of substituted *S*-ethynyl thio- and dithiophosphates with cytochrome P 450 leads to the cleavage of the P–S bond and the formation of 'killer' particles – thioketenes – which irreversibly damage isoforms of cytochrome P 450 carrying out this reaction.
2. The destruction of the isoforms of cytochrome P 450, which cleave the P–S bond, appears to decrease the degradation of substituted *S*-ethynyl thio- and dithiophosphates and to increase their toxicity.
3. Data about the destruction of cytochrome P 450 by the sulphur atom released by oxidative desulphuration point out that the reaction in arthropods is significantly faster than in warm-blooded animals.
4. These data suggest that the introduction of the acetylenic bond into a certain position of molecules of known industrial organophosphorus insecticides and their analogues can be used to increase their selectivity and effectiveness.

References

AGOSIN, M. (1985). Role microsomal oxidations in insecticide degradation. In *Comprehensive Insect Physiology, Biochemistry and Pharmacology*, Volume 12 (G.A. Kerkut and L.I. Gilbert, eds), pp. 647–712. Oxford, New York, Toronto.

BALASHOVA, E.K., BRESTKIN, AP. P., ZHUKOVSKIJ, JU.G., ROSENGART, V.I., SHERSTOBITOV, O.E., VICHREVAL, A., GODOVIKOV, N., BABASHEVA, K.K. and KABACHNIK, M.I. (1983). Synthesis and anticholinesterase activity of acetylenic organophosphorus compounds. *Dokl. Akad. Nauk. SSSR*, **272**(2), 503–506.

FURAKAWA, N., SATO, M. and SUZUKI, Y. (1986). Effects of alkyl phosophorothiates on the hepatic microsomal mixed-function oxidase system in rats. *Biochem. Pharm.*, **35**(6), 1019–1026.

LOWRY, O.H., ROSENBROUGH, N.J., FARR, A.L. and RANDALL, R.J. (1951). Protein measurement with the Folin phenol reagent. *Journal of Biological Chemistry*, **193**, 265–275.

OMURA, T. and SATO, R. (1964). The carbon monoxide binding

pigment of liver microsomes. Evidence for its hemoprotein nature. *Journal of Biological Chemistry*, **239**, 2373–2385.

TESTAI, E., CITTI, L., GERVASI, P. and TURCHI, G. (1982). Suicidal inactivation of hepatic cytochrome P 450 *in vitro* by some aliphatic olefins. *BBRC*, **107**(2), 633–641.

VICHREVA, L.A., GODOVIKOV, N.N., ROSENGART, V.I., SHERSTOBITOV, O.E. and KABACHNIK, M.I. (1988). High selectivity in the action of organophosphorus insectoacaricides, containing acetylene bond on arthropods. *Dokl. Akad. Nauk. SSSR*, **302**(3), 731–735.

WILKINSON, C.F. and MURRAY, M. (1984). Considerations of toxicologic interactions in developing new chemicals. *Drug Metabolism Reviews*, **15**, 897–917.

31
Isoenzymes of Glutathione *S*-transferases in Insects

U. SCHOKNECHT

Biological Research Centre Berlin, Stahnsdorfer Damm 81, 0-1532 Kleinmachnow, Germany

Introduction

Glutathione *S*-transferases (GSTs) belong to the enzymes which detoxify xenobiotics and can, therefore, cause insect resistance to insecticides. It is known that some insect species, *Periplaneta americana* (Usui *et al.*, 1977), *Costelytra zealandica* (Clark *et al.*, 1985), *Wiseana cervinata* (Clark and Drake, 1984), *Aedes aegypti* (Grant and Matsumura, 1989), *Musca domestica* (Clark *et al.*, 1984, 1986), possess sets of GST isoenzymes with different substrate specificity. In *M. domestica* isoenzymes with isoelectric points higher than pH 6.5 accept lindane, diazinon and methyl parathion as substrates, whereas in *P. americana* several isoenzymes are able to metabolize diazinon and methyl parathion.

There are, however, insect species which seem to express only a single GST, e.g. *Triatoma infestans* (Wood *et al.*, 1986), *Galleria mellonella* (Chang *et al.*, 1981) and *Ceratitis capitata* (Yawetz and Koren, 1984).

The separation of GST isoenzymes by isoelectrofocusing followed by the visualization of GST activity in a polyacrylamide gel with 1-chloro-2,4-dinitrobenzene (CDNB) is a suitable method for investigating the occurrence of GST isoenzymes. Using this method several insect species from different orders were examined.

Insecticides: Mechanism of Action and Resistance

Material and Methods

Adults of *P. americana, Myzus persicae, Apis mellifera, M. domestica* and *Drosophila melanogaster* and larvae from *Oncopeltus fasciatus, Leptinotarsa decemlineata, Heliothis virescens* and *Agrotis segetum* were used to prepare homogenates in 0.05 M Tris-CHl buffer pH 8.0. The homogenates were centrifuged for ten minutes at $6000 \times g$, fifteen minutes at $10\,000 \times g$ and finally for sixty minutes at $105\,000 \times g$ was used for all experiments. The specific GST activity was determined by spectrophotometric measurements according to Habig *et al.* (1974) under the following conditions: 0.04 mM CDNB, 4 mM GSH, 250 µg protein in 1 ml 0.05M Tris-HCl buffer pH 7.5, 22°C. The enzyme reaction was recorded at 344 nm. The protein content was determined by the method of Lowry.

Ampholine Page Plates pH 3.5–9.5 (Pharmacia) were used for isoelectrofocusing. The activity of GSTs in the polyacrylamide gels was visualized following the method of Kenney and Boyer (1981). The gels were incubated for ten minutes in 0.025 M potassium phosphate buffer pH 6.5 containing 5 mM GSH and 1 mM CDNB. The gels were then scanned at 344 nm with a Shimadzu Dual-Wavelength TLC-Scanner CS-910. This procedure was repeated after five and ten minutes. Positions with developing peaks represented GST activity.

Results

The results are shown in Figure 1. Table 1 summarizes our results and some selected literature data on the activity of GSTs with CDNB as well as the occurrence and isoelectric points of GST isoenzymes in insect species.

P. americana, M. persicae, A. mellifera, H. virescens, A. segetum, M. domestica and *D. melanogaster* belong to species with several isoenzymes. Their isoelectric points range from 3.8 to 8.2. For *O. fasciatus* and *L. decemlineata* only one GST, with isoelectric points of 4.3 and 4.6, respectively, could be detected.

Figure 1(a-i) Demonstration of glutathione *S*-transferase isoenzymes in several insect species according to the method of Kenney and Boyer (1981). (a) *Periplaneta americana*; (b) *Leptinotarsa decemlineata*; (c) *Oncopeltus fasciatus*; (d) *Heliothis virescens*; (e) *Agrotis segetum* (L); (f) *Musca domestica*; (g) *Drosophila melanogaster*; (h) *Apis mellifera*; (i) *Myzus persicae*.

Periplaneta americana (A)

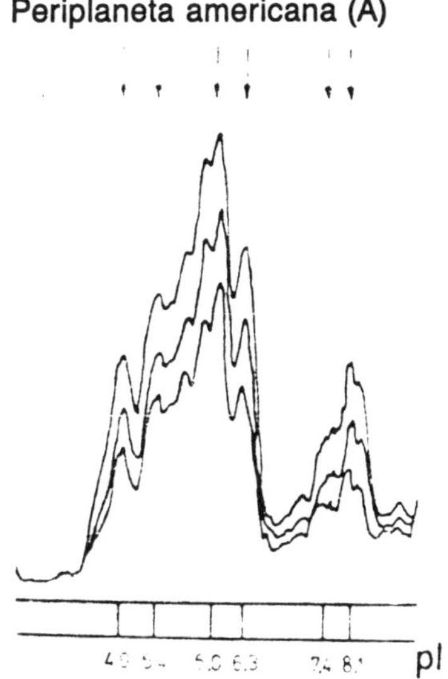

Figure 1a

Leptinotarsa decemlineata

Oncopeltus fasciatus

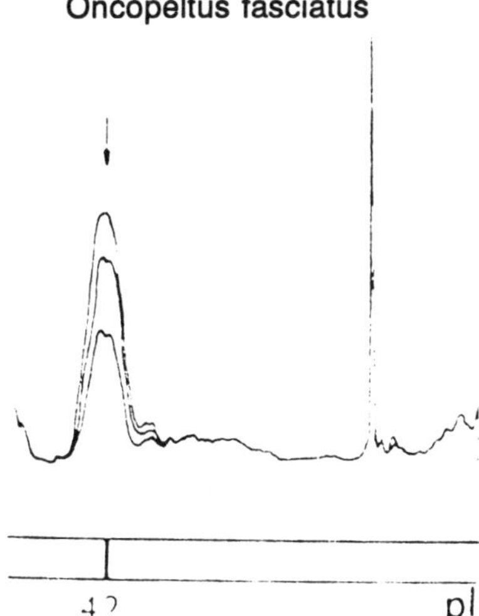

Figure 1c

Heliothis virescens

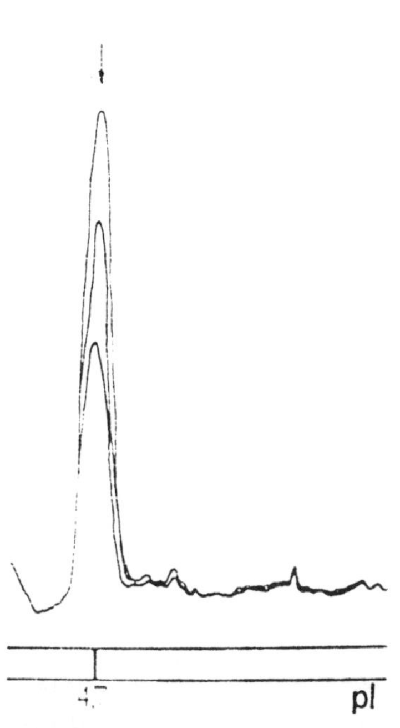

Figure 1b

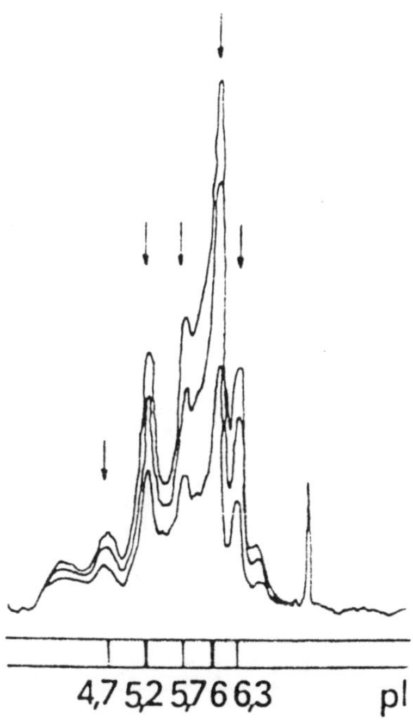

Figure 1d

Agrotis segetum (L)

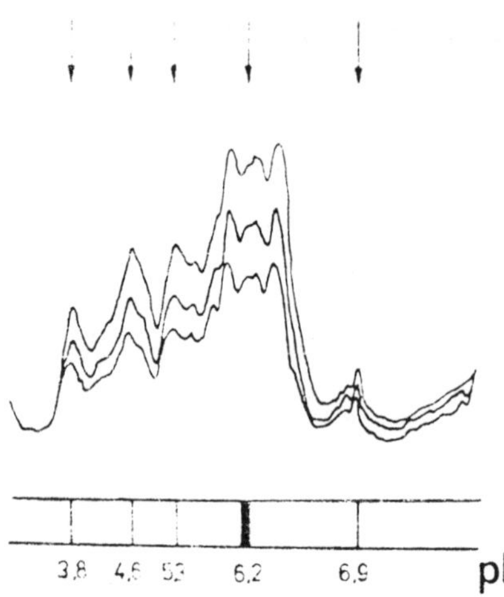

Figure 1e

Drosophila melanogaster

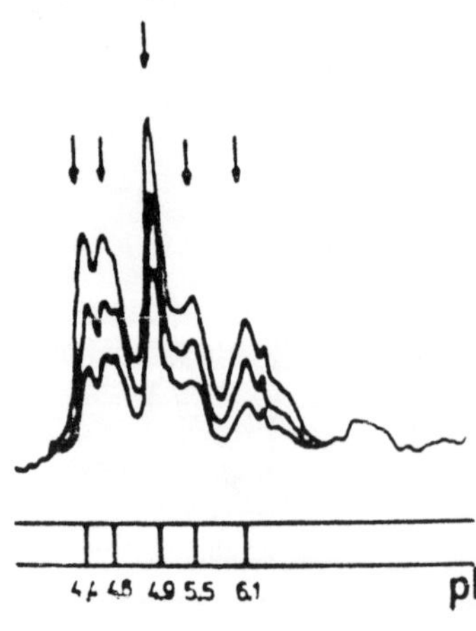

Figure 1g

Musca domestica

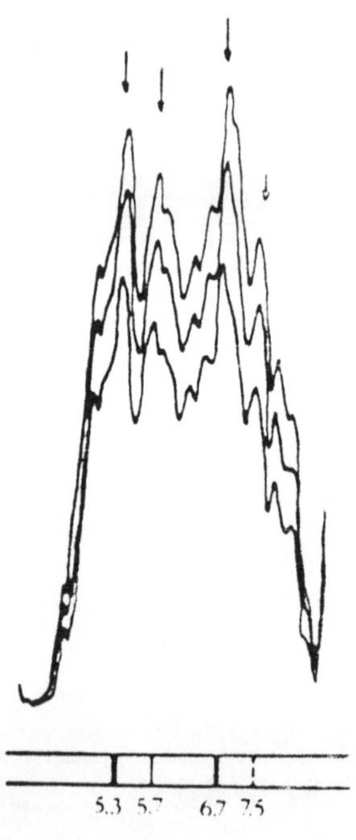

Figure 1f

Apis mellifera

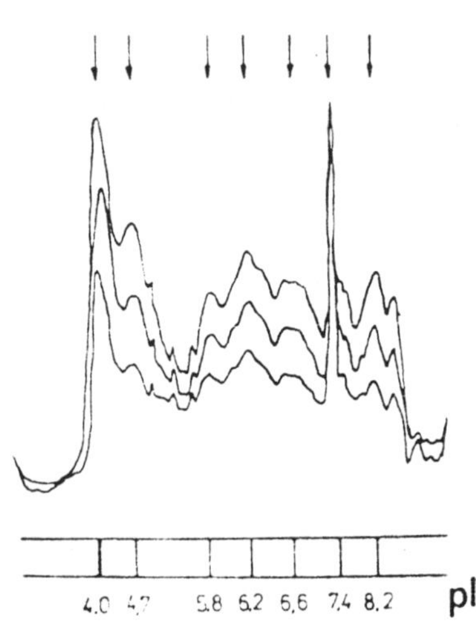

Figure 1h

Myzus persicae
Figure 1i

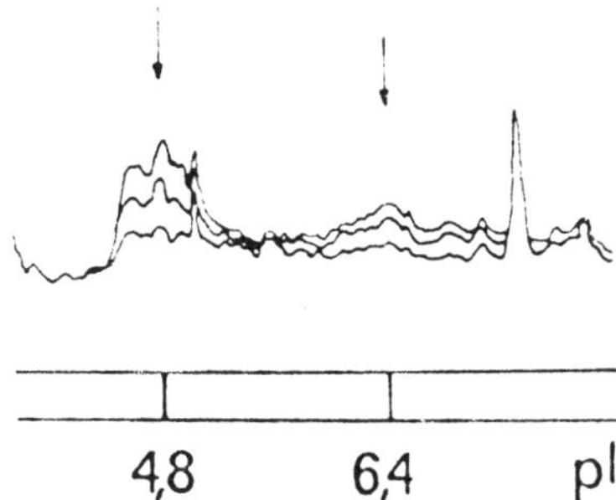

Table 1   Glutathione *S*-transferases in insects.

Order	Species (L = larvae, A = adults)	GST activity nmol CDNB conjugated / min × mg protein	Isoenzymes No.	Isoelectric points	Reference
Blattodea	*P. americana* (A)	154	6	4.9 5.4 6.0 6.3 7.4 8.1	—
	P. americana (A)	—	5		Usui *et al.* (1977)
Heteroptera	*O. fasciatus* (L)	38	1	4.3	
	T. infestans (A)	—	1	—	Wood *et al.* (1986)
Homonoptera	*M. persicae* (A)	297	2	4.7 6.4	—
Coleoptera	*C. zealandica* (L)	—	3	4.3 5.9 8.7	Clark *et al.* (1985)
	L. decemlineata (L)	56	1	4.6	—
Hymenoptera	*A. mellifera* (A)	47	7	4.0 4.7 5.8 6.2 6.6 7.4 8.2	—
Lepidoptera	*A. segetum* (L)	107	5	3.8 4.6 5.3 6.2 6.9	—
	G. mellonella (L)	—	1	8.25	Chang *et al.* (1981)
	H. virescens (L)	57	5	4.7 5.2 5.7 6.0 6.3	—
	W. cervinata (L)	—	4	—	Clark and Drake (1984)
Diptera	*A. aegypti* (L)	—	2	below 5.0 5.0	Grant and Matsumura (1989)
	C. capitata (A)	—	~1	5.7	Yawetz and Koren (1984)
	D. melanogaster (A)	149	5	4.4 4.6 4.9 5.5 6.1	—
	N. domestica (A)	230	4	5.3 5.7 6.7 7.5	—
	N. domestica (A) 3 strains	—	4	3.8–8.2 depending on the stain	Clark *et al.* (1984)

Conclusion

The knowledge that insect species may possess only a single GST or a series of GST isoenzymes conjugating CDNB raises two questions:

1. Is the occurrence of isoenzymes connected with a shift in substrate specificity and which isoenzymes conjugate insecticides of different chemical structures?
2. Is there a relationship between the expression of GST isoenzymes and evolutionary processes?

Detailed investigations of the substrate specificity as well as the molecular genetics of the GSTs in insects are necessary to answer these questions.

References

CHANG, C.K., CLARK, A.G., FIELDS, A. and POUND, S. (1981). Some properties of glutathione S-transferase from larvae of *Galleria mellonella*. *Insect Biochemistry*, **11**(2) 179–186.

CLARK, A.G. and DRAKE, B. (1984). Purification and properties of glutathione S-transferase (EC 2.5.1.18) from larvae of *Wiseana cervinata*. *Biochemical Journal*, **217**(1), 41–50.

CLARK, A.G., SHAMAAN, B.A., DAUTERMAN, W.C. and HAYAOKA, T. (1984). Characterization of multiple glutathione S-transferase from the housefly, *Musca domestica* (L.). *Pesticide Biochemistry and Physiology*, **22**(1), 51–59.

CLARK, A.G., DICK, G.L., MARTINSDALE, S.M. and SMITH, J.N. (1985). Glutathione S-transferase from the New Zealand grass grub, *Costelytra zealandica*: Their isolation and characterization and the effect on their activity of endogenous factors. *Insect Biochemistry*, **15**, 35–44.

CLARK, A.G., SHAMAAN, N.A., SINCLAIR, M.D. and DAUTERMAN, W.C. (1986). Insecticide metabolism by multiple glutathione S-transferase in two strains of the housefly, *Musca domestica* (L.). *Pesticide Biochemistry and Physiology*, **25**, 169–175.

GRANT, D.F. and MATSUMURA, F. (1989). Glutathione S-transferase 1 and 2 in susceptible and insecticide resistant *Aedes aegypti*. *Pesticide Biochemistry and Physiology*, **33**, 132–143.

HABIG, W.H., PABST, M.J. and JAKOBY, W.B. (1974). Glutathione S-transferase: The first enzymatic step in mercapturic acid formation. *Journal of Biological Chemistry*, **249**, 7130–7139.

KENNEY, W.C. and BOYER, T.D. (1981). Detection of glutathione S-transferase isoenzymes after isoelectrofocusing in polyacrylamide gels. *Analytical Biochemistry*, **116** 344–348.

USUI, K., FUKAMI, J.I. and SHISHIDO, T. (1977). Insect glutathione S-transferase: Separation of transferases from fat bodies of American cockroaches active on organophosphorus triesters. *Pesticide Biochemical Physiology*, **7**, 249–250.

WOOD, E., CASABE, N., MELGAR, F. and ZEBRA, E. (1986). Distribution and properties of glutathione S-transferase from *T. infestans*. *ComparativeBiochemistry and Physiology*, **84B**(4), 607–617.

YAWETZ, A. and KOREN, B. (1984). Purification and properties of the Mediterranean fruit fly *Ceratitis capitata* W. glutathione S-transferase. *Insect Biochemistry*, **14**(6), 663–670.

32

Drosophila Acetylcholinesterase and Housefly Glutathione Transferases: Structure and Involvement in the Resistance to Insecticides

D. FOURNIER, A. MUTERO, M. PRALAVORIO and
J.-M BRIDE

*INRA, Laboratoire de biologie des invertébrés, BO 2078, 06606 Antibes,
France*

Proteins known to be involved in the resistance of insects to insecticides are either specific target or metabolizing enzymes. Acetylcholinesterase (AChE, EC 3.1.1.7) is associated with cholinergic synapses where it rapidly terminates neurotransmission. This enzyme is the target site of organophosphorus and carbamate classes of pesticides which react as acetylcholine analogues to complex the enzyme. Following poisoning, the insect AChE is inhibited, the neurotransmitter is no longer metabolized and the neurotransmission is blocked. Glutathione transferases (GSTs, EC 2.5.1.1.8) are enzymes involved in the detoxification mechanisms of many molecules. They catalyse reactions in which the sulphur atom of glutathione provides electrons for a nucleophilic attack on a second electrophilic substrate. Variations in activity or in affinity of these enzymes in correlation to the resistance to insecticides have been reported for many years. We present here some data about the structure of these proteins and some of their variations associated with the resistance.

Drosophila Acetylcholinesterase

Structures of the gene and of the protein

In *Drosophila*, genetic and molecular studies (Hall and Kankel, 1976) have determined that AChE is encoded by a unique locus, *Ace*. It spreads over 34 kb and is composed of ten exons. In some strains, insertions in the introns enlarge the size of the gene above 40 kb (Figure 1). This gene is transcribed into two main mRNAs which originate from differential utilization of two polyadenylation sites (Fournier *et al.*, 1989). These mRNAs have been cloned by Hall and Spierer (1986) and present an unusual long 5′ non-coding sequence (1 kb) with several open reading frames. Hoffmann *et al.* (in preparation) recently succeeded in transforming and rescuing *Ace*⁻flies with a minigene corresponding to the cDNA driven by the *Ace* promoter. Deletion of the 5′ non-coding sequence does not seem to affect the expression of the protein. Thus the function of this 5′ leader and of its short open reading frames remains unknown. The open reading frame encodes a 70 kDa polypeptide (Figure 2). The NH_2-terminal peptide is hydrophobic enough to play the role of signal peptide as found in membrane associated and exported protein precursors. It is removed from the mature protein (Hass *et al.*, 1988). A hydrophobic COOH-terminal peptide is also removed and replaced by a glycolipid anchor. The exact site of the cut is still unknown but by analogy with Torpedo AChE the *C*-terminal amino acid of the mature protein is supposed to be cysteine 615. Several lines of evidence have contributed to prove the presence of this glycolipid anchor. First of all, Arpagaus and Toutant (1985) showed that insect AChE binds to detergent. Then, Gnagey *et al.* (1987) found that purified AChE contains two constitutive elements of glycolipid anchor, ethanolamine and glucosamine. Finally we found that *Drosophila* AChE anchor is sensitive to phosphatidyl-inositol phospholipases C and that these digestions uncover a specific epitope (Fournier *et al.*, 1988a). The role of this anchor to the membrane is still undetermined but by analogy with other proteins linked to membranes via a glycolipid anchor, it is unlikely that this anchor is responsible for the targeting of AChE into synapses. The externalization of the protein is related to the proteolytic cut of the precursor into two polypeptides of 16 and 55 kDa which remain non-covalently linked. The 16 kDa polypeptide corresponds to the *N*-terminus of the precursor while the 55 kDa polypeptide corresponds to the *C*-terminus (Fournier *et al.*, 1988b). This cut occurs in several sites inside an internal hydrophilic region (amino acids 148 to 180; Figure 2) which does not exist in sequences of uncut AChE found in vertebrates. This proteolytic cut does not correspond to an activation of the protein and can be inhibited by deletion of the hydrophilic region. The two polypeptides are *N*-glycosylated, on asparagine 126, 174 and 531 when the two other potential sites of *N*-glycosylation (331 and 569) does not seem to be used. In the mature protein, two 55 kDa polypeptides are covalently linked by a disulphide bond involving cysteine 615 (Mutero and Fournier, in preparation).

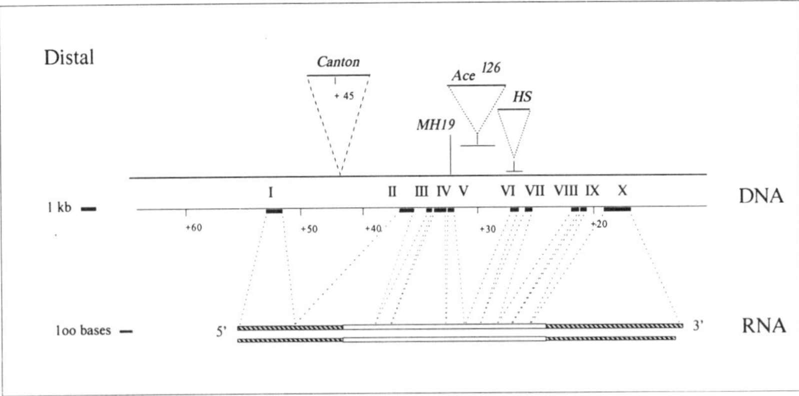

Figure 1   Molecular map of the acetylcholinesterase locus. The long horizontal line represents the DNA walk marked in kilobases. As the 5' end of the gene is on the left, the chromosomal polarity is drawn from distal to proximal. On top of the DNA line, triangles represent insertions, the short lines under the mutations indicate the limits of uncertainty. Vertical line indicates position of the point mutation in the resistant strain MH19. Note that the DNA scale has been established in Canton S. The black blocks on the genomic DNA show the ten exons of the transcription unit. The *Ace* transcript is drawn below the DNA line at a different scale. The hatched boxes represent the 5' and 3' untranslated regions and the open box symbolizes the AChE encoding region.

In conclusion, the mature *Drosophila*AChE has been characterized as an amphiphilic dimer linked to the membrane via a glycolipid anchor. Each of the two active subunits is composed of two polypeptides with 16 and 55 kDa as apparent molecular weight (Figure 3). Action of endogenous phosphatidyl-inositol phospholipases and proteases leads to several other isoforms: an amphiphilic monomer, an amphiphilic dimer with only one glycolipid anchor, a

```
                                           ▼
         maiscrqsrv lpmslplplt iplplvlvls lhlsgvcgVI DRLVVQTSSG    50

         PVRGRSVTVQ GREVHVYTGI PYAKPPVEDL RFRKPVPAEP WHGVLDATGL   100
                                    CHO
         SATCVQERYE YFPGFSGEEI WNPNTNVSED CLYINVWAPA KARLRHGrga   150

         nggehpngkq adtdhlihng npqnttnglp ILIWIYGGGF MTGSATLDIY   200
                                    CHO                *
         NADIMAAVGN VIVASFQYRV GAFGFLHLAP EMPSEFAEEA PGNVGLWDQA   250

         LAIRWLKDNA HAFGGNPEWM TLFGESAGSS SVNAQLMSPV TRGLVKRGMM   300

         QSGTMNAPWS HMTSEKAVEI GKALINDCNC NASMLKTNPA HVMSCMRSVD   350

         AKTISVQQWN SYSGILSFPS APTIDGAFLP ADPMTLMKTA DLKDYDILMG   400

         NVRDEGTYFL LYDLIDYFDK DDATALPRDK YLEIMNNIFG KATQAEREAI   450

         IFQYTSWEGN PGYQNQQQIG RAVGDHFFTC PTNEYAQALA ERGASVHYYY   500
                         *
         FTHRTSTSLW GEWMGVLHGD EIEYFFGQPL NNSLQYRPVE RELGKRMLSA   550
                                             CHO
         VIEFAKTGNP AQDGEEWPNF SKEDPVYYIF STDDKIEKLA RGPLAARCSF   600
                         ▼
         WNDYLPKVRS WAGTCdgdsg sasisprlql lgiaaliyic aalrtkrvf    649
                         |
                        SH
```

Figure 2   Primary structure of *Drosophila* acetylcholinesterase. The arrows indicate the *N*-terminal end and the presumed *C*-terminal amino acid of the mature protein. Cysteine at position 615 is involved in interchain disulphide bond. The 33 amino acids which are absent in vertebrate cholinesterases are in minuscule letters, a proteolytic cut occurs in this region. The stars mark the three amino acids supposedly involved in the catalytic triad. The CHOs symbolize the three Asn glycosylation sites.

hydrophilic monomer and a hydrophilic dimer.

Active sites of cholinesterases are composed of two moieties: the esterasic site and the anionic site. The anionic site contains a free carboxyl involved in the attraction of the substrate through an electrostatic interaction with the positive charge of the quaternary ammonium of acetylcholine. The esterasic site possesses a charge relay system involving the hydroxyl group of a serine, the imidazole group of a histidine and the carboxyl group of an aspartic acid (Rosenberry, 1975). It seems that the anionic site belongs to the 16 kDa polypeptide since mutagenesis of tyrosine 109 changes the affinity of the enzyme towards the substrate and the pH activity dependent pattern. The esterasic site is located on the 55 kDa polypeptide. Serine 276, histidine 518 and probably aspartic acid 248 seem to be the components of the catalytic triad.

Resistance to insecticides

Overproduction of acetylcholinesterase results in a higher resistance

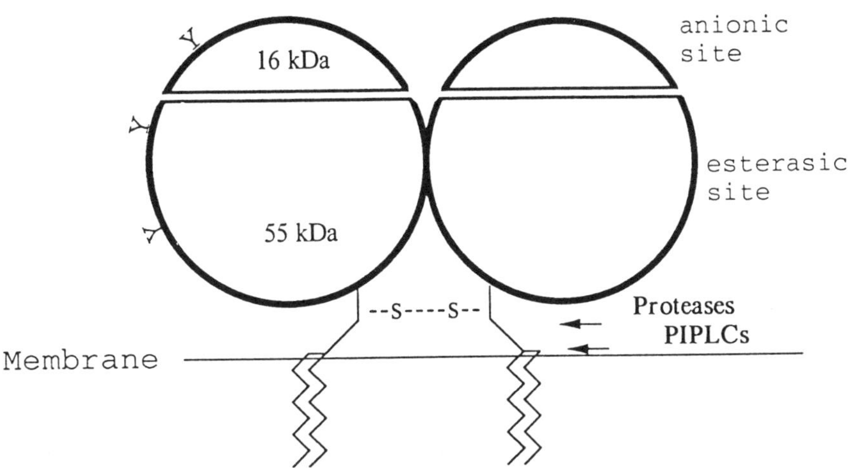

Figure 3   A model for *Drosophila* acetylcholinesterase structure. AChE has been characterized as a dimeric protein composed of two active units covalently associated. Each of them is composed of two polypeptides (55 kDA and 16 kDa) non-covalently associated. AChE is an amphiphilic protein linked to the membrane of the neuronal cholinergic synapses via a glycolipid anchor located at the *C*-terminal end of the 55 kDa polypeptide.

The two 55 kDa polypeptides are linked by disulphide bond while the 16 kDa polypeptides are not covalently linked.

to malathion. We succeeded in transforming wild type flies with a minigene corresponding to the cDNA driven by 1 kb of the promoter sequence. These flies display 120–130% activity of a normal fly due to the expression of the endogenous gene (100%) plus the expression of the minigene (20–30%). They are also more resistant to malathion. Conversely, hemizygous flies bearing one wild type gene and one lethal mutant gene display 50% of normal activity and they are more susceptible to malathion. In rescued flies, the two endogenous genes are lethally mutated. So the activity is only brought by the minigene. These flies display 20–30% of wild type activity and are the most susceptible to malathion. Thus AChE amount in the synapse is correlated to the resistance to insecticide (Figure 4). Nevertheless, we do not know yet if this phenomenon has been selected in natural conditions and what is the maximum expression of AChE in the synapse without affecting the fitness of the fly.

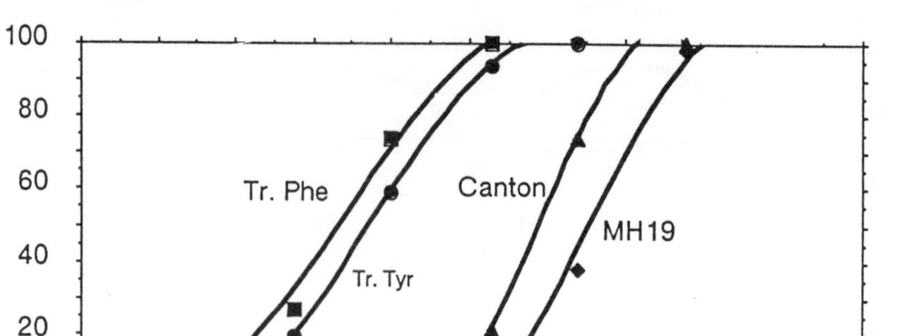

Figure 4   Effect of the quantity of AChE in synapses and effect of the Phe to Tyr mutation on the resistance. Malathion dose/mortality curves for the two wild strains (Canton S and MH19) and for the two rescued strains (Tr. Phe and Tr. Tyr). Flies were treated overnight by tarsal contact on filter paper soaked with malathion. The effect of AChE quantity in the synapse is shown by comparing Canton to Tr. Phe or MH19 to Tr. Tyr. It appears that rescued flies are more susceptible to malathion due to lower amounts of enzyme. Effect of mutation of phenylalanine 368 to tyrosine is shown by comparing Tr. Phe to Tr. Tyr or Canton to MH19.

From the work of Smissaert (1964), several instances of difference in inhibition of AChE by organophosphates have been reported in resistant insects. Among them, a *Drosophila* strain, MH19, was described by Morton and Singh (1982). Gene sequence comparison between this strain and a wild type showed only one mutation, namely the replacement of the phenylalanine 368 by a tyrosine. Mutagenesis of the wild type gene and rescue *Ace⁻*flies by P-element mediated transformation showed that this mutation is indeed responsible for the resistance in the MH19 strain (Figure 4). Furthermore, other *in vitro* mutagenesis at this site gave new patterns of resistance. Thus it appears that site 368 is a resistance site and that the pattern of resistance depends on the amino acid present at this position. Preliminary results of *in vitro* mutagenesis and expression in *Xenopus* oocyte indicate that other sites are potentially important for the resistance to insecticides. For example, mutations on tyrosine 109 produce slight differences in the pattern of resistance to insecticides.

In order to test if recombination can be a mechanism of resistance to insecticides, we mutated the same gene in two places, on tyrosine 109 and on phenylalanine 368. Preliminary results show that resistance results from the multiplication of the two patterns. Consequently, besides the occurrence of a new mutation, which produces a strong resistance to insecticides, we can hypothesize that strong resistance may originate from the combination of several weak mutations in the same gene and in the same protein.

Housefly Glutathione Transferases

Structure of the proteins

Two classes of glutathione transferases (GSTs) have been identified, GST1 and GST2. The first one migrates as a single band of apparent molecular weight 28 kDa in denaturing gel. In two-dimensional gel, this band is split in several spots with isoelectric points from 4 to 9. These charge isomers are serologically related. We prepared polyclonal antisera directed against the six major spots and each antiserum cross-related with all the other spots. We cloned the gene coding for GST1. The cDNA is 0.8 kb long and encodes a 208 amino acid protein (Figure 5). Our first hypothesis was that charge isomers originated from a multigene family but all the cDNAs we sequenced were identical within a strain. Furthermore, Southern experiments with DNA purified from a single fly show one or two bands by hybridization with the cDNA. These two results indicate that GST1 is encoded by only one gene. Thus charge isomers would originate from a post-translational modification. To test this hypothesis, we expressed the cDNA in *Escherichia coli*. In bacteria, GST1 is active and preliminary results indicate that catalytic properties of this protein are identical to what is found in the housefly. GST1 expressed in *E. coli* is also resolved in several charge isomers. In conclusion, it seems that GST1 is encoded by only one gene, and that variability results from post-translational modification that *E. coli* can process.

The second class of glutathione transferase (GST2) migrates as a single band of molecular weight 32 kDa in denaturing gel. In two-dimensional gels, this band is composed in several spots to each other at acidic pL. In gradient gel electrophoresis, the native protein migrates at 70 kDa, so GST2 is a dimer.

GST1 and GST2 belong to two different classes since they are serologically distinct and do not form heterodimeric association. An antiserum directed against one class of protein does not cross-react

```
MDFYYLPGSA  PCRSVLMTAK  ALGIELNKKL        30
   *          *

LNLQAGEHLK  PEFLKINPQH  TIPTLVDGDF        60
                         *     *

ALWESRAIMV  YLVEKYGKTD  SLFPKCPKKR        90
      *

AVINQRLYFD  MGTLYKSFAD  YYYPQIFAKA       120

PADPELFKKI  ETAFDFLNTF  LKGHEYAAGD       150
                                  *

SLTVADLALL  ASVSTFEVAS  FDFSKYPNVA       180
     *                        *

KWYANLKTVA  PGWEENWAGC  LEFKKYFG        208
```

Figure 5 Primary structure of glutathione transferase 1. This sequence is deduced from the cDNA isolated from the strain Cooper, a strain susceptible to insecticides. Stars indicate amino acids conserved in all GSTs so far sequenced. Variations found in the resistant strains Rutger and Kirokawa: IIe 116 to Val, Asp 135 to Glu and Lys 142 to Glu.

with the other one. Isolation of native enzymes by chromatofocusing results in the separation of the two polypeptides (Figure 6).

Resistance to insecticides

Insecticide resistance in insect pests of agricultural and medical importance has been attributed in many cases to GST because high levels of GST activity have been recorded in some resistant strains. One strain of *M. domestica* (Cornell R) is a good example of this augmentation of GST activity (Clark and Dauterman, 1982). Ottea and Plapp (1984) found that the difference was on the Vmax without effect on the Km. We showed that the increased activity found in Cornell R is correlated with an increased level of GST1 mRNA and protein. In contrast, we never saw any difference between the strains in Southern blot. Thus, it seems that resistance to insecticides in this strain is at least in part due to an overtranscription of the GST1 gene. On the other hand, when we compared sequences from two resistant strains, Rutger and Hirokawa, we found variations in the primary sequence with the susceptible strain Cooper (Figure 5). Both resistant strains harvested in the USA and in Japan show the same mutations, but the possibility that these mutations are indeed involved in increased GST activity and in insecticide metabolism remains to be tested.

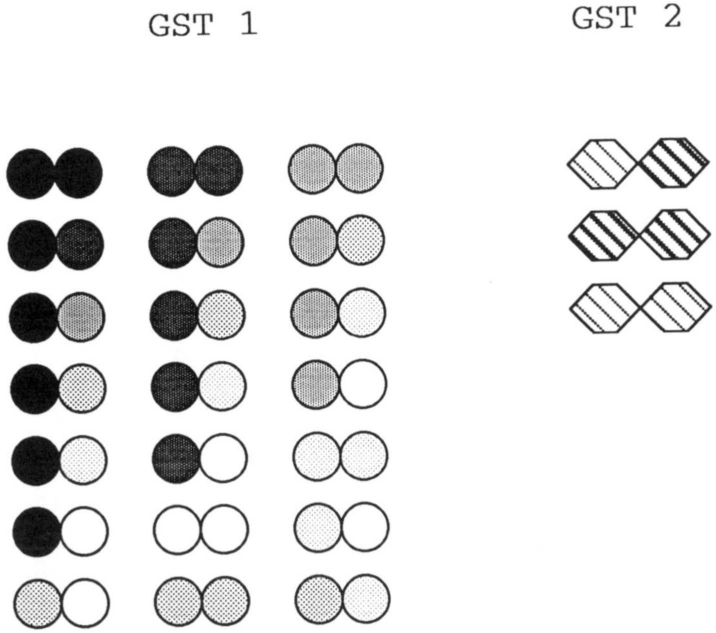

GST 1   GST 2

Figure 6   A model for housefly glutathione transferases. Two classes of glutathione transferase have been identified. GST1 are dimeric proteins composed of 28 kDa polypeptides with isoelectric points from 4 to 9. GST2 are dimeric proteins composed of 32 kDa acidic polypeptides. Antisera prepared against each class have no immunological cross-reactivity. Heterodimeric associations between the two classes have not been detected.

References

ARPAGAUS, M. and TOUTANT, J.P. (1985). Polymorphism of acetylcholinesterase in adult *Pieris brassicae* heads. Evidence for detergent-insensitive and Triton X-100-interacting forms. *Neurochem. Int.*, **7**, 793–804.

CLARK, A.G. and DAUTERMAN, W.C. (1982). The characterization by affinity chromatography of glutathione S-transferase from different strains of housefly. *Pesticide Biochemical Physiology*, **17**, 307–314.

FOURNIER, D., BERGE, J.B., CARDOSO DE ALMEIDA, M.L. and BORDIER, C. (1988a). Acetylcholinesterase from *Musca domestica* and *Drosophila melanogaster* brain are linked to membranes by a glycolipid anchor sensitive to an endogenous phospholipase. *J. Neurochem.*, **50**, 1158–1163.

FOURNIER, D., BRIDE, J.M., KARCH, F. and BERGE, J.B. (1988b). Acetylcholinesterase from *Drosophila melanogaster*, identification of two subunits encoded by the same gene. *FEBS Lett.*, **238**, 333–337.

FOURNIER, D., KARCH, F., BRIDE, J.M., HALL, J.M.C., BERGE, J.B. and SPIERER, P. (1989). *Drosophila melanogaster* acetylcholinesterase gene: structure, evolution and mutations. *Journal of Molecular Biology*, **210**, 15–22.

GNAGEY, A., FORTE, M. and ROSENBERRY, T. (1987). Isolation and characterization of acetylcholinesterase from *Drosophila*. *Journal of Biological Chemistry*, **262**, 13290–13298.

HALL, J.C. and KANKEL, D. (1976). Genetics of acetylcholinesterase in *Drosophila melanogaster*. *Genetics*, **83**, 517–535.

HALL, L.M.C. and SPIERER, P. (1986). The *Ace* locus of *Drosophila melanogaster*: structural gene for acetylcholinesterase with an unusual 5′ leader. *EMBO Journal*, **5**, 2949–2954.

HASS, R., MARSHALL, T.L. and ROSENBERRY, T. (1988). *Drosophila* acetylcholinesterase: demonstration of a glycoinositol phospholipid anchor and an endogenous proteolitic cleavage. *Biochemistry*, **27** 6453–6457.

MORTON, R.A. and SINGH, R.S. (1982). The association between malathion resistance and acetylcholinesterase in *Drosophila melanogaster*. *Biochemical Genetics*, **20**, 179–198.

OTTEA, J.A. and PLAPP, F.W.Jr (1984). Glutathione transferase in housefly: biochemical and genetic changes associated with induction and insecticide resistance. *Pesticide Biochemistry and Physiology*, **22**, 203–208.

ROSENBERRY, T. (1975). Acetylcholinesterase. *Advances in Enzymology*, **43**, 103–208.

SMISSAERT, H.R. (1964). Cholinesterase inhibition in spider mites susceptible and resistant to organophosphate. *Science*, **143**, 129–131.

33
Molecular Biology of Insecticide Resistance: Coamplification of Transposon-like Elements with an Esterase Gene Responsible for Organophosphate Resistance in *Culex* Mosquitoes

G. MOUCHÈS, M. AGARWAL and K. CAMPBELL

Laboratoire d'Ecologie Moléculaire, Université de Pau et des Pays de l'Adour, 6400 Pau, France

Introduction

Gene amplification seems to be a fundamental and widely occurring mechanism of overproduction of proteins for counteracting environmental stress. Insecticide resistance in several insect species is associated with increased esterase activity resulting from amplification of the corresponding structural gene (Devonshire and Field, 1991). In *Culex pipiens quinquefasciatus* from California, high levels of organophosphate (OP) resistance (800 times) are due to the esterase B1 gene, which is amplified at least 250-fold (Mouchès *et al.*, 1986, 1987). Many populations of several species of mosquitoes of the genus *Culex* have become resistant to a variety of OP due to such an increased production of detoxifying esterases. Other genes very similar to esterase B1 gene such as B2 and B3 were shown to be amplified respectively in *C. p. quinquefasciatus* and *C. tarsalis* strains resistant to OP (Beyssat-Arnaouty *et al.*, 1989; Raymond *et al.*,

1989). Evidence indicates that amplification of each esterase B gene occurs as a single and independent event (Raymond *et al.*, 1990), suggesting that it may be induced by a specific mechanism.

Preliminary characterization of the structure of the amplification unit or amplicon encompassing the structural esterase B1 gene in the OP-resistant *C. p. quinquefasciatus* TEM-R strain demonstrates that two repetitive DNA sequences dispersed in the mosquito genome are coamplified with the esterase gene in insecticide-resistant insects (Mouchès *et al.*, 1990). This suggests the OP-insecticides have selected amplification not only of the esterase B1 gene in *Culex* field populations, but also of DNA sequences structurally related to transposable elements which may play a role in the amplification process. Indeed, sequencing of these repetitive elements demonstrates that they are functionally related to transposons.

Structure of the Esterase B1 Amplicon

In the genome of the *C. p. quinquefasciatus* TEM-R strain, the amplicon responsible for the insecticide resistance covers at least 30 kilobases (kb) and contains a constant and highly conserved 'core' of 25 kb (Mouchès *et al.*, 1990) (Figure 1). This core is the common part of the various amplification units that may have a much longer sequence. Each core carries a singly copy of the esterase B1 gene which represents only 2.8 kb of the 25 kb of the core. Flanking sequences extending on either side beyond the core are heterogeneous in their structure, probably as a result of rearrangements that have occurred during the amplification process.

As a consequence of their coamplification with the esterase B1 gene, all DNA sequences within the amplicon core are present in a large number of copies of TEM-R genomic DNA. In contrast, the number of copies of most of these sequences in the genome of OP-susceptible mosquitoes is considerably lower: they are present as single or low number copies in the genomes of mosquitoes lacking overproduction of the esterase B1 protein. Exceptions to this were found in CE1 and CE2 (Figure 1), which contain sequences belonging to families of repetitive elements scattered through the genome of both susceptible and OP-resistant mosquitoes and furthermore coamplified with esterase B1.

The Esterase B1 Gene

The esterase B1 gene is 2773 bp long and from the 5′ to 3′-end it contains four exons of 279, 138, 882 and 564 bp and three introns of

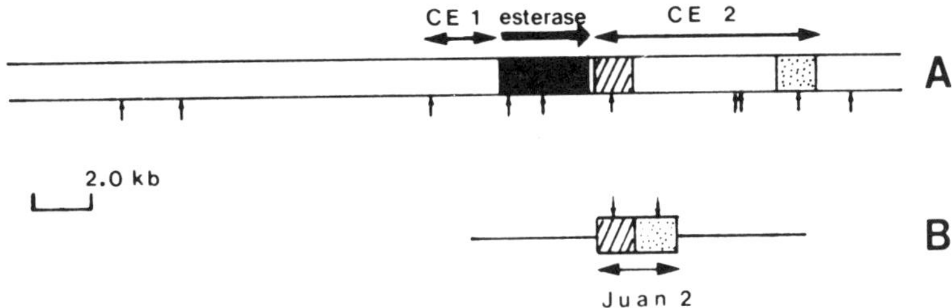

Figure 1 Organization of the core of the amplicon-containing esterase B1 gene and of one truncated copy of the LINE-retroposon Juan element. The long double line represents the genomic DNA core. The small vertical arrows indicate Eco RI restriction sites. Each core carries a single copy of the esterase B1 gene (black box); the gene is framed by two sequences, CE1 and CE2, which contain transposon-like elements. The two ends of CE2 are parts of the highly repetitive LINE-retroposon element Juan represented below the amplicon. Flanking sequences on either side of the core are heterogeneous. These maps are improved versions of the one published previously (Mouchès *et al.*, 1990).

775, 52 and 61 bp (Mouchès *et al.*, 1990) (Figure 2). The inferred polypeptide encoded by the exons' open reading frames contain 540 residues and has a calculated M_r of 59 000. The deduced amino acid sequence of the enzyme presents, from residues 189 to 196, the characteristics of the consensus 8-residue polypeptide shared at the active site of eukaryotic serine esterases. In addition, esterase B1 contains regions strongly similar to human butyrylcholinesterase, acetylcholinesterases from *Torpedo* and *Drosophila melangaster*, esterase 6 from *D. melanogaster*, *Heliothis* juvenile hormone esterase, rabbit liver esterase and rat thyroglobulin. Regions of homologies with these proteins are found mainly between the NH_2-terminal amino acid and the esterasic site (Mouchès *et al.*, 1990, Pasteur *et al.*, 1990). These relationships are also strengthened by alignment of the esterase B1 nucleotide sequence with those of acetylcholinesterase and esterase 6 from *D. melanogaster*. Areas of similarities among these sequences were found upon dot matrix analysis when the coincidence of 30 out of 50 bp was taken as a criterion of homology.

The complete cDNA specific of esterase B1 was inserted, under the control of the strong heat-shock inducible promoter hsp, into the *D. melangaster* P-transposon. Transgenic insects transformed by this chimaeric DNA expressed the esterase B1 gene and were found to be resistant to the OP insecticide Temephos. This demonstrates without ambiguity that the esterase B1 gene is indeed responsible for insecticide resistance.

The CE1 Elements Family

One *Culex* repetitive sequence, CE1, lies upstream of the esterase B1 gene 5'-end (Figure 1) (Mouchès *et al.*, 1990). Other copies of this sequence are present in the heterogeneous 5' flanking edge of the amplicon. The coamplification of CE1 along with the esterase B1 gene drastically increases the abundance of this element in the genome of TEM-R insects. The nucleotide sequence of CE1 shows that it indeed belongs to a family of transposable-like elements which are scattered through the genome of *Culex pipiens* mosquitoes, whatever their insecticide resistance status is. However, the fact that

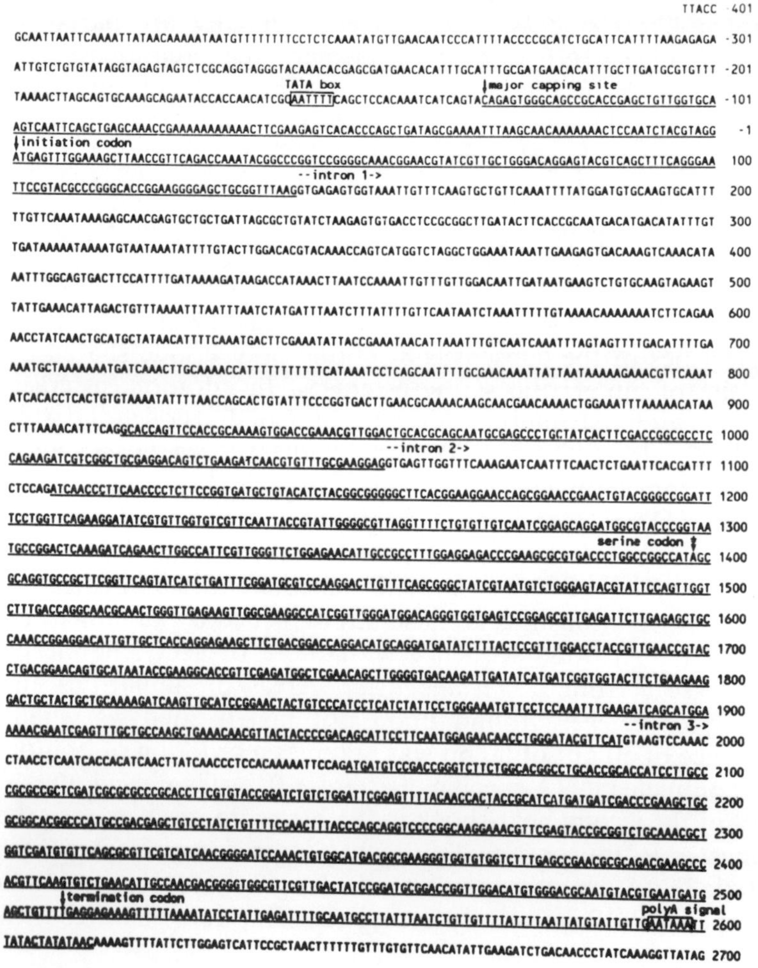

```
                                                                            TTACC  401
GCAATTAATTCAAAATTATAACAAAAATAATGTTTTTTTTCCTCTCAAATATGTTGAACAATCCCATTTTACCCCGCATCTGCATTCATTTTAAGAGAGA  -301
ATTGTCTGTGTATAGGTAGAGTAGTCTCGCAGGTAGGGTACAAACACGAGCGATGAACACATTTGCATTTGCGATGAACACATTTGCTTGATGCGTGTTT  -201
                                      TATA box                      |major capping site
TAAAACTTAGCAGTGCAAAGCAGAATACCACCAACATCGGAATTTTCAGCTCCACAAATCATCAGTACAGAGTGGGCAGCCGCACCGAGCTGTTGGTGCA  -101
AGTCAATTCAGCTGAGCAAACCGAAAAAAAAAAAACTTCGAAGAGTCACACCCAGCTGATAGCGAAAATTTAAGCAACAAAAAAACTCCAATCTACGTAGG    -1
|initiation codon
ATGAGTTTGGAAAGCTTAACCGTTCAGACCAAATACGGCCCGGTCCGGGGCAAACGGAACGTATCGTTGCTGGGACAGGAGTACGTCAGCTTTCAGGGAA   100
                                               --intron 1->
TTCCGTACGCCCGGGCACCGGAAGGGGAGCTGCGGTTTAAGGTGAGAGTGGTAAATTGTTTCAAGTGCTGTTCAAATTTTATGGATGTGCAAGTGCATTT   200
TTGTTCAAATAAAGAGCAACGAGTGCTGCTGATTAGCGCTGTATCTAAGAGTGTGACCTCCGCGGCTTGATACTTCACCGCAATGACATGACATATTTGT   300
TGATAAAAATAAAATGTAATAAATATTTTGTACTTGGACACGTACAAACCAGTCATGGTCTAGGCTGGAAATAAATTGAAGAGTGACAAAGTCAAACATA   400
AATTTGGCAGTGACTTCCATTTTGATAAAAGATAAGACCATAAACTTAATCCAAAATTGTTTGTTGGACAATTGATAATGAAGTCTGTGCAAGTAGAAGT   500
TATTGAAACATTAGACTGTTTAAAATTTAATTTAATCTATGATTTAATCTTTATTTTGTTCAATAATCTAAATTTTTGTAAAACAAAAAAATCTTCAGAA   600
AACCTATCAACTGCATGCTATAACATTTTCAAATGACTTCGAAATATTACCGAAATAACATTAAATTTGTCAATCAAATTTAGTAGTTTTGACATTTTGA   700
AAATGCTAAAAAAATGATCAAACTTGCAAAACCATTTTTTTTTTTCATAAATCCTCAGCAATTTTGCGAACAAATTATTAATAAAAGAAACGTTCAAAT   800
ATCACACCTCACTGTGTAAAATATTTTAACCAGCACTGTATTTCCCGGTGACTTGAACGCAAAACAAGCAACGAACAAAACTGGAAATTTAAAAACATAA   900
CTTTAAAACATTTCAGGCACCAGTTCCACCGCAAAAGTGGACCGAAACGTTGGACTGCACGCAGCAATGCGAGCCCTGCTATCACTTCGACCGGCGCCTC  1000
                                               --intron 2->
CAGAAGATCGTCGGCTGCGAGGACAGTCTGAAGATCAACGTGTTTGCGAAGGAGGTGAGTTGGTTTCAAAGAATCAATTTCAACTCTGAATTCACGATTT  1100
CTCCAGATCAACCCTTCAACCCCTCTTCCGGTGATGCTGTACATCTACGGCGGGGGCTTCACGGAAGGAACCAGCGGAACCGAACTGTACGGGCCGGATT  1200
TCCTGGTTCAGAAGGATATCGTGTTGGTGTCGTTCAATTACCGTATTGGGGCGTTAGGTTTTCTGTGTTGTCAATCGGAGCAGGATGGCGTACCCGGTAA  1300
                                                                          serine codon |
TGCCGGACTCAAAGATCAGAACTTGGCCATTCGTTGGGTTCTGGAGAACATTGCCGCCTTTGGAGGAGACCCGAAGCGCGTGACCCTGGCCGGCCATAGC  1400
GCAGGTGCCGCTTCGGTTCAGTATCATCTGATTTCGGATGCGTCCAAGGACTTGTTTCAGCGGGCTATCGTAATGTCTGGGAGTACGTATTCCAGTTGGT  1500
CTTTGACCAGGCAACGCAACTGGGTTGAGAAGTTGGCGAAGGCCATCGGTTGGGATGGACAGGGTGGTGAGTCCGGAGCGTTGAGATTCTTGAGAGCTGC  1600
CAAACCGGAGGACATTGTTGCTCACCAGGAGAAGCTTCTGACGGACCAGGACATGCAGGATGATATCTTTACTCCGTTTGGACCTACCGTTGAACCGTAC  1700
CTGACGGAACAGTGCATAATACCGAAGGCACCGTTCGAGATGGCTCGAACAGCTTGGGGTGACAAGATTGATATCATGATCGGTGGTACTTCTGAAGAAG  1800
GACTGCTACTGCTGCAAAAGATCAAGTTGCATCCGGAACTACTGTCCCATCCTCATCTATTCCTGGGAAATGTTCCTCCAAATTTGAAGATCAGCATGGA  1900
                                                                          --intron 3->
AAAACGAATCGAGTTTGCTGCCAAGCTGAAACAACGTTACTACCCCGACAGCATTCCTTCAATGGAGAACAACCTGGGATACGTTCATGTAAGTCCAAAC  2000
CTAACCTCAATCACCACATCAACTTATCAACCCTCCACAAAAATTCCAGATGATGTCCGACCGGGTCTTCTGGCACGGCCTGCACCGCACCATCCTTGCC  2100
CGCGCCGCTCGATCGCGCGCCCGCACCTTCGTGTACCGGATCGTGTCTGGATTCGGAGTTTTACAACCACTACCGCATCATGATGATCGACCCGAAGCTGC  2200
GCGGCACGGCCCATGCCGACGAGCTGTCCTATCTGTTTTTCCAACTTTACCCAGCAGGTCCCCGGCAAGGAAACGTTCGAGTACCGCGGTCTGCAAACGCT  2300
GGTCGATGTGTTCAGCGCGTTCGTCATCAACGGGGATCCAAACTGTGGCATGACGGCGAAGGGTGGTGTGGTCTTTGAGCCGAACGCGCAGACGAAGCCC  2400
ACGTTCAAGTGTCTGAACATTGCCAACGACGGGGTGGCGTTCGTTGACTATCCGGATGCGGACCGGTTGGACATGTGGGACGCAATGTACGTGAATGATG  2500
                 |termination codon                                              polyA signal
AGCTGTTTTGAGGAGAAAGTTTTTAAAATATCCTATTGAGATTTTGCAATGCCTTATTTAATCTGTTGTTTTTATTTTAATTATGTATTGTTGAATAAAT  2600
TATACTATATAACAAAAGTTTTATTCTTGGAGTCATTCCGCTAACTTTTTTGTTTGTGTTCAACATATTGAAGATCTGACAACCCTATCAAAGGTTATAG  2700
```

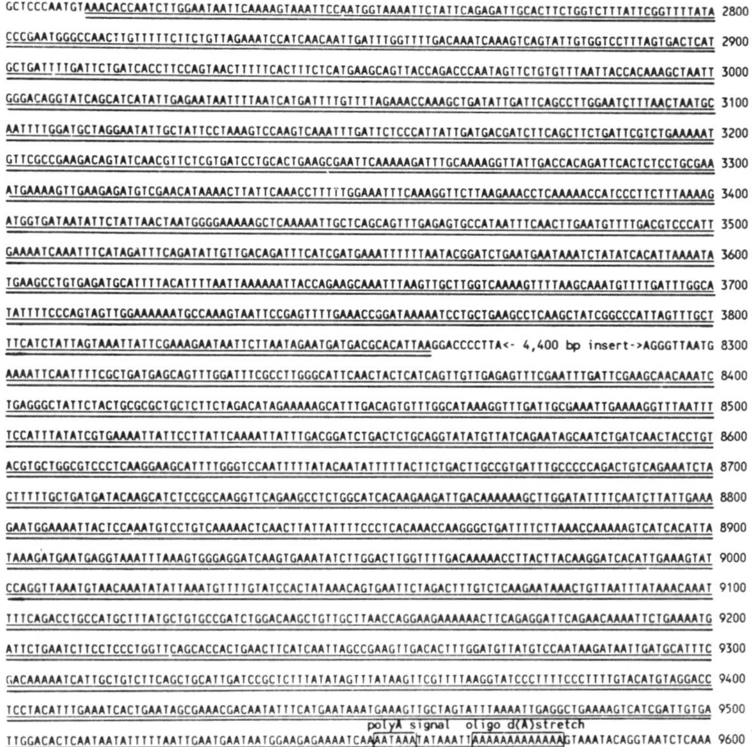

Figure 2 Nucleotide sequence of the esterase B1 amplicon core from the beginning of the esterase B1 gene to the end of the truncated LINE-retroposon Juan element coamplified with the structural gene. The esterase B1 exons are underlined by a single line whereas the Juan specific sequence is doubly underlined. The esterase B1 sequence previously reported (Mouchès *et al.*, 1990) is reproduced here (nucleotides −405 to 2700) with some minor corrections. The putative TATA box (AATTTT) and the poly-adenylation signal are boxed; the major capping site, the initiation and termination codons and the codon encoding the serine typical of the esterasic site are indicated by arrows. The Juan copy coamplified with esterase B1 into the amplicon is split by the intercalation of a 4.4 kb sequence and indirect evidence suggests that this composite structure, called CE2, may play a role in the amplification process. The polyadenylation signal and oligo dA stretch at the 3′-end of the Juan element are boxed.

CE1-like sequences are not present in the amplicon core of the overproduced esterases B2 and B3 makes it unlikely that this element family plays any active role in the esterase B1 amplification process.

Furthermore, as hybridization experiments revealed the presence of reiterated sequences homologous to CE1 in the genome of *C. p. quinquefasciatus* and *C. pipiens*, but not of *C. tarsalis*, CE1 elements have probably invaded the genome of *C. pipiens* mosquitoes only recently.

The CE2 Sequence

The CE2 sequence contains a truncated copy of the Juan LINE-retroposon element. The CE2 element coamplified into the amplicon along with esterase B1 contains sequences which belong to a family of highly conserved repeats, Juan elements (Figure 1). Truncated Juan elements represent about 3% of the total *Culex* DNA and seem to be scattered throughout the genome of every *C. pipiens* and *C. tarsalis* strain analysed so far, regardless of insecticide resistance status (Mouchès *et al.*, 1990). Furthermore, Juan-like elements have been found to be conserved and reiterated in other mosquito species such as the yellow fever vector *Aedes aegyptii*, the malaria vector *Anopheles stephensi* and the plant pest *Ceratitis capitata* (Mouchès *et al.*, unpublished results).

We have sequenced three representative copies of the Juan family from *Culex* mosquitoes, one coamplified with the esterase B1 gene (Juan 1) and two others reiterated in the mosquito genome, but outside the amplicon (Juan 2 and Juan 3) (Mouchès *et al.*, 1991). The three Juan copies have different flanking edges, but were found to be nearly the same with the exception of some minor base substitutions near their 3′-end. Juan elements have an oligo (A) stretch at their 3′-end, with a preceding polyadenylation signal (Figure 2). Furthermore, comparison of both their nucleotide and deduced amino acid sequences with those present in the GenBank and NRF databases revealed significant homologies of Juan with some LINEs, a superfamily of retrotransposable ubiquitous DNA elements (Di Nocera and Sasaki, 1990). Areas of similarities were found upon dot matrix analysis between Juan and the LINEs *D. melangaster* sequences Jockey (Priimägi *et al.*, 1988), F and G (but not *D. teisseiri* I factor) when the coincidence of 30 out of 50 bp was taken as a criterion of homology. The alignments of the Juan specific sequences and the short ORF they encode, with those of Jockey, F and G show that the areas of similarities are located in the ORF 2 of these elements, which codes protein sequences homologous to the characteristic domains for the reverse transcriptase. All these features demonstrate that Juan elements are structurally related to

LINEs retroposons, even though they appear to be truncated at their 5′-end and carry only degenerated ORFs. In the esterase B1 amplicon core, the Juan copy lies downstream of the 3′-end of the structural gene and is split by the intercalation of a 4.4 kb sequence which is not often repeated in the genome of OP-susceptible strains (Figures 1 and 2). Indirect evidence suggests that this composite CE2 structure may play a role in the amplification process of esterase B1 gene (Mouchès *et al.*, unpublished results).

Concluding Remarks

Our results show that selection of *Culex* field populations by OP insecticides has resulted in amplification not only of the detoxifying esterase B1 gene but also of a large amount of DNA surrounding this gene, part of which are DNA sequences structurally related to transposable elements. The finding of such elements functionally related to transposons in an amplicon responsible for insecticide resistance may be of great importance in understanding the biology of insecticide-resistant strains.

As pointed out by Devonshire and Field (1991), the generation of many copies of esterase genes in resistant mosquitoes and aphids could be mediated by mobile genetic elements, and insecticide selection of insects with amplified esterase genes might then play a role in the dispersion and spreading of transposable elements in natural populations. The Juan retroposon-like element coamplified with esterase B1 is truncated, contains degenerated ORFs and an insertion. Thus, its role in the amplification process cannot be as an active retroposon. However, the existence in the mosquito genome of a large number of very similar Juan copies indicates that the retroposition of these defective LINEs occurs via a passive mechanism and suggests that they can use reverse transcriptase in trans. Thus, Juan elements may be involved in the passive retroposition of the esterase B1 gene and sequences coamplified with this structural gene.

Moreover, as some transposons are associated in insects with hybrid dysgenesis, the identification of such an element in the mosquito genomes may be useful in the future for genetic control of these human-disease vectors. Active units of the Juan family are now being researched in mosquitoes, using strategies allowing the selective isolation of recently transposed LINE elements (Sakaki *et al.*, cited in Di Nocera and Sasaki, 1990). Also, genetic engineering and transformation offer the possibility of modifying the genetic material of insects. These techniques will make it possible, for example, to develop genetic methods such as autocidal control for combating insect pests and vectors. We have shown that two cloned genes encoding proteins that detoxify insecticides are good selection markers for the purpose of developing transformation techniques. The first one is esterase B1 from *Culex* mosquitoes; the second has been cloned from bacteria which break down various organic phosphates (Mouchès, 1989). Furthermore, the reiterated DNA elements associated with gene amplification in insecticide-resistant *Culex* may be good candidates for the construction of a gene vector in dipterous insects and we are looking into the use of the CE1, CE2 and Juan elements as vectors for developing gene transformation systems in dipterous insects as well as other species.

Acknowledgements

We would like to thank Karima Bergeroo and Solange Karama for technical assistance and Dr Thierry Candresse (INRA Bordeaux) and Dr Olivier Hyrien (CNRS Paris) for help in sequence analysis. This work was supported by grants from Ministère de l'Education Nationale and Conseil Régional d'Aquitaine to Claude Mouchès. Keith Campbell was a recipient from Conseil Général des Pyrénées-Atlantiques and Centre d'Etudes de l'Université de Californie à Pau.

References

BEYSSAT-ARNAOUTY, V., MOUCHÈS, C., GEORGHIOU, G.P. and PASTEUR, N. (1989). Detection of organophosphate detoxifying esterases by dot blot immunoassay in *Culex* mosquitoes. *Journal of American Mosquito Control Association*, **5**, 196–200.

DEVONSHIRE, A.L. and FIELD, L.M. (1991). Gene amplification and insecticide resistance. *Annual Review of Entomology*, **36**, 1–23.

DI NOCERA, P.P. and SASAKI, Y. (1990). LINEs: a superfamily of retrotransposable ubiquitous DNA elements. *Trends in Genetics*, **6**, 29–30.

MOUCHÈS, C. (1989). Génie génétique et transgénèse des insectes pour améliorer les techniques de lutte biologique. *C. R. Acad. Agric. Fr.*, **75**, 27–31.

MOUCHÈS, C., PASTEUR, N., BERGE, J.B., HYRIEN, O., RAYMOND, M., ROBERT DE SAINT VINCENT, B., DE SILVESTRI, M. and GEORGHIOU, G.P. (1986). Amplification of an esterase gene is responsible for insecticide resistance in a California *Culex* mosquito. *Science*, **233**, 778–780.

MOUCHÈS, C., MAGNIN, M., BERGE, J.B., DE SILVESTRI, M., BEYSSAT, V., PASTEUR, N. and GEORGHIOU, G.P. (1987). Overproduction of detoxifying esterases in organophosphate resistant *Culex* mosquitoes and their presence in other insects. *Proceedings of the National Academy of Sciences, USA*, **84**, 2113–2116.

MOUCHÈS, C., PAUPLIN, Y., AGARWAL, M., LEMIEUX, L., HERZOG, M., ABADON, M., BEYSSAT-ARNAOUTY, V., HYRIEN, O., ROBERT DE SAINT VINCENT, B., GEORGHIOU, G.P. and PASTEUR, N. (1990). Characterization of amplification core and esterase B1 gene responsible for insecticide relationship in *Culex*. *Proceedings of the National Academy of Sciences, USA*, **87**, 2574–2578.

MOUCHÈS, C., AGARWAL, M., LEMIEUX, L., ABADON, M. and CAMPBELL, K. (1991). Sequence of a truncated retroposon-like element scattered through the genome of *Culex* mosquitoes and coamplified with a gene responsible for insecticide resistance. *Gene* (submitted).

PASTEUR, N., RAYMOND, M., PAUPLIN, Y., NANCE, E., HEYSE, D. and MOUCHÈS, C. (1990). Role of gene amplification in insecticide resistance. In *Pesticides and Alternatives: Innovative Chemical and Biological Approaches to Pest Control* (J.E. Casida, ed.), pp. 439–447. Elsevier.

PRIIMÄGI, A.F., MIZROKHI, L.V. and ILYIN, Y.V. (1988). The *Drosophila* mobile element jockey belongs to LINEs and contains coding sequences homologous to some retroviral proteins. *Gene*, **70**, 253–262.

RAYMOND, M., BEYSSAT-ARNAOUTY, V., SIVASUBRAMANIAN, N., MOUCHÈS, C. GEORGHIOU, G.P. AND PASTEUR, N. (1989). Amplification of various esterase B's responsible for organophosphate resistance in *Culex* mosquitoes. *Biochemical Genetics*, **27**, 417–423.

34
Possible Genetic Start Points and End Points of Insecticide Resistance Evolution

P. RICHTER

Biological Research Centre Berlin, Stahnsdorfer Damm 81, 0-1532 Kleinmachnow, Germany

Introduction

There is increasing evidence that a large number of mechanisms of pesticide resistance in insects and spider mites are inherited as a single autosomal gene (Oppenoorth, 1985; Roush and McKenzie, 1987). Hardy–Weinberg relations are therefore often used to predict resistance development (Taylor and Headley, 1975; Otto, 1988). In contrast to many mathematical models, the selection process of an R-allele cannot be explained only by the increase in its frequency. There are different genetic events that occur during the first selection of an appropriate allele and after its establishment in a finite population. Although commonly not distinguished, the prevention and management of resistance may therefore require very different strategies.

Possible States of Resistance Prior to the Introduction of a New Insecticide

Increasing amounts of data indicate that the genetic change of morphological and physiological characters is caused by a few major genes and that the selective advantage conferred by a mutation is generally quite large with a rather small number of gene substitutions required for a particular event of adaptive evolution

---

Insecticides: Mechanism of Action and Resistance

(Nei, 1988). Available data suggest that this is also largely true for the evolution of insecticide and acaricide resistance. Relatively small changes may cause an insensitivity of the target site (Soderlund *et al.*, 1989), the production of a detoxifying enzyme with increased substrate specificity to the insecticide (Brown, 1990) or the production of a greater quantity of enzyme during the translational process (Paigen, 1989). The following considerations are therefore restricted to single-gene-inherited resistance, although with the amplification of esterase genes in *Myzus persicae* (Field *et al.*, 1988) and *Culex* mosquitoes (Mouchès *et al.*, 1990) some important exceptions have been discovered.

Deterministic models often used to predict resistance are of limited value for the development of resistance prevention strategies because they contain a number of restrictions and simplifications that are not adequate at very low allele frequencies. Table 1 shows that for a finite population a critical allele frequency exists below which the occurrence not only of homozygote-resistant individuals but also of heterozygotes is questionable. Alleles of this character rather behave stochastically. Additionally, changes in the genotype surrounding the R-allele cannot be neglected throughout the selection process, and possible fitness disadvantages of the resistant individuals in selection-free intervals may not be constant.

Table 1 Number of susceptible (SS), heterozygous (SR) and homozygous (RR) resistant individuals expected in a population of 1 million insects assuming a frequency of q of the resistance allele R

q	SS	SR	RR
1×10^{-3}	998 001	1998.0	1
1×10^{-4}	999 800	199.9	1×10^{-2}
1×10^{-5}	999 980	19.9	1×10^{-4}
1×10^{-6}	999 998	1.99	1×10^{-6}
1×10^{-7}	999 999.8	0.199	1×10^{-8}
.	.	.	.
.	.	.	.
.	.	.	.
1×10^{-10}	999 999.9	1.9×10^{-5}	1×10^{-14}

The possible starting points of resistance evolution are discussed with respect to the initial frequency of the R-allele, the state of its fixation and distribution in the species and the possibilities of its detection in a resistance risk assessment programme. Note that in terms of molecular genetics 'the *R*-allele' is not necessarily one allele, but summarizes all alleles at the same locus that are able to confer a certain degree of resistance.

Proceeding on these attributes the possible states of resistance alleles prior to selection can roughly be divided into four groups: stable alleles, rare variant alleles, extremely rare variant alleles, and new mutations.

Stable alleles

Traditional biotesting involves up to 100 individuals and allows the detection of resistance at a level of 3–5% (Roush and Miller, 1986). Survival rates of this order clearly indicate the existence of cross-resistance to a compound that had been used in the past. To detect possible cross-resistance at early stages of product development, a collection of purified strains of the target species with resistance to different pesticides is very useful.

Stable alleles have already been fixed by selection. The higher their frequency in one population, the more widely they are distributed in neighbouring populations. Their behaviour under selection is largely deterministic.

Rare variant alleles

The term is adopted from Kimura (1983a) keeping in mind that resistance alleles are selectively neutral or only slightly deleterious when they are not selected for by a pesticide. Kimura found it appropriate to take the frequency q of a rare variant allele not exceeding 0.01. In the examples he used to apply his theory to estimate the fraction of selectivity neutral mutations, sample sizes were between 10^2 and 10^4. In a resistance risk assessment programme, resistance alleles at frequencies of 10^{-2}–10^{-4} may appear by testing large samples or pre-selecting ('pressing') representative populations from the chosen target area with low doses to concentrate them.

Rare variant alleles are fixed in populations at random. Until selection their behaviour is stochastic according to Kimura's neutral theory of molecular evolution (Kimura, 1983b). Due to their, perhaps, long history of neutral drift, they may be very widespread among populations as has, for instance, been shown by Singh and Rhomberg (1987) for a number of loci of *Drosophila melanogaster*.

Extremely rare variant alleles

I would like to identify them as a part of the rare variant allele fraction only for the one pragmatic reason that the possibility to detect them even in very large samples is nearly nil. In most cases their frequency is expected to be much below 10^{-4}. Nevertheless they are fixed at random, and with great probability they are also distributed very widely.

New mutations

I doubt that the actual mutation process is a very important source of resistance alleles in insects and spider mites. I would rather agree with Nei (1988) that the present genetic variability of a population is a product of its evolution in the past. The main reason is that the required mutations have evidently to occur at selectively relevant sites representing only a part of the whole sequence needed for the translation of an enzyme or receptor protein.

In contrast to stable and rare alleles, mutations are local and not fixed. There is also no possibility to detect them. Even large laboratory selection experiments involving the application of mutagens cannot make sure that alleles eventually selected are similar to those in the field.

If, despite an intense search, no resistance allele is found, there is much reason to presume the extremely rare variant allele fraction to be the main source of the expected resistance. As with mutations, any finite population will then contain only single resistant individuals.

The Faith of Rare Resistance Alleles under Selection

When extremely rare variant alleles are exposed to high selection pressure, they are, notwithstanding their selective advantage, threatened by accidental extinction.

The aim of pesticide treatment is to reduce the target population substantially. The population recovers then from three sources of founder individuals: immigrants, survivors in refugia, and resistant individuals. There is a short period of time after treatment during which the resistant survivors are solitary. A resistant male may not find a female to reproduce with, previously mated females having been killed as susceptibles. A resistant virgin may remain unmated, or some offspring carrying the resistance allele may be destroyed during their development, especially due to an increase in the relative importance of density-independent mortality factors. These events correspond to the principle of Allee (1931) who supposed the rate of increase of a population to be less than optimum not only in cases of overcrowding, but also when the population density is very small.

Additionally, the small number of resistant individuals can more likely be killed by another pesticide to which they are not cross-resistant.

A rapid influx of susceptible immigrants, a considerable amount of survivors in refugia or survivors of a low-dose strategy will greatly enhance the chance of the trait to be conserved. Once a certain quantity of resistant insects escaped the threat of extinction, the allele can quickly increase in frequency during following treatments with the same selective compound.

Resistance selection is diversity-reducing, and the chromosome carrying the resistance allele would no longer contribute to the genetic variability of the population of origin. The genetic diversity is restored by recombination to susceptibles. As has been described by Baker (1982), two relatively independent processes of recombination are involved.

Consecutive backcrossing continuously eliminates the specific genetic environment of the resistant allele. During very few generations all chromosomes not carrying the resistance allele are replaced by chromosomes from susceptibles. A more time-consuming process is the exchange of genes jointly inherited with the resistance allele by crossingover. The longer the crossover distance to the allele, the shorter the time needed for an exchange. So some genetic environment in the vicinity of the resistance allele is hitchhiked, perhaps over years.

Resistance alleles do not necessarily result in fitness disadvantages in the absence of selection. Slightly disadvantageous alleles are adjusted by the selection of fitness modifiers, and sometimes resistance is improved by the selection of enhancer genes. Appropriate examples are, for instance, known for the housefly (Whitehead *et al.*, 1985; Sawicki *et al.*, 1986) and the sheep blowfly (McKenzie and Purvis, 1984; Clarke and McKenzie, 1987).

The Prevention of Multiple Resistance

The relationship between two or more resistance genes depends on their chromosomal location. Resistance genes located on different chromosomes are independently inherited. As soon as the selection pressure is ceased, their frequency can decrease either by a relatively lower fitness compared to susceptibles or simply by dilution. The risk of translocating the resistance to another chromosome is small.

On the contrary, resistance genes on homologous chromosomes are competing. They are selected against each other even in the case of dominant inheritance. There is a considerable chance for recombination of two resistance genes on to one chromosome by crossingover, especially when both genes are present in high frequencies and the selecting pesticides are used in rotation. Both genes are then jointly inherited and concomitantly selected. This seems to be the most ultimate form of multiple resistance because all the compounds concerned fall out of the group of alternative pesticides.

Conclusions

Strategies of the prevention or management of insecticide and acaricide resistance have to consider the state of fixation of probably existing resistance alleles. Without selection appropriate alleles can be fixed in populations only at random. They have very little chance to improve their adjustment by the selection of modifier genes (Wright, 1951). It is the aim of resistance prevention to disturb or interrupt the fixation of these alleles by selection.

Contrariwise, resistance management is concerned with established mechanisms, and its aim is to reduce the frequency of established alleles and to suppress the genetical recombination of different genes, thus preventing the occurrence of multiple resistance caused by the selection of several resistance genes on to one chromosome.

References

ALLEE, W.C. (1931). *Animal Aggregations. A Study in General Sociology*. Chicago.

BAKER, J. (1982). Selective effects of insecticides on within-species variation: the lessons to be learnt when considering the environmental effects of pollutants. *Agriculture and Environment*, **7**, 187–198.

BROWN, T.M. (1990). Biochemical and genetic mechanisms of insecticide resistance. In *Managing Resistance to Agrochemicals: From Fundamental Research to Practical Strategies* (M.B. Green, W.K. Moberg and H. Le Baron, eds). *ACS Symposium Series*, **421**, 61–76.

CLARKE, G.M. and McKENZIE, J.A. (1987). Developmental stability of insecticide resistant phenotypes in blowfly: a result of canalizing natural selection. *Nature*, **325**, 345–346.

FIELD, L.M., DEVONSHIRE, A.L. and FORDE, B.G. (1988). Molecular evidence that insecticide resistance in peach potato aphids (*Myzus persicae* SULZ.) results from amplification of esterase gene. *Biochemical Journal*, **251**, 309–312.

KIMURA, M. (1983a). Rare variant alleles in the light of the neutral theory. *Molecular and Biological Evolution*, **1**, 84–93.

KIMURA, M. (1983b). *The Natural Theory of Molecular Evolution*. Cambridge University Press, Cambridge.

McKENZIE, J.A. and PURVIS, A. (1984). Chromosomal localization of fitness modifiers of diazinon resistance genotypes of *Lucilia cuprina*. *Heredity*, **56**, 625–634.

MOUCHÈS, C., PAUPLIN, Y., AGARWAL, M., LEMIEUX, L., HERZOG, M., ABADON, M., BEYSSAT-ARNAOUTY, V., HYRIEN, O., ROBERT DE SAINT VINCENT, B., GEOR-GHIOU, G.P. and PASTEUR, N. (1990). Characterization of amplification core and esterase B1 gene responsible for insecticide relationship in *Culex*. *Proceedings of the National Academy of Sciences, USA*, **87**, 2574–2578.

NEI, M. (1988). Relative roles of mutation and selection in the maintenance of genetic variability. *Philosophical Transactions of the Royal Society of London*, **319**, 615–629.

OPPENOORTH, F.J. (1985). Biochemistry and genetics of resistance. In *Comprehensive Insect Physiology, Biochemistry and Pharmacology* (G.A. Kerkut and L.I. Gilbert, eds), Volume 12, *Insect Control*, pp. 731–773.

OTTO, D. (1988). Die Ausbildung von Resistenz in Insektenpopulationen nach Insektizideinwirkung. *Biologische Rundschau*, **26**, 135–149.

PAIGEN, K. (1989). Experimental approaches to the study of regulatory evolution. *The American Naturalist*, **134**, 440–458.

ROUSH, R.T. and McKENZIE, J.A. (1987). Ecological genetics of insecticide and acaricide resistance. *Annual Review of Entomology*, **32**, 361–380.

ROUSH, R.T. and MILLER, G.L. (1986). Considerations for design of insecticide resistance monitoring programs. *Journal of Economic Entomology*, **79**, 293–298.

SAWICKI, R.M., FARNHAM, A.W., DENHOLM, I. and CHURCH, V.J. (1985). Potentiation of super-kdr resistance to deltamethrin and other pyrethroids by an intensifier (factor 161) on autosome 2 in the housefly (*Musca domestica* L.). *Pesticide Science*, **17**, 483–488.

SINGH, R.S. and RHOMBERG, L.R. (1987). A comprehensive study of genetic variation in natural populations of *Drosophila melanogaster*. I. Estimates of gene flow from rare alleles. *Genetics*, **115**, 313–322.

SODERLUND, D.M., BLOOMQUIST, J.R., WONG, F., PAYNE,

L.L. and KNIPPLE, D.C. (1989). Molecular neurobiology: implications for insecticide action and resistance. *Pesticide Science*, **26**, 359–374.

TAYLOR C.R. and HEADLEY, J.C. (1975). Insecticide resistance and the evaluation of control strategies for an insect population. *Canadian Entomology*, **107**, 237–242.

WHITEHEAD, J.R., ROUSH, R.T. and NORMENT, B.R. (1985). Resistance stability and coadaptation in diazinon-resistant house-flies (Diptera: Muscidae). *Journal of Economic Entomology*, **78**, 25–29.

WRIGHT, S. (1951). Fisher and Ford on 'the Sewall Wright Effect'. *American Science*, **39**, 452–459.

35
A Critical Comment on the Evaluation of the Resistance Level in Field Populations by the Resistance Index R_i

D. OTTO, E. MOLL and P. RICHTER

Biological Research Centre Berlin, Stahnsdorfer Damm 81, 0-1532 Kleinmachnow, Germany

The resistance level in an insect strain or field population is commonly expressed by the resistance index R_i. This is a comparison of the LD_{50} value of this population with the LD_{50} of a 'normal susceptible' reference strain:

$$R_i = \frac{LD_{50} \text{ of the field population}}{LD_{50} \text{ of a 'normal susceptible' reference strain}}$$

LC_{50}, LD_{95} or LC_{95} can be used likewise.

This chapter attempts to show that a R_i estimated in this way is generally not sufficient to evaluate the toxicological situation in field populations because:

- the denominator is a subjectively chosen term, influencing substantially R_i;
- the numerator, the LD_{50} of a heterogeneous field population with mixed toxicological phenotypes, cannot be calculated by common probit analysis.

Insecticides: Mechanism of Action and Resistance

Relativity of R_i

One of the disadvantages of R_i is that it is not an objective and real parameter of the evaluated population because it completely depends on the LD_{50} of the reference strain, which is chosen subjectively. This means that R_i data published by different authors using different standard strains are not comparable, as shown in Table 1.

Table 1 Response of *Tetranychus urticae* to dimethoate. R_i values of a greenhouse population in comparison to different reference strains

Susceptible standard strain	LD_{50} (ppm a_i)	TuGG Greenhouse at Groß Gaglow $LD_{50} = 2400$ ppm a_i
TuS Kleinmachnow, kept since 1972 in laboratory	15	$R_i = 160$
TuA Kleinmachnow, collected in 1988 in the nature preserve Aken	1.5	$R_i = 1600$
TuL Leningrad, laboratory standard	2.0	$R_i = 1200$

For some species the WHO has recommended and introduced standard strains. This is a suitable way of overcoming the problem, but our laboratory observations on the *Musca domestica* SRS strain from Padua/Italy have shown that such a strain can change its susceptibility during long-term laboratory rearing.

Brent (1986) pointed out that the resistance index or resistance factor is '. . . often used, but the basis of its calculation needs careful consideration. The choice of susceptible reference strains and any shift in their response in time can effect greatly the value of this index. . . . If the reference strain has been kept away from all chemical treatments for years in a laboratory culture, it may be abnormally susceptible.' Some laboratory strains kept without contact with insecticides may have the significance of a museum piece. They are necessary for biochemical and genetical comparison experiments, but they are hardly suitable to describe the resistance situation in the field.

Today it is virtually impossible to find a pest population with its original susceptibility. Which one, for instance, is the 'normal natural susceptible population' of the Colorado potato beetle after more than forty years of control by pesticides?

To avoid this problem, a resistance ratio is sometimes calculated comparing the highest and the lowest LD_{50} found for an insecticide in different areas within a region (Huang and Smilowitz, 1990).

We attempted to find an inherent quantitative characterization of a population independent of a separate population or laboratory strain. As a field population is composed of susceptible and resistant genotypes it is more promising to compare its resistant part to its own susceptible part.

The Problem of Mixed Populations

There is evidence that resistance of a practical significance is usually controlled by a single gene (Whitten and McKenzie, 1982; Roush and Croft, 1986; Roush, 1989) and that the toxicology of resistance results from the fact that a field population is a mixture of SS, SR, and RR genotypes at the frequencies p^2, $2pq$, and q^2, respectively, differing phenologically in their susceptibility to insecticides (Georghiou, 1983). In such a heterogeneous sample the calculation of only one straight log dosage–probit mortality regression line (ld–pm line) for a dosage–mortality relationship analysis is not possible (Tsukamoto, 1963, 1983).

The following example is based on only one R-allele conferring a high degree of resistance. Such a model ought to be appropriate to field populations. It may consist of approximately 50% susceptible individuals and 50% carriers of the dominant R-gene (Figure 1a). In tests with increasing concentrations, using a standard biotest as recommended by the FAO, WHO or regional institutions, mortality is investigated (Figure 1b). Starting with a low dosage, mortality increases gradually until the dosage is reached at which all susceptible individuals, i.e. 50% of the sample are killed. Despite a further increase in dosage mortality remains at 50%. Only when the dosage reaches the amount at which the resistant part of the sample begins to die is a further increase in the dosage–mortality line established (Figure 1b).

Usually one straight ld–pm line is calculated which is adjusted to all the experimental points regardless of to which phenological group

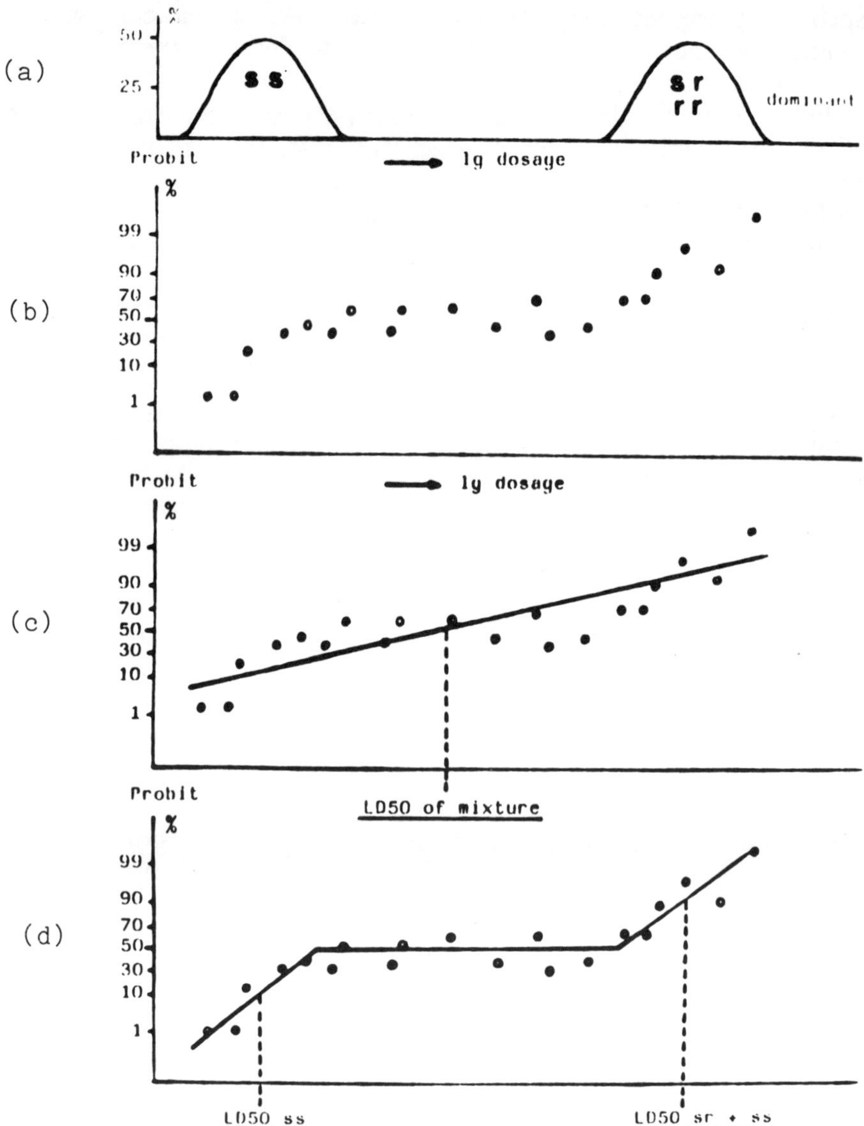

Figure 1a–d   Illustration of the distinction between common probit analysis and spline analysis in evaluating resistance (for explanation see text).

they belong. The deviation of these points from the regression line is so high that it is more than only random as compared to the random deviation revealed by χ^2-test. The LD_{50} derived from this diagram has thus no relevance to the real situation (Figure 1c).

Spline regression analysis in bioassays and evaluation of the resistance level

A mathematical method has been developed which splits the concentration–mortality points in the probit diagram into two or three groups belonging to the susceptible or to the resistant part of the population (Heinig *et al.*, 1989).

The method of spline regression with one or two variable break points ideally gives two separate ascending lines, one for the susceptible part and one for the resistant part (Figure 1d). Both are connected by a plateau, which reveals the frequency of susceptible individuals in a sample (Figure 1d).

Transforming the two probit lines each to a full probit scale, it is possible to calculate the LD_{50} separately for both the susceptible and the resistant phenotypes of the population (Figure 1d). Comparing these two LD_{50} values, an objective and real R_i is obtained without reference to a laboratory strain. This way it is possible to determine:

- the frequency of the resistant phenotypes in a field population; and
- the level of resistance when compared to the susceptible ones of the same population.

Mathematical background of spline regression

To simplify calculations, we elucidated the elementary case of a population containing a number of susceptible individuals and resistant individuals of one genotype, as for instance in the dominant inheritance model. There are three linear splines connected by two break points (Figure 1d). The first and third splines represent the response of the susceptible and resistant individuals, respectively. The spline function connecting them is believed to represent the responses of all susceptible individuals and also the non-response of the resistant part of the population.

Taking into account that in many cases the number of resistant or susceptible individuals may be too small to identify, the analysis can be reduced to two splines connected by one break point (Figure 5).

The spline function for one break point z is

$$y_1 = f_1(z, c, x) = c_1 + c_2 x + c_3 (x - z)$$

where
$$(x - z) = \begin{cases} x - z & \text{when } x - z > 0 \\ 0 & \text{when } x - z \leqslant 0 \end{cases}$$

and for two break points z_1, z_2

$$y_2 = f_2(z, c, x) = c_1 + c_2 x + c_3(x - z)_1 + c_4(x - z_2)$$

where

$$(x - z_j) = \begin{cases} x - z_j & \text{when } x - z_j > 0; \ j = 1, 2 \\ 0 & \text{when } x - z \leqslant 0; \ j = 1, 2 \end{cases}$$

In the transformed lc–pm system (x, y), any c_i with $i = 1, 2, 3, (4)$ is the coefficient of the spline function and z_j with $j = 1, 2$ and z are the values of the break points on the x axis (log dosages).

The break points are located by the least squares method so that for all transformed data pairs (x_l, y_l) with $l = 1 \dots n$:

$$\sum_{l=1}^{n}(y_1 - f(z, c, x_1))^2 \longrightarrow \begin{array}{c} \text{Minimum} \\ \text{z, c} \end{array}$$

Although the spline functions are linear in the coefficients c_i, they are not linear in their break points z_j. This problem is solved by the following two minimizations:

$$\sum_{l=1}^{n}(y_1 - f(z, c, x_1))^2 \longrightarrow \begin{array}{c} \text{Minimum} \\ z; c \text{ fixed} \end{array}$$

and

$$\sum_{l=1}^{n}(y_1 - f(z, c, x_1))^2 \longrightarrow \begin{array}{c} \text{Minimum} \\ c; z \text{ fixed} \end{array}$$

For this purpose the transformed interval between the least and the highest concentration is divided into N sections.

Each of these raster points could be a break point $x_1 < z_1 < z_2 < x_n$ or $x_1 < z < x_n$. For every pair of raster points for two break points and for every raster point for one break point the coefficients c_i are calculated. They are the result of a linear equation system consisting of n equations with four or, for one break point, three variables according to the Gauss normal equation and the modified or square-root-free Cholesky method (Schwetlick, 1979; Engeln-Müllges and Reutter, 1988). The function with the lowest mean square deviation or residuum is the spline function.

Due to the discretization of the interval the result cannot be unique. For this reason both the coefficient c_i and the break points z_j can numerically differ depending on N. The spline function is, therefore, calculated approximately. Having determined one or two of the $N-1$ raster points as break points, the interval segment of each break point is divided into N segments to improve the spline function. The problem is thus solved by a suitable combination of the number of segments N and the number of improvement of the spline function.

Application of Spline Regression in Practice

Figure 2 shows the application of the spline regression method in practice. A *Tetranychus urticae* population from a greenhouse producing cucumber near Erfurt/Thuringia was analysed for its resistance to the acaricide dicarzol (active ingredient, a_i = formetanate). The computer print shows a line with two break points, the equations for the splines representing the susceptible and the resistant parts and the coordinates of the break points.

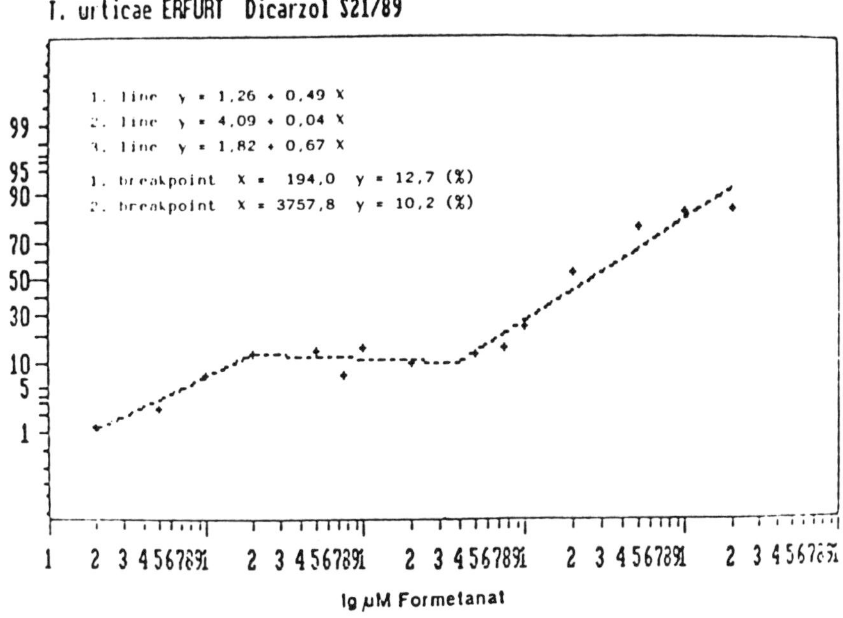

Figure 2 Computer print of a spline regression analysis of the dicarzol toxicology of a greenhouse population of *Tetranychus urticae* from Erfurt.

The frequency of the resistant individuals and LC_{50} of the susceptible and the resistant part of the sample were calculated in three steps (Figure 3):

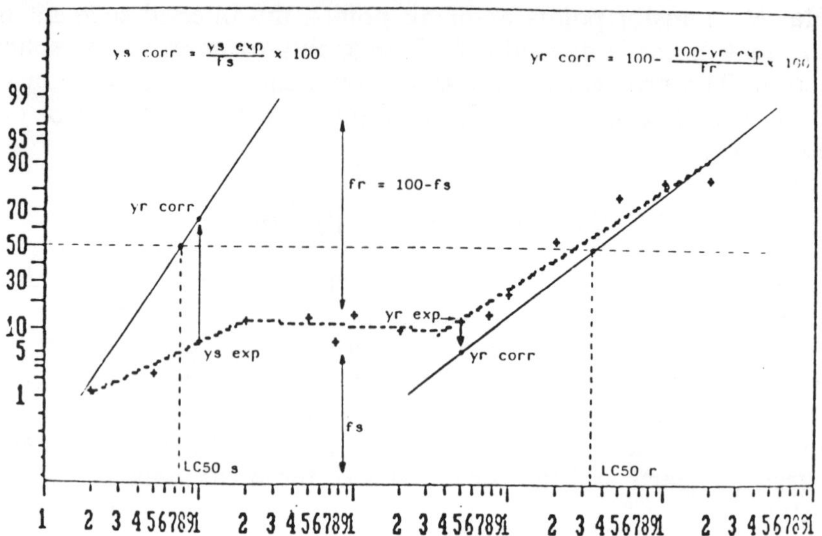

Figure 3   Method of calculating the frequency of the susceptible (f_s) and of the resistant (f_r) individuals and LC_{50} values of the susceptible ($LC_{50\,s}$) and the resistant ($LC_{50\,r}$) part of the populations.

Step 1   Determination of the frequency of the susceptible and the resistant phenotypes by calculating the average ($\pm$ standard deviation) of all experimental mortality values between the two break points after probit transformation. Retransforming this average, the frequency of the susceptible individuals f_s and the frequency of the resistant individuals $f_r = 100 - f_s$ are obtained. When 100% = 1, f_r equals $1 - f_s$.

Step 2   Transformation of the line for the susceptible part ($0 \ldots f_s$) to a full probit scale ($0 \ldots 100\%$) by the equation

$$Y_{s\ corr} = \frac{Y_{s\ exp}}{f_s} \times 100 \qquad\qquad [1a]$$

or when 100% = 1

$$Y_{s\ corr} = \frac{Y_{s\ exp}}{f_s} \qquad\qquad [1b]$$

where $Y_{s\ exp}$ is the experimental value of the percentage mortality of the susceptible phenotypes at each concentration and $Y_{s\ corr}$ is the corrected value of the percentage mortality of the susceptible phenotypes at each concentration.

Having transformed each experimental dosage–mortality point ($Y_{s\ exp}$) to $Y_{s\ corr}$, the LC_{50} of the susceptible part can be calculated by common probit analysis.

Step 3   Transformation of the line for the resistant part (f_s... 100) to a full probit scale (0 ... 100%) by equation

$$Y_{r\ corr} = 100 - \frac{100 - Y_{r\ exp}}{100 - f_s} \times 100 \qquad [2a]$$

or when 100% = 1

$$Y_{r\ corr} = \frac{Y_{r\ exp} - 1 + f_r}{f_r} \qquad [2b]$$

where $Y_{r\ exp}$ and $Y_{r\ corr}$ are defined in Step 2, but related to the resistant phenotypes.

Having transformed each experimental dosage–mortality point ($Y_{r\ exp}$) to $Y_{r\ corr}$, the LC_{50} of the resistant part can be calculated by common probit analysis.

Applying these equations to the above-mentioned example, the greenhouse population of *T. urticae* in Erfurt, the LC_{50} value of the susceptible and resistant phenotypes and resistance frequency are calculated (Figure 4) as

$$fr = 79\%$$
$$LC_{50\ s} = 76\ \mu M\ a_i$$
$$LC_{50\ r} = 33\ 000\ \mu M\ a_i$$

and

$$R_i = \frac{LC_{50}\ r}{LC_{50}\ s} = \frac{33\ 000\ \mu M}{76\ \mu M} = 434$$

The former R_i method would have given only one probit line and one LC_{50} of about 20 300 μM a_i. Comparing this LC_{50} with that of our laboratory standard strain, the R_i would have been 23.

Outlook

The application of the spline regression method is limited to a certain extent, for the following reasons:

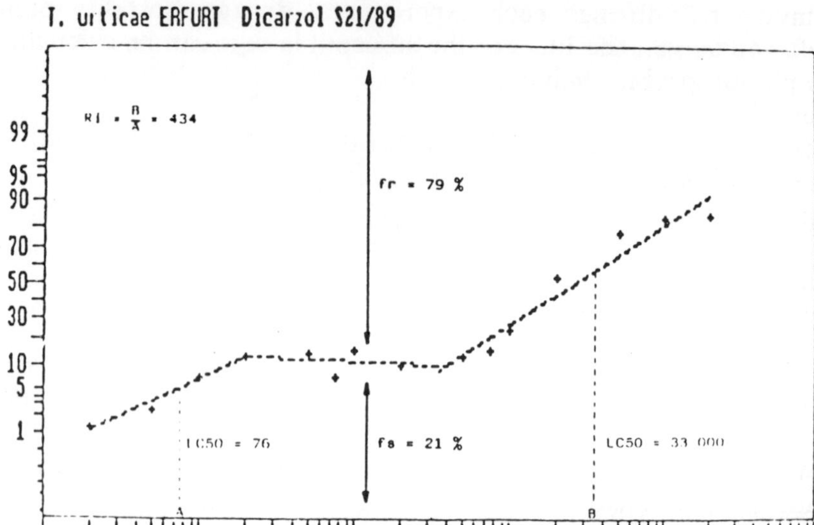

Figure 4 Results from calculating f_r, $LC_{50\ s}$, $LC_{50\ r}$, and R_i related to the mixed dicarzol-resistant *Tetranychus urticae* population in Erfurt.

1. It is appropriate only when resistance is mainly determined by one factor and when resistance and susceptibility of individuals clearly differs. If pesticide resistance is polygenic, the resistant phenotype as expressed in the ld–pm line will be continuous (Via, 1986), not split by a plateau, and break points cannot be established.

2. It gives information only when the resistant phenotype has already reached a frequency in the range from 10^{-2} to 10^{-1} as bioassays seldom involve samples of more than 100 individuals for one dosage test (Roush and Miller, 1986; Martinson *et al.*, 1991). At lower degrees of resistance (frequency $< 10^{-3}$) biochemical and related tests may be helpful. They can provide separate data for a large number of individuals so that the research worker can create a characteristic diagram of the distribution of the phenotypes or even of the genotypes in a field sample. The earliest occurrence of *R*-genes in the gene pool, when they are very rare, can be considered only theoretically (Richter, this volume Chapter 33).

3. If resistance frequency equals 90% and more, the mortality in the susceptible part is not different from the random mortality

among the untreated control (Figure 5). In this case there is no left-hand ascending line and no left-hand break point, the plateau is placed in the lower random range, and the LC_{50} of the rare susceptible individuals cannot be determined. The R_i of the resistant individuals cannot, therefore, be calculated comparing the LC_{50} with the susceptible part of the same population. In spite of this, spline regression has a remarkable advantage in this situation: The LD_{50} of the spline of the population can be calculated more precisely, because the regression line is not burdened with many values from the non-resistant part as it is in a normal probit analysis.

Despite these restrictions we consider the spline regression a good tool for the resistance analysis of field populations giving more adequate information on the population resistance than does the common probit analysis.

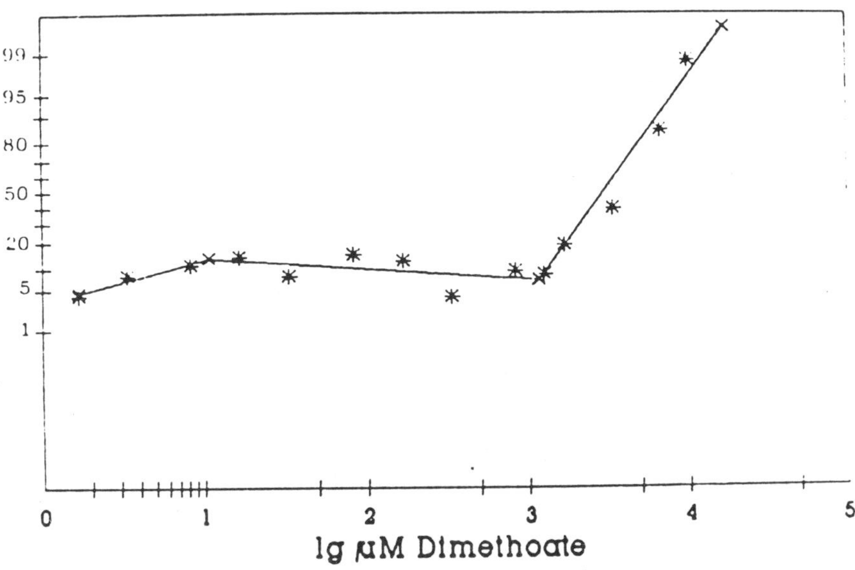

Figure 5 Spline regression of a greenhouse population of *Tetranychus urticae* in Erfurt with a high frequency of individuals resistant to Bi 58 (a_i dimethoate).

References

BRENT, K.J. (1986). Detection and monitoring of resistant forms – an overview. In *Pesticide Resistance: Strategies and Tactics for Management*, pp. 298–312. National Academic Press, Washington DC.

ENGELN-MÜLLGES, G. and REUTTER, F. (1988). *Formelsammlung zur Numerischen Mathematik mit Standard-Fortran 77-Programmen*. BI-Wissenschaftsverlag, 6th edn.

GEORGHIOU, G.P. (1983). Management of resistance in arthropods. In *Pesticide Resistance to Pesticides* (G.P. Georghiou and T. Saito, eds), pp. 769–792. Plenum Press, New York and London.

HEINIG, P., OTTO, D. and MOLL, E. (1989). Spline regression as a tool for analysis of population genetics in resistant populations. In *Proceedings of the First International Symposium 'Insecticides – Mechanisms of Action and Resistance'. Reinhardsbrunn, 1988. Tagungsbericht der Akademie der Landwirtschaftswissenschaften der DDR*, **274**, 215–218.

HUANG, J. and SMILOWITZ, Z. (1990). Insecticide resistance in Colorado potato beetle – Pennsylvania. In *Pest Resistance Management* (Pesticide Research Center Michigan State University), **2**(1), 22.

MARTINSON, T.E., NYROP, J.P., DENNIHY, T.J. and REISSING, V.H. (1991). Temporal variability in repeated bioassay of field populations of European red mite (Acari: tetranychiade): Implications for resistance monitoring. *Journal of Economic Entomology*, **84**, in press.

ROUSH, R.T. (1989). Designing resistance management programs: How can you choose. *Pesticide Science*, **26**, 423–441.

ROUSH, R.T. and CROFT, B.A. (1986). Experimental population genetics and ecological studies of insecticide resistance in insects and mites. In *Pesticide Resistance Strategies and Tactics for Management*, pp. 257–270. National Academic Press, Washington DC.

ROUSH, R.T. and MILLER, G.L. (1986). Considerations for design of insecticide resistance monitoring programs. *Journal of Economic Entomology*, **79**, 293–298.

SCHWETLICK, H. (1979). *Numerische Lösung nichtlinearer Gleichungen*. VEB Deutscher Verlag der Wissenschaften, Berlin.

TSUKAMOTO, M. (1963). The log dosage–probit mortality curve in genetic researches of insect resistance to insecticides. *Botyu-Kagaku*, **28**, 91–98.

TSUKAMOTO, M. (1983). Methods of genetic analysis of

insecticide resistance. In *Pest Resistance to Pesticides* (G.P. Georghiou and T. Saito, eds), pp. 71–98. Plenum Press, New York and London.

VIA, S. (1986). Quantitative genetic models and the evolution of pesticide resistance. In *Pesticide Resistance Strategies and Tactics for Management*, pp. 222–235. National Academic Press, Washington DC.

WHITTEN, M.H. and McKENZIE, J.A. (1982). The genetic basis for pesticide resistance. In *Proceedings of the Third Australian Conference on Grassland Invertebrate Ecology* (K.E. Lee, ed.), pp. 1–16. Adelaide S.A. Government Printer. Cited in KEIDING, J. (1986). Prediction or resistance risk assessment. In *Pesticide Resistance Strategies and Tactics for Management*, pp. 279–297. National Academic Press, Washington DC.

36
Present Status, Mechanisms and Countermeasures of Insecticide Resistance in Vegetable Pests in China

Z. TANG

Shanghai Institute of Entomology, Academia Sinica, Shanghai 200025, China

Introduction

In the 1960s, organophosphorus insecticides (OPIs) such as trichlorfon, dichlorvos and dimethoate were recommended in Shanghai for control of vegetable pests, followed by acephate in the 1970s and the pyrethroid insecticides fenvalerate and deltamethrin in the early 1980s. These insecticides played a vital role in the production of good quality vegetables. Continuous application of the same insecticides has resulted in the rapid sequential development of resistance to commonly used insecticides and has affected pest control programmes.

By the end of 1990, resistance was reported in at least eight species of vegetable pests. This chapter deals with the status and mechanisms of insecticide resistance in vegetable pests, and the countermeasures adopted in China.

Resistance in Vegetable Pests

At present there are eight species of vegetable pests known to resist insecticides in China. These species are as follows: the melon-cotton aphid (*Aphis gossypii*), the willow aphid (*Cavariella salicicola*), the mustard aphid (*Lipaphis erysimi*), the cabbage flea beetle (*Phyllotreta*

Insecticides: Mechanism of Action and Resistance
© 1992   Intercept Ltd, P.O. Box 716, Andover, Hants SP10 1YG, UK

cittata), the greenhouse whitefly (*Trialeurodes vaporariorum*), the green peach aphid (*Myzus persicae*), the cabbage worm *Pieris rapae* and the diamondback moth *Plutella xylostella*. The cases and levels of insecticide resistance in these pests are given in Table 1. In general, resistance levels of vegetable pests in northern China are higher than in eastern China, except for the diamondback moth.

Table 1 Cases and levels of resistance of vegetable pests to insecticides in China, 1990

species	Insecticide and resistance level	Method tested	Areas and Years occurred
Melon-cotton aphid (*Aphis gossypii*)	Fenvalerate ($\times 223$) Deltamethrin ($\times 520$) Dimethoate ($\times 18$) Omethoate ($\times 13$) Pirimicarb ($\times 81$)	Dipping	Beijing and Hebei (1983–86)
	Fenvalerate ($\times 14$) Deltamethrin ($\times 14$) Flucythrinate ($\times 19$) Cypermethrin ($\times 14$) Permethrin ($\times 4$) Acephate ($\times 4$) Dimethoate ($\times 6$)	Dipping	Shanghai (1979–90)
Willow aphid (*Cavariella salicicola*)	Malathion ($\times 35$) Deltamethrin ($\times 300$) Fenvalerate ($\times 230$)	Topical application	Shanghai (1986)
Mustard aphid (*Lipaphis erysimi*)	Fenvalerate ($\times 8$) Flucythrinate ($\times 16$) Cypermethrin ($\times 5$) Deltamethrin ($\times 7$)	Dipping	(1981–88)
Greenhouse whitefly (*Trialeurodes vaporariorum*)	Deltamethrin ($\times 6290$) Fenvalerate ($\times 1942$) Malathion ($\times 10$)	Dipping	Beijing (1987–88)
Green peach aphid (*Myzus persicae*)	Dimethoate ($\times 56$) Omethoate ($\times 217$) Fenvalerate ($\times 1088$) Deltamethrin ($\times 929$)	Topical application	Beijing (1983–88)
	Fenvalerate ($\times 7$) Deltamethrin ($\times 8$)	Topical application	Shanghai (1981–87)
Cabbage worm (*Pieris rapae*	DDT ($\times 5$–55) Trichlorfon ($\times 100$) Acephate ($\times 6$)	Topical application	Shandong, Hubei, Heilonjiang, Sichuan, Jiangsu

Table 1 (continued)

Species	Insecticide and resistance level	Method tested	Areas and years occurred
	Deltamethrin ($\times$ 5–28)		and Shanxi (1981)
	Fenvalerate ($\times$ 6)		Shanghai (1988)
	Trichlorfon ($> \times$ 100)		
Diamondback moth (*Plutella xylostella*	Deltamethrin ($\times$ 287)	Leaf residue	Shanghai (1979–86)
	Fenvalerate ($\times$ 435)		
	Cypermethrin ($\times$ 411)		
	Flucythrinate ($\times$ 925)		
	Permethrin ($\times$ 58)		
	Acephate ($\times$ 29)		(1979–89)
	DDT ($> \times$ 3)	Topical application	Shanghai and Guangzhou (1988)
	Carbaryl ($> \times$ 3)		
	Deltamethrin ($> \times$ 10414)		
	Fenvalerate ($\times$ 2102–$>$3569		
	Permethrin ($\times$ 245–1533)		
	Malathion ($\times$ 27–144)		
Cabbage flea beetle (*Phyllotreta cittata*)	DDT ($\times$ 3)	Spray	Zhejiang (1964)

Of the eight species of resistant pests, resistance in the diamondback moth (DBM) is particularly serious in the eastern and southern coastline of China where the climate is good for this pest and there are about twenty generations per year. The DBM collected from Shanghai, Fuzhou, Xiamen and Gongchu have become very strongly resistant to all classes of available insecticides such as OPIs, carbamates and pyrethroids. Among them, the resistance levels are particularly high for pyrethroids, deltamethrin (1014-fold), fenvalerate (2103- and 3569-fold) and permethrin (245- and 1533-fold) as shown by the topical application method recommended by FAO (Busvine, 1980).

Monitoring for resistance of DBM to pyrethroids and acephate from 1979 to 1986 in Shanghai suburbs showed that the resistance ratios at LC_{50} were 645 for fenvalerate, 287 for deltamethrin, 411 for cypermethrin, 925 for flucythrinate, 58 for permethrin and 29 for acephate, respectively (Wu and Gu, 1986).

Monitoring for resistance in the melon-cotton aphid (MCA) showed resistance to be $\times 521$ for deltamethrin, $\times 223$ for fenvalerate, $\times 18$ for dimethoate, $\times 13$ for omethoate and $\times 81$ for pirimicarb respectively in Beijing and Hebei Provinces during 1983 to 1986 (Zheng *et al.*, 1988). Resistance in MCA, however, in Shanghai was much lower than that in Beijing and Hebei (see Table 1).

Myzus persicae showed higher resistance to pyrethroids such as fenvalerate ($\times 1088$) and deltamethrin ($\times 929$) and to OPIs omethoate ($\times 217$) and dimethoate ($\times 56$) when the LC_{50}. tested in 1988 were compared with the LC_{50}s from 1983 in Beijing (Zheng *et al.*, 1989) and low resistance to fenvalerate ($\times 7$), deltamethrin ($\times 8$) and dimethoate ($\times 4$) from 1981 to 1987 in Shanghai.

Cavariella salicicola has also shown resistance to fenvalerate ($\times 230$), deltamethrin ($\times 300$) and malathion ($\times 35$) in Shanghai.

Trialeurodes vaporariorum has been recorded as highly resistant to deltamethrin ($\times 6290$), fenvalerate ($\times 1942$), but less resistant to malathion ($\times 10$) in Beijing (Zhen and Rui, 1988).

Pieris rapae appeared to be resistant to trichlorfon and DDT in certain areas in 1981 (Mu *et al.*, 1984). In Shanghai suburbs this pest has been controlled by fenvalerate, acephate and clorfluazuron. Surveys during 1983–88 showed that resistance to these insecticides increased ($\times 31$, $\times 18$ and $\times 2$, respectively).

Mechanisms of Resistance

As resistance in the DBM has become a serious problem in Shanghai suburbs, much work has been done on the mechanisms of resistance to find more effective ways of overcoming or delaying resistance.

The *in vitro* studies of synergism showed:

1. that TBPT (*S, S, S*-tribytyl phosphorotrithioate, an inhibitor of esterases) at 0.2082 µg per larva synergized the effect of malathion ($\times 59.6$ and $\times 32$, respectively) against the resistant (R) strain but had no effect on deltamethrin and permethrin;
2. that PB, an inhibitor of mixed function oxidases (MFO), at a dosage of 0.2082µg per larva, synergized the effect of deltamethrin ($\times 5$) and permethrin ($\times 13$);
3. that DMC [1, 1-bis(*p*-chlorophenyl)ethanol], an inhibitor of dehydrochlorase, and PB did not synergize the effects of DDT.

These results suggest that the DBM resistance was associated with MFO, malathion carboxylesterase (MCE) and kdr (knockdown resistance).

The *in vivo* studies with [^{14}C]permethrin and [^{14}C]fenvalerate indicated:

1. that penetration of permethrin into the R strain was not one of the resistance mechanisms, because 50% of [^{14}C]permethrin penetrated into the body of both R and S strains within one hour after application;
2. that PB obviously inhibited metabolisms of [^{14}C]permethrin and fenvalerate in the R strain, but TBPT and TPP did not. This suggests that MFO was one of the mechanisms of pyrethroid resistance (Zhou *et al.*, in press).

The studies of esterases showed:

1. that in esterase activity and V_{max} there were no significant differences between R and S strains;
2. that inhibition by paraoxon was very strong in both R and S larvae, but very weak by TBPT and TPP. I_{50} determined for these three inhibitors were similar in the R and S strains;
3. that K_m with α-NA (naphthyl acetate) in the R strain was 1.75 times as high as that in the S strain but there was no clear difference with α-NA alone. The phosphatase activity was very low and the difference between the two strains was not significant.

These results suggest that resistance in the DBM to malathion may be related only to a highly specific MCE, and not to general non-specific esterases. This is consistent with the synergism results and [^{14}C]permethrin and fenvalerate metabolism (Tang and Zhou, in press).

Glutathione *S*-transferase (GST) in both R and S strains of the DBM was also investigated by *in vitro* studies. The R strain was found to contain about two times more GST activity towards both DCNB (1,2-dichloro-4-nitrobenzene) and CDNB (1-chloro-2,4-dinitrobenzene) than the S strain. The GST from both strains had similar K_m values in the S strain. The GST from R and S strains had an optimum pH 8.7 for CDNB. These results appear to show a quantitative rather than qualitative difference between the R and S strains and thus indicate an important mechanism of resistance in the DBM (Tang and Zhou, in press).

Acetylcholinesterase (AChE) from R and S larvae was characterized and examined to determine whether insensitive AChE (AChE-R) contributed to insecticide resistance. The relative activity of cholinesterase (ChE) of R and S strains toward acetylthiocholine (ATCh), propynylthiocholine (PrTCh) and butyrylthiocholine (BuTCh) was 0.3-, 0.7- and 6.8-fold, respectively. The kinetic parameters, V_{max} and K_m, were also determined with the above three substrates. V_{max} ratio with S strain was 0.4, 1.4 and 7.0, K_m ratio was 21.9, 87.0 and 37.8, respectively. A qualitative difference in AChE was shown between the R and S strains. In order to confirm insensitive AChE to inhibition, I_{50} ratio and bimolecular rate constant, K_i, were measured in R and S strains. The I_{50} ratios of R and S were 333.3 for eserine, 66.7 for carbaryl, 17.2 for paraoxon, 12.5 for DDVP and 10.0 for propoxur and the K_i ratios of insensitive AChE to paraoxon, DDVP, propoxur and carbaryl were 126.4, 86.5, 32.9 and 12.3, respectively. It is concluded that resistance to OPIs and carbamates R strain larvae is associated with a mutant form of AChE of broad insensitivity (Tang and Zhou, in press).

Based on the above *in vivo* and *in vitro* results, the mechanisms responsible for resistance to DDT, OPIs, carbamates and pyrethroids in DBM involve several factors: AChE insensitivity, higher MFO, GST and MCE activity and, probably, nerve insensitivity (kdr).

Since resistance in other vegetable pests is not as serious as in DBM, their mechanisms have not yet been researched in detail and are summarized in Table 2.

Table 2   Mechanism of resistance in vegetable pests in China

Species	Resistance to insecticide	Mechanism
Aphis gossypii	Pyrethroids, OPIs and carbamates	AChE-R, Est and MFO
Lipaphis erysimi	Pyrethroids and OPIs	Est and MFO
Myzus persicae	Pyrethroids and OPIs	Est MFO and AChE-R
Cavariella salicicola	Pyrethroids and OPIs	Est and MFO
Plutella xylostella	Pyrethroids, OPIs, carbamates and DDT	AChE-R, MFO GST, kdr and MCE

OPIs, organophosphorus insecticides; AChE-R, insensitive acetylcholinesterase; Est, esterase; MFO, mixed function oxidase; GST, glutathione *S*-transferase; kdr, knockdown resistance; MCE, malathion carboxylesterase.

Countermeasures to Resistance

The number of species of insects and mites in which resistance has been reported is over 500 species (Brattsten, 1989), of which at least sixty-six are pyrethroid-resistant. In addition to pyrethroids, induction of resistance has been demonstrated to the juvenile hormone, to the chitin synthesis inhibitor, diflubenzuron, in houseflies (Pimprikar and Georghiou, 1979), to chlorfluazuron and teflubenzuron in the DBM (Perng *et al.*, 1988), to *Bacillus thuringiensis* (Bt) in the Indian meal moth and the almond moth (McGaughey, 1985; McGaughey and Beeman, 1988) and in the DBM (Tabashnik *et al.*, 1990). It should be noted that a permethrin-resistant strain of the housefly is also resistant to avermectin isolated from the actinomyces *Streptomyces avermitilis* which holds a great deal of promise for the control of many key pests (Scott, 1989).

From the above information, it is obvious that the risk of resistance will continue to exist regardless of the chemical nature of the pesticide or even biological agents used. Therefore, approaches to the problem including delaying or prevention of resistance development are urgently needed. During the last few years, much has been said and written about theoretical aspects of management (Georghiou 1983; NRC, 1986, Tang *et al.*, 1988). But practical approaches to managing resistance have received very scant attention, with the exceptional case of *Heliothis armigera* in Australia (Forrester and Cahill, 1987). The following measures have been considered as possible methods for delaying resistance in vegetable pests in China:

1. choosing potential replacement insecticides and, at the same time, improving the use of methods;
2. limiting a potential insecticide to areas/seasons where/when application threshold has been reached, with dosages high enough to ensure killing resistant heterozygotes;
3. using synergists;
4. using a mixture of insecticides, in rotation, in a mosaic rotation.

Alternation of insecticides is the most commonly used strategy to combat resistance. It is most important that the alternative insecticide must have no cross-resistance to previous insecticides used. For example, as described above, the DBM has developed resistance to conventional insecticides. The resistance mechanisms of this pest involved MFO, GST, MCE, AChE-R as well as kdr. Fortunately, no cross-resistance to dimethypo [disodium S,S'-2-dimethylaminotrime-thylene bis(thiosulphate)], insect growth regulators (IGRs) such as chlorfluazuron and teflubenzuron or Br was found in the resistant DBM. Thus, these insecticides have been recommended to control the resistant DBM in Shanghai. It must be mentioned, however, that DBM has already developed resistance to IGRs in Thailand (Perng *et al.*, 1988) and to Bt in Hawaii (Tabashnik *et al.*, 1990). Because of this we recommended the use of IGRs in Shanghai only once during the peak population period of DBM. At other times, Bt and dimethypo are used in a mixture, in rotation and in mosaic control.

For the insects with population mobility, mosaic control may be used with two or several unrelated insecticides. In this case migrants that have not been killed in their sectors of origin by insecticide A will be killed upon exposure to insecticide B used in the adjoining sectors (Georghiou, 1983; Byford *et al.*, 1987). The effectiveness of such a strategy will depend on a reasonably high rate of exchange of migrants between sectors. This strategy, however, is not suitable for pests such as apterous adult aphids, reproduced by means of parthenogenesis and with little mobility. Because this situation usually produces a uniform resistance with no dilution of R-gene(s), an attempt was made to expand the concept of mosaic control by first spraying a 'mosaic' with unrelated insecticides A and B, then subsequently interchanging treatments in different sectors to counteract migration between adjacent sectors. This strategy is known as mosaic rotation.

In Meiton, Shanghai County, where vegetables have been grown for at least forty years, dimethoate is alternated with fenvalerate every season in small trial areas. Since 1983 control has improved, and the aphids are more susceptible to these insecticides than those from Jinwei, Jinshan County, Shanghai, where vegetables have been grown for only fifteen years. The aphids are resistant to dimethoate ($\times 6$) and fenvalerate ($\times 12$) (Tang *et al.*, 1988).

Estimates of esterase activity in single aphids, which provide a rapid and accurate way of detecting resistance and estimating the proportion of OP-resistant aphids in the population, have shown that aphids with low (OD < 0.3) or moderate (OD 0.3–0.5) activity were more frequent in Meiton (30% and 68%) than in Jingwei (8% and 85%). The aphids with most resistance (OD > 0.6) were more common in Jingwei (7%) than in Meiton (2%). The results obtained from bioassays and estimates of esterase activity in single aphids show that mosaic rotation offers prospects for delaying insecticide resistance in *Lipaphis erysimi*, as the continuous use of dimethoate followed by the continuous use of fenvalerate has caused serious resistance problems (Tang *et al.*, 1988).

Conclusion

By the end of 1990, there were at least eight species of vegetable pests reported as resistant to insecticides in China. Among them resistance in the DBM to insecticides is particularly serious. This pest has already developed resistance to all classes of insecticides available such as DDT, OPIs, carbamates and pyrethroids. *In vivo* and *in vitro* studies, and metabolism of [^{14}C]permethrin and fenvalerate indicated that mechanisms of resistance in the DBM were associated with MFO, GST, MCE, AChE-R as well as possibly kdr.

Designing a scientifically valid, economically acceptable resistance management programme for a particular pest often appears to be a complex task, but present strategies have demonstrated that resistance is no longer a 'one-way street'. It is evident that insecticide resistance is neither inevitable nor invincible.

References

BRATTSTEN, L.B. (1989). Insecticide resistance, research and management. *Pesticide Science*, **26**, 329–332.

BUSVINE, J.R. (1980). *Recommended methods for measurement of pest resistance to insecticides*. FAO plant production and protection paper **21**. Rome.

BYFORD, R.L., LOCKWOOD, J.A. and SPARKS, T.C. (1987). A novel resistance management strategy for horn flies (Diptera: Muscidae). *Journal of Economic Entomology*, **80**, 291–369.

FORRESTER, N.W. and CAHILL, M. (1987). Management of insecticide resistance in *Heliothis armigera* (Hubner) in Australia. In *Combating Resistance in Xenobiotics: Biological and Chemical Approaches* (M. Ford *et al.*, eds), pp. 127–137. Ellis Horwood, Chichester, England.

GEORGHIOU, G.P. (1983). Management of resistance in arthropods. In *Pest Resistance to Pesticides* (G.P. Georghiou and T. Saito, eds), pp. 769–792. Plenum Press, New York.

McGAUGHEY, W.H. (1985). Insect resistance to the biological insecticide *Bacillus thuringiensis*. *Science*, **229**, 193–195.

McGAUGHEY, W.H. and BEEMAN, R.W. (1988). Resistance to *Bacillus thuringiensis* in colonies of Indian meal moth and almond moth (Lepidoptera: Pyralidae). *Journal of Economic Entomology*, **81** 28–33.

MU, L.Y., WANG, X.Y., LUO, W.C. and ZHAO, Y. (1984). A study of resistance to cabbage worm (*Pieris rapae*) to insecticides. *Acta Phytophylacica Sinica*, **11**, 267–273. (In Chinese with English abstract.)

NATIONAL RESEARCH COUNCIL (1986). *Pesticide Resistance: Strategies and Tactics for Management*. National Academy Press,

Washington DC.

PERNG, F.S., YAO, M.X., HUNG, C.F. and SUN, C.N. (1988). Teflubenzuron resistance in diamondback moth (Lepidoptera: Plutellidae). *Journal of Economic Entomology*, **81**, 1277–1282.

PIMPRIKAR, G.D. and GEORGHIOU, G.P. (1979). Mechanisms of resistance to diflubenzuron in the housefly, *Musca domestica* (L.). *Pesticide Biochemistry and Physiology*, **12**, 10–22.

SCOTT, J.G. (1989). Cross-resistance to the biological insecticide abamectin in pyrethroid-resistant houseflies. *Pesticide Biochemistry and Physiology*, **34**, 27–31.

TABASHNIK, B.E., CUSHING, N.L., FINSON, N. and JOHNSON, M.W. (1990). Field development of resistance to *Bacillus thuringiensis* in diamondback moth (Lepidoptera: Plutellidae). *Journal of Economic Entomology*, **83**, 1671–1676.

TANG, Z.H. and ZHOU, C.L. (in press,a). Acetylcholinesterase and its insensitivity to inhibition in resistant *Plutella xylostella*. *Acta Entomologica Sinica*. (In Chinese with English abstract.)

TANG, Z.H. and ZHOU, C.L. (in press,b). The role of detoxication esterases in insecticide resistance of diamondback moth, *Plutella xylostella*. *Acta Entomologica Sinica*. (In Chinese with English abstract.)

TANG, Z.H. and ZHOU, C.L. (in press,c). Glutathione *S*-transferases in resistant and susceptible strains of diamondback moth, *Plutella xylostella*. *Acta Entomologica Sinica*. (In Chinese with English abstract.)

TANG, Z.H., GONG, K.Y. and YOU, Z.P. (1988). Present status and countermeasures of insecticide resistance in agricultural pests in China. *Pesticide Science*, **23**, 189–198.

WU, S.C. and GU, X.Z. (1986). Toxicity of the diamondback moth to pyrethroids. *Plant Protection*, **12**, 19–20. (In Chinese.)

ZHEN, B.Z. and RUI, C.H. (1989). Insecticide resistance and mechanism of *Trialeurodes vaporariorum* (West) inBeijing. In *A Compilation of Abstracts for Second Symposium on Pesticide Toxicology, Nanjing*, p. 31, 7–10 November.

ZHEN, B.Z., GAO, L.W., WANG, Z.G., LIANG, T.T., GAO, B.J. and GAO, H. (1988). Resistance mechanism of organophosphorus and carbamate insecticides in *Aphis gossypii. Glov. Acta Phytophylacica Sinica*, **16**, 131–138. (In Chinese with English abstract.)

ZHOU, C.L., TANG, Z.H. and ZHANG, L.M. (in press). Resistance of diamondback moth to synthetic pyrethroids and its relationship with microsomal mixed function oxidase. *Acta Phytophylacica Sinica*. (In Chinese with English abstract.)

37
Cotton Pest Resistance to Pesticides and its Management in the USSR

G.T. SUKHOROCHENKO

All-Union Institute of Plant Protection, 188 620 Leningrad, Pushkin-6. Shosse Poldbelskogo, 3, USSR

Introduction

Cotton pests have been subjected to intense chemical pressure in all cotton-growing regions of the world. As a result, resistance has become very widespread.

By the late 1980s populations of at least thirty-nine cotton pest species from all over the world had been registered to be resistant. They represent about 15% of the total number of resistant phytophages. Among them 49% are sucking pests: mites (18%); phytophagous bugs (18%); and leafhoppers and aphids (13%). Of the remaining 51%, 39% belong to Lepidoptera and 12% to others. Resistance has been registered in all cotton regions and to all groups of pesticides applied.

In the USSR, as in other countries, cotton is one of the most intensively cultivated crops. Therefore, first cases of resistance in this crop were registered in the early 1960s. Presently, resistance occurs in populations of the four major pests of cotton: two-spotted spider mite (*Tetranychus urticae* Koch), cotton bollworm (*Heliothis armigera* Hbn.), cotton aphid (*Aphis gossypii* Glov.) and large cotton aphid (*Acyrthosyphon gossypii* Mordv.). Resistance development in these arthropods is discussed below.

Insecticides: Mechanism of Action and Resistance
© 1992   Intercept Ltd, P.O. Box 716, Andover, Hants SP10 1YG, UK

Spider Mites

The two-spotted spider mite is one of the most harmful cotton pests in the Soviet Union. Its economic significance strongly increased during the late 1950s when the area colonized by this pest increased to 65% of the total area occupied by cotton from only 30% in the 1930s (Uspenkij, 1960). This coincided with world-wide mass reproduction of spider mites on many crops as a consequence of activities instrumental in increasing their efficiency such as transition to monocultures, intense application of fertilizers, which improved the quality of the mites' food and the use of universal pesticides killing predators and stimulating mite development.

In the 1950s spider mite control in the Soviet Union was widely performed with organophosphates (OPs), like Systox (merkaptofos), Schradan (shradan) and Intration (thiometon). Later Metasystox (methylmerkaptofos) and Rogor (dimethoate) were used instead of these compounds because they were very harmful to warm-blooded animals and man. In 1961 the first stage of spider mite resistance to Metasystox was noted in the Yangiyul region of Uzbekistan (Ivanova and Kornilov, 1964) and in 1965 a 350-fold resistance level was reached in the Fergana Valley. It was in this period when hard losses in the efficiency of OPs to control the above-mentioned pest species were first reported (Smirnova, 1968).

The situation became very serious in southern Tadzhikistan where the number of Metasystox treatments to control spider mites reached six to eight per year and raw cotton losses caused by this pest increased by up to 25–50%. Resistance levels observed were 485-fold to Metasystox and 540-fold to Rogor. High levels of cross-resistance were found to Kilval (vamidothion) and phenkapton although they had never been applied in that zone (Smirnova and Kornilov, 1968).

In 1968 mapping of pest resistance to Metasystox in cotton-growing regions revealed its emergence in areas with intense application of OPs, namely the Fergana region in Uzbekistan and Kulyab region of Tadzihikistan. Although rotation of specific acaricides (Acrex (dinobuton), Kelthane (dicofol), milbex, Neoron (brompropylate), Tedion (tetradifon)) has decreased mite resistance in these regions since 1971, resistance has generally increased and expanded (Table 1).

In the 1970s susceptibility of spider mites to the acaricides applied in rotation revealed that a high level of OP resistance was preserved (Table 2). This can be accounted for by the use of OPs to control other pests (aphids, cotton bollworm). Nevertheless rotations considerably delayed the development of resistance to other compounds.

Table 1 The degree of spider mite resistance to Metasystox in different cotton-growing zones of the Soviet Union (1968–71)

Republic, zone	District	Level of resistance	
		1968	1971
Azerbaidzhan	Saatlin	1.5	2.5
Mugan	Sabirabad	1.7	3.7
	Pushkin	0.3	8.8
Shirvan	Salyan	7.2	3.0
Kirgiziya	Oshsk	0.3	0.5
Turkmenistan			
Maryisk	Murgab	–	3.0
Tashauz	Tashauz	–	7.0
Uzbekistan			
Fergana	Bagdad	325.0	120.0
	Papsk	325.0	78.0
Namangan	Namangan	16.6	–
Tashkent	Yangiyul	5.7	–
Surkhandaryin	Denaus	2.5	36.0
	Sariasiisk-Uzun	6.5	42.0
Tadzhikistan			
Vakhsh	Kolkhozobad	0.7	50.0
	Kurgantyubinsk	3.8– 6.0	20.0
	Kuibyshev	0.9– 6.4	26.0
	Pyanzha	6.4– 9.0	33.0
Gissar	Lenin	11.8	200.0
	Ordzhonikidzeabad	0.9	9.0
	Tursun-zade	66.4	70.0
	Yavan	0.9	0.9
Kulyab	Vosey	24.0– 38.0	75.0
	Kulyab	36.00–189.0	86.0
	Moskva	37.0– 67.00	3.0–21.0
	Parkhar	97.0–485.00	0.0–11.0
Leninabad	Ashta	0.7	7.0
	Naus	0.9	8.0
	Insfarin	0.9	4.0
	Khozhent	281.0	36.0

Table 2 The susceptibility of Parkhar spider mite population to acaricides and insecticides applied in rotation

Acaricide	LC$_{50}$ (% a$_i$)		Level of resistance	
	before treatment	at the end of the season	before treatment	at the end of the season
1978				
Roger	0.135 (0.078–0.191)	0.05 (0.048–0.052)	675.0	250.0
Phosalone	0.082 (0.042–0.122)	0.063 (0.050–0.076)	51.0	39.0
Dursban	0.0041 (0.0033–0.005)	0.0015 (0.0014–0.0017)	4.1	1.5
Kelthane	0.022 (0.015–0.028)	0.005 (0.0034–0.060)	33.0	5.0
Acrex	0.0013 (0.011–0.0015)	0.009 (0.008–0.01)	4.3	3.0
Omite	0.0056 (0.0036–0.007)	0.0048 (0.004–0.0056)	4.6	4.0
Plictran	0.0012 (0.008–0.0016)	0.001 (0.0008–0.0012)	2.0	1.7
1979				
Roger	0.150 (0.140–0.160)	0.090 (0.070–0.011)	750.0	450.0
Phosalone	0.027 (0.0214–0.0326)	0.026 (0.0162–0.0368)	17.0	16.2
Dursban	0.006 (0.005–0.0071)	0.005 (0.004–0.006)	6.0	5.0
Kelthane	0.120 (0.01–0.014)	0.0016 (0.0013–0.0019)	12.0	1.6
Acrex	0.0005 (0.00048–0.0006)	0.0002 (0.00016–0.00024)	1.8	0.6
Omite	0.0009 (0.0008–0.001)	0.0009 (0.0006–0.00012)	0.8	0.8
Plictran	0.0013 (0.0004–0.0022)	0.0052 (0.0042–0.01)	2.1	4.1

Cotton Aphids

Aphids had been successfully controlled by organophosphorus insectoacaricides during the late 1950s. However, during the late 1960s the number of treatments to control spider mites increased substantially and resulted in the development of OP resistance in cotton aphid populations. Between 1971 and 1973 resistance of cotton aphids and large cotton aphids in the Kulyab region of Tadzhikistan increased from 1.5-fold to 10.5-fold. Taking into account an expansion of the cotton-growing area and its colonization by aphids, the efficiency of insecticides decreased from 99% to 73.6–87.5% (Ivanova, 1975). In the mid-1960s aphids colonized 40 000–120 000 ha of cotton or about 20–30% of the cropped area. In the first half of the 1970s, 131 000 ha of the cropped areas were colonized by aphids and treatment was applied to 116 000–

125 000 ha. Banning organophosphorus compounds from spider mite control, the development of resistance to them was delayed in the southern zone of the republic's cotton region. Despite this resistance to phosalone, which was applied under the regional control scheme against either aphids or cotton bollworm, aphicides were applied in rotation (Table 3).

Table 3 Response of cotton aphids to aphicides applied in a rotation system in south Tadzhikistan between 1977 and 1981

Aphicide	Susceptible line	Population in 1977		Population in 1981	
	LC_{50} (%a$_i$)	LC_{50} (%a$_i$)	Level of resistance	LC_{50} (%a$_i$)	Level of resistance
Alfalfa aphid					
Rogor	0.001	0.0012 (0.0009–0.00015)	1.2	0.00035 (0.00027–0.00043)	0.35
Phosalone	0.0002	0.00024 (0.00018–0.0003)	1.2	0.00017 (0.00013–0.00021)	0.85
Croneton	0.0011	–	–	–	–
Cotton aphid					
Rogor	0.0002	0.00012 (0.00009–0.00015)	0.6	0.0001 (0.00008–0.00012)	0.50
Phosalone	0.00005	0.0001 (0.00008–0.00011)	2.0	0.00017 (0.00013–0.00021)	3.4
Croneton	0.0006	–	–	0.00075 (0.00073–0.00077)	
Large cotton aphid					
Rogor	0.0001	0.00035 (0.00027–0.00043)	3.5	–	–
Phosalone	0.00002	0.00017 (0.00013–0.00021)	8.5	–	–

Cotton Bollworm

Emergence of bollworm resistance to DDT was first noted in the late 1960s in Azerbaidzhan and southern Tadzhikistan (Sukhorochenko, 1976; Ragimov, 1981). In the mid-1970s group resistance to organochlorine insecticides emerged but sensitivity to the DDT-replacements phosalone, Gardona (tetrachlorvinphos) and Sevin (carbaryl) was observed as well (Table 4). In the 1970s the development of resistance to phosalone and Sevin was reported from southern Tadzhikistan (Table 5). Until that time the development of multiple resistance to organochlorine, organophosphorus and carbamate compounds took place in pest populations in the regions with the most intensive cultivation.

Table 4 Responses of cotton bollworm populations from three cotton growing republics of the Soviet Union to insecticides (1975)

Insecticide	LD_{50} (mg g^{-1} a$_i$)		larvae 3rd instar		Level of resistance		
	Laboratory line	Azerbaidzhan	Tadzhikistan	Turkmenistan			
			populations				
		1	2	3	1	2	3
Sevin	73.3 + 7.6	120.0 + 60.0	97.6 + 3.1	73.6 + 10.0	1.6	1.3	1.0
Gardona	79.2 + 9.1	80.0 + 17.4	72.8 + 8.1	79.5 + 3.2	1.1	1.0	1.0
Phosalone	234.0 + 12.0	374.0 + 51.0	237.0 + 0.36	232.0 + 8.5	1.6	1.0	0.9
Dilor	235.0 + 23.0	1350.0 + 151.0	1453.0 + 90.0	151.0 + 15.0	5.8	6.2	0.6
Toxaphen	148.0 + 13.0	510.0 + 11.0	552.0 + 37.0	145.0 + 13.0	3.4	3.7	1.0
DDT	105.0 + 12.0	560.0 + 80.0	1176.0-70.0	113.0 + 13.0	5.3	11.0	1.1

Thus, by the end of the 1970s the situation in southern Tadzhikistan made it necessary to overcome resistance or to retard its development to certain compounds from different chemical groups in populations of the major pests. It was clear that pesticide pressure had caused resistance and that a resistance control system without loss in yield had to be worked out integrating a considerable decrease in pesticide application and avoidance of yield loss.

Management of Cotton Pest Resistance

According to technical probabilities and our conception of the problem the following tactical recommendations have been developed:

- Application of pesticides by economic thresholds, which provide reduction of treated areas and accordingly diminish the number of treated populations; resistance monitoring to exclude resisted products from control programmes as early as possible; selection of replacement products on the basis of cross-resistance studies; inclusion of compounds from different chemical groups in the rotation scheme due to the knowledge of cross-resistance patterns and resistance mechanisms.

- Reduction of pesticide pressure using alternative means, e.g. microbiological preparations.
- Adjustment of pesticide application according to their effect on the respective predominant harmful and beneficial arthropods and according to the population parameters of the key pests; preservation of beneficial entomophauna.

Table 5   Response of cotton bollworm populations from Tadzhikistan cotton-growing zones to insecticides (1977)

Insecticide	LC_{50} (a_i) for 1st instar larvae	Level of resistance
	Central zone **Tursun-zade district**	
DDT	0.0284 (0.0221–0.0265)	9.5
Sevin	0.0041 (0.0023–0.0071)	1.6
Phosalone	0.0204 (0.0165–0.0251)	5.5
Thiodan	0.0142 (0.0098–0.0204)	2.8
	Southern zone **Parkhar district**	
DDT	0.2020 (0.2670–0.3180)	97.3
Sevin	0.0260 (0.0230–0.03180)	10.0
Phosalone	0.0150 (0.0130–0.0170)	4.0
Thiodan	0.0050 (0.0040–0.0060)	–
Ambush	0.00021 (0.00017–0.00025)	–
	Kulyab district	
DDT	0.0905 (0.0666–0.1240)	30.2
Sevin	0.0159 (0.0103–0.0245)	6.1
Phosalone	0.0189 (0.0167–0.0214)	5.1
Thiodan	0.0123 (0.0089–0.0171)	2.5
Ambush	0.00013 (0.00011–0.00015)	–

A control system was developed in which all treatments were carried out strictly according to economic thresholds (Table 6). Usually control measures start with aphid control at the stage of one or two leaf pairs followed by cotton bollworm control. DDT preparations were excluded and the use of Bi 58 (dimethoate) was limited to reduce the danger of resistance to compounds of other chemical groups.

Table 6   Pesticide rotation for zones with spider mite, cotton aphid and cotton bollworm resistance to preparations of different chemical groups.

Crop development stage	Object of control	Treatment
Seedling	Aphid complex (alfalfa aphid, cotton aphid)	High resistance ($>$ 50-fold) – Ethaphos; moderate resistance ($<$ 50-fold) – phosalone
Vegetation	Spider mite emergence,	Ringe or local treatment with Acrex or Omite
	Repeated treatment of aphids necessary	High resistance – spraying with pyrethroids; moderate resistance – Croneton or Pirimor
	Further control of spider mite necessary	Treatment of separate fields with halogen organic acaricides (Kelthane, Mithran, Neoron)
Early budding	Cotton bollworm 1st generation	Two treatments of separate fields with BTB (last one to field populated with spider mites, too)
	Spider mite	Acrex (Isophen)
Mass budding, ovary formation	Cotton bollworm 2nd generation	1st treatment – Thiodan, 2nd – pyrethroids (Ambush, Anomethrin-N, Cymbush, Ripcord, Decis, Sumicidin)
	Spider mite control necessary	Plictran, Omite, Tedion or sulphur preparations
Pod formation	Cotton bollworm 3rd generation, control necessary	Sevin

In aphid control a rotation system was established containing organophosphorus and carbamate compounds whereas against spider mites specific acaricides were used.

To prevent cotton bollworm resistance, microbiopreparations or combinations of phosalone and Gardona were applied against the first generation, Thiodan (endosulfan) and pyrethroids against the second generation and Sevin against the third generation. The possible development of resistance to pyrethroids and Thiodan and phosalone was also considered. Therefore, field populations of the cotton bollworm treated with microbiopreparations during the first generation were treated with pyrethroids during the second generation. Phosalone or Gardona against the first generation rotated with Thiodan against the second generation. If necessary, Sevin was applied against the third generation. These generations can result in cross-resistance to OPs; however, they are separated by Thiodan during the season.

In all seasonal programmes the demand for pesticides to be as safe as possible for the main beneficial zoophages in cotton fields was strictly taken into account.

Pesticides were divided into the following toxicity classes according to their influence on beneficial arthropods (Sukhorochenko and Nedirov, 1985):

1. Slightly toxic pesticides causing 20% decrease in population density ten days after treatment under field conditions.
2. Moderately toxic pesticides causing 21–50% decrease in population density ten days after treatment.
3. Highly toxic pesticides causing more than 50% decrease in population density twenty days after treatment.

Pesticides selected for rotation were classified according to this scale (Table 7).

The capability of broad-spectrum insecticides to stimulate the development of populations of sucking phytophages, especially spider mites, was considered to determine their optimum place in the seasonal rotations scheme. Pesticides slightly or moderately toxic to zoophages and not causing an increase in sucking pest populations such as microbiopreparations, phosalone, Dilor, and Thiodan may be used in the first half of the vegetation cycle up the middle of July. From the middle of July to the end of August more toxic pesticides may be used (pyrethroids, Sevin). Although they may result in an increase in sucking pest populations, this is less dangerous to the crop than during the first half of the season. Specific acaricides can be used safely during the whole vegetation period.

Table 7 Classes of pesticide hazard to beneficial arthropods

Hazard class

Pesticide	Adonia variegata	Orius niger	Nabis palifer	Deraeocoris punctulatus	Aeolothrips intermedius	Chrisopa carnea	Araneina spp.	Trichogramma evanescens	Zoophages
Roger	3	2	3	3	3	3	1	3	2.6
Phosalone	1	2	2	2	3	2	1	0	1.8
Gardona	3	1	1	2	3	0	1	0	1.4
Dilor	1	1	1	1	1	0	1	0	0.8
Thiodan	3	0	3	2	3	1	2	1	1.8
Kelthane	2	0	0	0	0	0	0	1	0.6
Milbex	1	0	2	0	0	2	1	1	0.9
Sevin	3	3	3	3	3	2	3	3	2.9
Croneton	0	1	1	2	1	1	0	2	1.0
Pirimor	1	2	1	2	2	2	2	2	1.8
Ambush	3	2	3	2	3	3	3	2	2.6
Cymbush	3	2	3	1	3	2	3	3	2.5
Decis	3	2	3	2	3	3	3	3	2.8
Sumicidin	3	2	3	2	3	1	3	3	2.5
Biopreparation	3	1	2	0	0	2	2	0	1.3

The position of insecticides in the rotation system against cotton bollworm has been corrected regarding their efficacy in inhibiting pest populations. Some population indices of survival at pre-imaginal stages substantially decreased against a background of Thiodan and pyrethroids compared to Gardona and phosalone both in the treated generation and in the following generation. Application had to be performed during the period of maximum infestation of the cotton plants (Table 8). This period coincides with the development of the second generation of the pest resulting in a reduction of the number of insecticide treatments in the third generation.

In southern Tadzhikistan a six-year approbation of control schemes based on the above-mentioned approaches showed a high biological effect of all included products resulting in a 1.5- to 2-fold reduction of the total volume of the pesticides used. The total number of treatments did not exceed five per season and was reduced to three or four treatments in years when the number of mites or bollworms was low. The control of highly resistant species was favoured and the development of multiple resistant forms excluded.

Table 8 Demographic characteristics of populations of *Heliothis armigera* Hbn. (treated and untreated)

Pesticide	Generation	Population parameter		
		R_o	T	r_m
Thiodan	2	9.32	44.5	0.0503
	3*	0	0	0
Untreated	2	37.30	33.5	0.1090
	3*	51.58	41.4	0.0950
Phosalone	1	27.23	41.3	0.0818
	2*	2.24	43.5	0.0190
Untreated	1	99.78	41.7	0.1111
	2*	2.67	46.4	0.0250
Gardona	1	95.94	39.4	0.1168
	2*	24.98	34.5	0.0935
Untreated	1	101.40	36.3	0.1261
	2*	102.71	34.2	0.1326

*Daughter generation.

As a result bollworm resistance to DDT reversed and susceptibility to pyrethroids increased. Spider mite resistance to Rogor became stable on the tolerance level and a slow decrease in mite susceptibility to specific acaricides was in progress, whereas aphid resistance to OPs was inhibited (Table 9).

The reduction of treatments also improved living conditions of predators and other zoophagous arthropods. Therefore, increase in full biological effect of treatments was furthered. The average annual economic effect of the programme runs at 60 roubles per ha (Sukhorochenko and Smirnova, 1985).

Since 1985 the developed scheme of pest control has been used by cotton-growing farms of southern Tadzhikistan on an area of about 20 000–25 000 ha. Considerable changes have occurred during the last years. The use of numerous products, e.g. Kelthane, Acrex,Tedion and Sevin, is now prohibited. New preparations have been integrated into the control scheme, in particular Danitol, a pyrethroid with acaricide effect, and Nurelle D, an insectoacaricide. These products are effective not only in controlling cotton bollworm but also against spider mites, aphids and bugs. The sumicidin isomer 'Sumi-Alpha' has been integrated into the system instead of Sumicidine.

Table 9 Reversion and retarding of resistance in cotton pest populations after rotations of pesticides

Pest	Pesticide	Resistance level	
		1977-1978	1986-1987
Tetranychus	Rogor	250.0	21.0
urticae Koch	Phosalone	39.0	20.0
	Kelthane	22.0	32.0
	Acrex	4.3	2.3
	Omite	4.0	0.7
	Plictran	1.7	14.5
Aphis	Rogor	0.5	5.7
gossypii	Phosalone	3.4	22.0
Glov.	Croneton	1.3	1.0
	Pirimor	1.0	1.0
Acyrthosyphon	Rogor	3.5	3.6
gossypii	Phosalone	8.5	8.5
Mordv.	Croneton	1.0	1.0
Heliothis	DDT	97.3	7.0
armigera	Sevin	29.9	0.7
Hbn.	Phosalone	4.0	5.4
	Gardona	1.0	1.2
	Thiodan	1.0	2.0
	Ambush	1.0	0.1
	Decis	1.0	0.1
	Cymbush	1.0	0.1
	Sumicidin	1.0	0.3

Examinations of pesticide assortments of different companies showed the integration of the acaricide Nissorun, the plant growth inhibitor chlorfluazuron and some others into the system in the future.

In our opinion the basic principles of the control system outlined provide opportunities for further improvements in plant protection.

References

GAPLEVSKAYA, L.N. (1970). Vliyanie primenyaemykh akaricidov na chislennost pautinnogo klesha v usloviyakh Tadzhikistana. In *Tezisy Dokladov. Vtoroe soveshchanie po resistentnosti vreditelej k khimichskim sredstvam zashchity rastenij* (in Russian), pp. 58–62. Leningrad.

IVANOVA, G.P. (1975). Ustoichivost khlopkovykh tlej k fosfororganicheskim insekticidam i preduprezhdenie ee razvitiya (in Russian). *Khimiya v selskom khozyaistvii*, **13**, 47–50.

IVANOVA, N.I. and KORNILOV, N.G. (1964). O privykanii pautinnogo klesha na khlopchatnike k merkaptofozy (in Russian). In *Trudy VNII zashchity rastenij*, **20**, 12–17.

RAGIMOV, Z.A. (1981). Integrirovannaya zashchita. *Zashchita rastenij*, **7**, 24–26.

SMIRNOVA, A.A (1968). Sostoyanie voprosa ob ustoichivosti pautinnogo klesha na khlopchatnike k fosfororganicheskim akaricidam (in Russian). In *Tezisy Dokladov. Soveshchanie po rezistentnosti kleshchej k akaricidam*, pp. 3–5. Leningrad.

SMIRNOVA, A.A and KORNILOV, V.G. (1972). O razvitii ustojchivosti i DDT khlopkovoi sovki (*Heliothis armigera*) v juzhnom Tadzhikistane. In *Kratkie Tezisy Dokladov Tretego Soveshchaniya po Resistentnosti Vreditelej k Khimicheskim Sredstvam Zashchity Rastenij*, pp. 87–89. Leningrad.

SUKHOROCHENKO, G.I. (1976). Ustoichivost khlopkovoi sovki k insekticidam. *Byulleten VIZR*, **37**, 34–37.

SUKHOROCHENKO, G.I. and NEDIROV, D.O. (1985). Vliyanie sovremennykh insekticidov na vrednuyu i poleznuyu entomofaunu khlopchatnika. *Byulleten VIZR*, **60**, 7–12.

SUKHOROCHENKO, G.I. and SMIRNOVA, A.A. (1985). Preduprezhdenie rezistentnosti u vreditelej khlopchatnika. *Zashchita rastenij*, **10**, 20–21.

USPENKIJ, F.M. (1960). *Obyknovenny pautinny klesh v oroshaemykh rayonakh sredney azii*. Tashkent.

38
Mechanisms of Permethrin Resistance of the Housefly (*Musca domestica*) in Japan

Y. TAKADA,[1] M. HIRANO[1] and T. HIROYOSHI[2]

[1]*Takarazuka Research Center, Sumitomo Chemical Cl., Ltd,
2-1, 4-Chome Takatsukasa, Takarazuka-Shi, Hyogo-Ken, 665*
[2]*Department of Genetics, Medical School, Osaka University, Japan*

Introduction

Ahn *et al.* (1986a,b) demonstrated that reduced nerve insensitivity was a major resistance mechanism to permethrin in the pyrethroid-resistant 228_{e2b} strain of the housefly in Denmark. On the other hand, Motoyama (1984) reported that an increased metabolism by microsomal cytochrome P-450-dependent mono-oxygenase system might be responsible for a large part of the pyrethroid resistance in Mashiko colony of the housefly. We evaluated the mechanisms of permethrin resistance in two other colonies collected in Japan, Miyakonojo colony and Akagi colony.

Material and Methods

Houseflies used

The houseflies *Musca domestica* L. used were as follows: Miyako-nojo colony was collected in Miyakonojo, Miyazaki Prefecture in 1983. Akagi colony was collected in Akagi, Gunma Prefecture in 1984. Akagi PP5 and Akagi PP15 were selected by permethrin for five generations and fifteen generations, respectively. The 228_{e2b} strain was originally established by Keiding in 1976 from Danish

---

Insecticides: Mechanism of Action and Resistance
© 1992   Intercept Ltd, P.O. Box 716, Andover, Hants SP10 1YG, UK

flies, and CSMA strain is a standard susceptible strain. Some visible marker strains were also used for linkage group analysis study.

Chemicals

The following chemicals were used: permethrin, cypermethrin, fenvalerate, fenitrothion, piperonyl butoxide and S,S,S-tributyl phosphorotrithioate (DEF).

Toxicity test

The appropriate dosages of test chemicals in 0.5 µl of acetone were topically applied to the thoracic dorsa of four-day-old female adult flies. In cases of synergist testing, 10 µg of the synergist was applied four hours prior to the application of insecticides.

Analysis of recessive resistant factor on the third chromosome

The susceptible $B_\chi{}^2$ strain, possessing one dominant marker on the third chromosome, was used in this study. Crossing method was based on that described by Takada *et al.* (1988).

Electrophysiological study

Permethrin (1 µl) was applied to the dorsal thorax of the female adult flies. The electric record was obtained from the femur of the leg as shown in Figure 1.

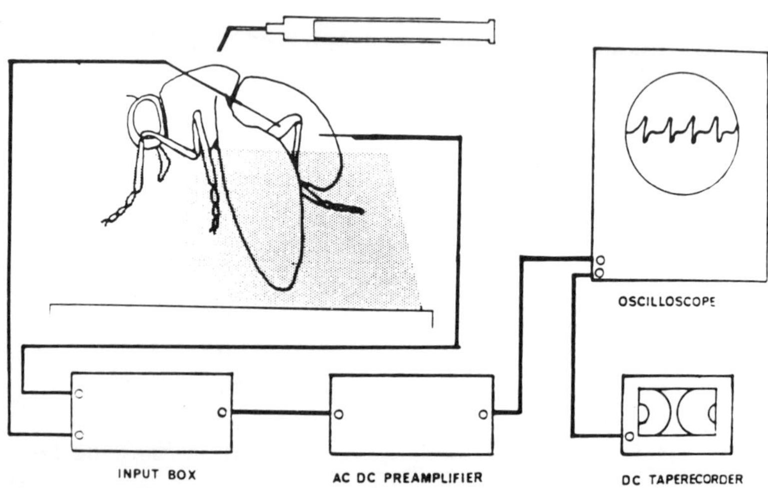

Figure 1   Schematic drawing of the recording methods.

Linkage map study on the third chromosome

This method was based on that described by Takada *et al.* (1990).

Results and Discussion

Miyakonojo colony showed the moderate resistance to pyrethroids tested and fenitrothion (Table 1). In order to know the mechanism of permethrin resistance in this colony, the effect of metabolism was investigated using two kinds of synergists, piperonyl butoxicide (PB) and DEF. The effect of DEF was small. PB, however, showed significant synergistic activity, and the LD_{50} value of permethrin to this colony with PB was almost equal to that of the CSMA strain (Table 2). Furthermore, no resistant factor was detected on the third chromosome where *pen* and *kdr* genes are located by the study using the B_χ^2 strain (Table 3). That is, the main resistance mechanism of Miyakonojo colony to permethrin seems to be metabolic by mixed function oxidase.

Table 1   Susceptibility of Akagi, Akagi PP15, Miyakonojo, 228_{e2b} and CSMA strains to permethrin, cypermethrin, fenvalerate and fenitrothion

	LD_{50} (μg per female fly)				
Insecticide	CSMA	Akagi	Akagi PP15	Miyakonojo	228_{e2b}
Permethrin	0.023	7.8	73	0.27	4.4
Cypermethrin	0.010	5.8	51	0.19	2.7
Fenvalerate	0.028	>10	>10	0.40	20
Fenitrothion	0.064	2.3	2.2	2.0	1.1

Table 2   The additional effect of piperonyl butoxide (PB) and *S,S,S*-tributyl phosphorotrithioate (DEF) against permethrin on Akagi, Akagi PP15, Miyakonojo, 228_{e2b} and CSMA strains

	LD_{50} (μg per female fly)				
	CSMA	Akagi	Akagi PP15	Miyakonojo	228_{e2b}
Permethrin only	0.023	7.8	73	0.27	4.4
with PB	0.0045	0.19	0.45	0.033	0.20
with DEF	0.0063	4.7	19	0.10	2.6

Table 3   Susceptibility of backcrossed progeny to insecticides

Insecticide	Genotype of III chromosome	LD_{50} (µg per female fly)			
		Akagi	Akagi PP15	Miyakonojo	228_{e2b}
Permethrin	$+/+$	7.3	37	0.14	2.5
	$+/B_\chi^2$	0.11	0.19	0.13	0.14
Fenvalerate	$+/+$	–	–	0.32	12
	$+/B_\chi^2$	–	–	0.33	0.59
Fenitrothion	$+/+$	3.0	1.3	1.6	0.51
	$+/B_\chi^2$	3.0	0.81	1.6	0.59

Akagi colony showed high resistance to pyrethroids testedand moderate resistance to fenitrothion. Akagi PP15 strain was almost ten times as resistant to permethrin and cypermethrin in comparison with the original Akagi colony (Table 1). Next, we evaluated the mechanisms of permethrin resistance in this colony. The effect of PB on Akagi and Akagi PP15 was recognized, but the effect of DEF on these two strains was small. That is, the metabolism caused by mixed function oxidase is one of the resistant mechanisms to permethrin. These LD_{50} values, however, were still larger than that of CSMA strain (Table 2), so a further mechanism must exist. In these strains, LD_{50} values to permethrin between $+/+$ and $+/B_\chi^2$ were different like the 228_{e2b} strain. This result shows that the resistance factor to permethrin is on the third chromosome (Table 3). No penetrating effect was recognized in Akagi PP15 strain by the study using ^{14}C- labelled permethrin (Takada *et al.*, in press). On the other hand, Akagi PP15 was resistant to permethrin not only for lethal activity but also for knockdown activity (Takada *et al.*, in press). Electrophysiological study was conducted *in vivo* to know the effect of permethrin on the nervous system of the housefly. Spontaneous activity of Akagi PP15 was decreased in comparison with that of the CSMA strain (Figure 2). This result shows that one of the resistant mechanisms to permethrin is the reduced sensitivity of the nervous system. Next, the locus for a recessive resistance gene to permethrin on the third chromosome was determined . The resistance gene was calculated to be located at the right side of 40.8 unit from the locus of visible marker pointed wing in Akagi PP5 strain and 39.8 unit in 228_{e2b} strain, respectively. Thus, the resistance genes in both strains seemed to be allelic with each other (Figure 3). Judging from their location on the third chromosome and from some physiological properties to insecticides, they might also be allelic to *kdr* gene already known.

Permethrin 1 μg

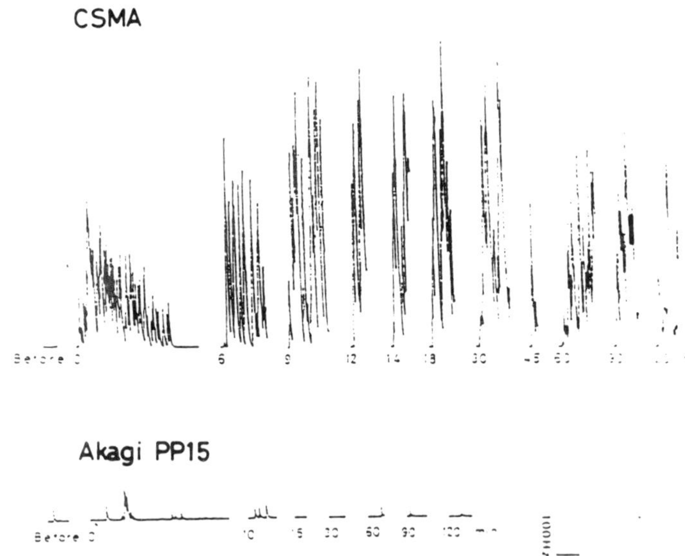

Figure 2 Changes of firing frequency.

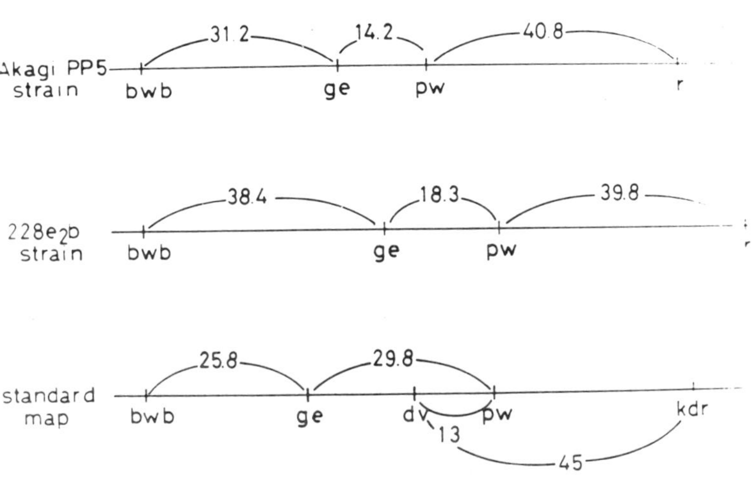

Figure 3 Linkage maps of the third chromosomes.

In conclusion, the resistance mechanisms to permethrin in Akagi colony was the metabolism by mixed function oxidase and low sensitivity of the target site. On the other hand, the main resistance mechanism in Miyakonojo colony is the metabolism by mixed function oxidase.

References

AHN, Y.J. *et al.* (1986a). *Pesticide Biochemistry and Physiology*, **26**, 231.
AHN, Y.J. *et al.* (1986b). *Journal of Pesticide Science*, **11**, 591.
KEIDING, J. (1976). *Pesticide Science*, **7**, 283.
MOTOYAMA, N. (1984). *Journal of Pesticide Science*, **9**, 523.
TAKADA, *et al.* (1988). *Applied Entomological Zoology*, **23**, 144.
TAKADA, *et al.* (1990). *Applied Entomological Zoology*.
TAKADA, *et al.*, in press.

39

World Distribution of Esterases Involved in Organophosphate Insecticide Resistance in *Culex* Mosquitoes, and Methods for Detection

G.P. GEORGHIOU

Department of Entomology, University of California, Riverside, CA 92521, USA

Several esterases are implicated in organophosphate (OP) insecticide resistance in *Culex* mosquitoes. These esterases are designated as type A or B depending on their preferential hydrolysis of α- or β-naphthyl acetate in the presence of both substrates at equal concentrations, and are numbered consecutively based on the order in which they were described. At least eight esterases have been recognized by starch gel electrophoresis in *Culex pipiens* or *C. quinquefasciatus* (A_1, A_2, A_4, B_1, B_2, B_4) and *C. tarsalis* (A_3, B_3). Each esterase confers a broad spectrum of OP resistance at varying levels of intensity. In general, highest resistance is conferred by esterase B_1 followed in decreasing order by the combinations of A_3B_3; A_2B_2, A_1, and A_4B_4. In at least the case of B_1, B_2, B_3 and B_4 the genes encoding these enzymes are highly amplified. Esterase B_1 was shown to be overproduced at levels corresonding roughly to the degree of resistance observed in larvae toward chlorpyrifos.

We have investigated the world distribution of these esterases through field collections made personally or received from collaborators, and through reports in the available literature. Our collections were examined electrophoretically for the esterase(s)

Insecticides: Mechanism of Action and Resistance
© 1992   Intercept Ltd, P.O. Box 716, Andover, Hants SP10 1YG, UK

present and were bioassayed as fourth-instar larvae with chlorphyrifos, temephos, fenthion and malathion, using either a single diagnostic concentration or a full range of dosages, depending on numbers available. The data obtained to date (to be published in detail elsewhere) show that

1. esterase B_1 is the principal esterase in North America (USA, Mexico, Guatemala, Caribbean islands) and is also found in Hawaii, South Korea and China;
2. esterase A_1 is present in southern Europe (southern France, Italy, Greece, Cyprus);
3. esterases A_2B_2 are the principal esterases in Africa and in Asia (Pakistan, Sri Lanka, Thailand, China, South Korea, Japan) and also occur in Greece and Cyprus. They have also been detected in the vicinity of major U.S. ports and in Marseille, France, suggesting recent introductions;
4. esterases A_4B_4 have been detected at low frequencies in collections from the Mediterranean area (France, Greece, Cyprus) and in a single collection from Tennessee, USA;
5. Finally, esterases A_3B_3 are known only in *C. tarsalis* in the USA and appear to be the principal esterases responsible for OP resistance in this species.

40
Selection of the Red Flour Beetle (*Tribolium castaneum* Herbst) for Resistance to Phosphine in the Laboratory and Biological Observations on the Resistant Strain

F.A. EL-LAKWAH, S.M. AHMED, M.M. KHATTAB and A.M. ABDEL-LATIEF

Faculty of Agriculture at Moshtohor, Plant Protection Department, Tukh, Kalyubia, Egypt

Introduction

Phosphine can be considered as one of the most effective fumigants in recent years to control insects in grains, flour, agricultural products and animal foods.

While the resistance of stored grain insects to contact insecticides has appeared in many parts of the world (Dyte and Blackman, 1970), resistance to fumigants has not become a practical problem. Fumigation with phosphine to control stored product insects infesting a variety of commodities has gained momentum over the past thirty years and there have been numerous published accounts of success or failure since Heseltine and Thompson (1957) and Lindgren *et al.* (1958) recommended it use.

Phosphine was most effective at higher temperatures, while long exposures at low concentrations were more effective than short exposures at high concentrations (Barbara *et al.*, 1976).

Howe and Hole (1967) emphasized the need for complete control

Insecticides: Mechanism of Action and Resistance
© 1992   Intercept Ltd, P.O. Box 716, Andover, Hants SP10 1YG, UK

in practical fumigations and the desirability of finding the reasons for failure.

One of the major problems that has developed in many insect control programmes in recent years concerns the resistance that is acquired when successive generations are exposed to toxic agents. Survivors from progressive selections can develop characteristics that will make the toxicant ineffective or uneconomical to use. Resistance to the fumigants methyl bromide, phosphine (Champ and Dyte, 1976) and ethylene dibromide (Bond, 1973) has been found in field populations of stored products insects in recent years.

Halliday *et al.* (1983) demonstrated in laboratory studies on stored grain in the United Kingdom that considerable resistance to phosphine had developed in *Sitophilus oryzae* (L.), *Cryptolestes* and *Rhyzopertha dominica* (F.) from Bangladesh. This had been caused by persistent fumigation of bagged grain without ensuring that adequate concentration–time products of phosphine were achieved to ensure effective control of all species of pests.

In Egypt, fumigation of stored products has been widely practised since 1960. Under field conditions, complaints of incomplete kill after fumigations with phosphine and methyl bromide in some localities were reported. It was believed that such failures might be attributed to development of resistance among stored grain insects to the fumigants used.

The aim of this work was to study the development of resistance in the red flour beetle (*Tribolium castaneum* Herbst) to phosphine in the laboratory and the biological characteristics of the phosphine-resistant strain (F_4) in comparison to the parental strain.

Materials and Methods

Generation of phosphine

Throughout this work phostoxin® pellets were used to obtain phosphine gas (product of DEGESCH Co., Germany). One phosphine pellet contains approximately 56% aluminium phosphide which reacts with water according to the equation:

$$AlP + 3HOH \rightarrow Al(OH)_3 + PH_3$$

A pellet weights 0.6 g and produces approximately o.2 g PH_3.

For generation of the gas, a glass tube (1.5×3 cm) containing 1 pellet + 2 ml water was introduced into a Dreshel flask. This flask was connected on one side to the gas reservoir and on the other side to a recirculatory pump, which was also connected to the gas reservoir. The pump was operated for half an hour. Then the flask was separated and the reservoir was closed.

Fumigation procedure

Fumigation experiments were performed at $26\pm1°C$ and $6\pm1°C$. The relative humidity during the fumigations ranged between 55 and 65%.

A recirculatory multi-flask apparatus was constructed to provide a fumigation chamber suitable for the concentrations used. Phosphine was recirculated through a series of Dreshel flasks with ground-glass joints (Dresmarchelier, 1984). The flasks with a volume of 0.55 l were connected to each other with PVC tubing. The joints were greased. The flasks were connected to two gas reservoirs of 8- l volume to enable the measurement of the gas concentration. The fumigations were performed with constant concentrations and at various exposure periods.

Measuring of phosphine concentrations

Concentrations of PH_3 were determined in the gas reservoir using Draegar gas detector tubes (50/a). The required gas concentration was then obtained by dilution of the gas using the Dreshel flask and the recirculatory pump.

$$1 \text{ ppm } PH_3 = 1 \text{ vpm} = 1.413 \text{ µg/l}$$

Pre-fumigation procedure

Adults of *Tribolium castaneum* (seven to fourteen days old) were used in the fumigation tests. Batches of thirty insects were introduced into wire gauze cages (diameter 14 mm, height 45 mm), which were filled with about 2 g of wheat flour and then covered with rubber stoppers. Three replicas were used in each treatment.

Post-fumigation procedure

After exposure, insects were transferred to glass petri dishes with about 3 g wheat flour and kept at $26\pm1°C$, $60\pm5\%$ r.h. for assessment of the mortality. Mortality was determined three days following fumigation. The percentage mortality was corrected using Abbott's formula (1925).

Selection procedure

To study the development of resistance in *T. castaneum* to phosphine, fumigation experiments were conducted in the laboratory at $26\pm1°C$, $60\pm5\%$ r.h. and $6\pm1°C$, $60\pm5\%$ r.h., using 7–14-

day-old adults, at a fixed concentration of phosphine (28.26 μg/l) and varying exposure periods.

The original culture of *T. castaneum* laboratory strain used in these tests was reared in the laboratory for three years without subjection to any chemical pressure. Once the regression line of the laboratory strain (parent stock) was drawn, the LT_{50} of that generation was extrapolated from the line. Then a number of insects amounting to 240 were exposed to the median lethal dose of phosphine ($LC.T_{50}$) which caused approximately 50% kill. Adult insects surviving this selection procedure were reared at $26 \pm 1°C$, $60 \pm 5\%$ r.h. to produce a successive generation. The same procedure of selection was applied to each generation – all of them reared under the same condition as the laboratory strain. Six selections were done and the susceptibility of the adult populations of all six generations was determined.

Biological investigations

Experiments were carried out in the laboratory at $30 \pm 1°C$, $75 \pm 5\%$ r.h. to study some biological aspects for the fourth generation of phosphine-selected strain of *T. castaneum* in comparison with those obtained for the parental stock (laboratory strain). In this respect the following were recorded:

- average number of daily eggs laid per female:
- average total number of eggs per female laid during fourteen days;
- pre-oviposition period;
- incubation period of eggs;
- percentage of hatching and developmental periods for the various insect stages.

Number of eggs

Tests were carried out to assess number of eggs laid by females during an observation period of fourteen days. Freshly emerged unmated males and females were paired in a glass tube (3 × 5 cm) containing a little wheat flour and covered with muslin. Every day the number of eggs laid in the flour was counted and the flour was replaced. The total number of eggs laid by a female during fourteen days was recorded. Hatching of eggs was also recorded for determining the viability. Each treatment was replicated four times.

The incubation period and the developmental span

To observe the incubation period of eggs and the developmental periods for the various stages, one-day-old eggs were taken and only one egg was put with a little wheat flour in a glass tube (3 × 5 cm) and then covered with muslin – ten replicas were used for each laboratory and phosphine-resistant strain. The incubation period for every egg was recorded. The developmental stages were observed and their developmental periods were also registered. Furthermore, the mortality rate for the larvae was obtained.

Statistical analysis

The toxicity data obtained were subjected to probit analysis (Finney, 1971), using a computer program of Noack and Reichmuth (1978).

The biological data were subjected to statistical analysis using the F-test.

Results and Discussion

Laboratory selection of *T. castaneum* adults for resistance to phosphine

Tables 1 and 3 show the lethal times and parameters of probitregression line estimates for adults of different generations of *T. castaneum*, selected for resistance to phosphine at 26 ± 1°C and 6 ± 1°C. The resistance ratios (RR) for the various generations at the two test temperatures are summarized in Tables 2 and 4.

Data indicate that the lethal times needed for a certain mortality with a fixed concentration of phosphine (28.26 µg/l) were obviously shorter at 26 ± 1°C than at 6 ± 1°C for the different strains of *T. castaneum*. The resistance ratio at the LT_{99} level obtained of the adults of the different generations of the insect at 26 ± 1°C was 1.42, 5.53 and 10.49 for the second, fourth and sixth generations respectively.

The corresponding ratio at 6 ± 1°C was 1.39, 4.32 and 5.65 for the second, fourth and sixth generations respectively.

Table 5 indicates the lethal response ($LC.T_{50}$, $LC.T_{90}$, $LC.T_{99}$ and $LC.T_{99.9}$) and the resistance ratio at $LC.T_{99}$ and $LC.T_{99.9}$ levels for adults of the parent and the phosphine-selected strains of *T. castaneum* at 26 ± 1°C and 6 ± 1°C. The lethal dosages needed to kill 99% of this insect species were increased from 0.684 mg h · l^{-1} in F_6 at 26 ± 1°C and from 2.883 mg h · l^{-1} in F_6 at 6 ± 1°C. Accordingly, the adults of the sixth generation showed at the $LC.T_{99}$ levels 10.5-fold resistance to the fumigant at 26 ± 1°C and 5.65-fold resistance at 6 ± 1°C, when compared with the parent strain. At the $LC.T_{99.9}$ levels, adults of the sixth generation showed 13.85-fold resistance to phosphine at higher temperature and 7.21-fold resistance at lower temperature.

Table 1   Lethal times and parameters of probit regression line estimates for adults of different generations of *Tribolium castaneum* (Herbst), exposed to a fixed concentration of phosphine (28.26 µg/l) at $26\pm1°C$ and $60\pm5\%$ r.h.

Generation	Lethal times (h) parameters of regression line							95% confidence limits at					
	LT_{50}	LT_{90}	LT_{99}	Slope$\pm$SE	b	F	R	LT_{50}(h)		LT_{90}(h)		LT_{99}(h)	
								Lower	Upper	Lower	Upper	Lower	Upper
Parent (Lab. strain)	5.2	10.0	17.0	4.525 ± 0.762	1.75	4	0.999	4.61	5.95	8.52	11.73	12.72	22.81
Second generation (F_2)	7.9	14.6	24.2	4.744 ± 0.472	0.75	4	0.999	7.23	8.53	12.53	16.94	18.87	31.03
Fourth generation (F_4)	19.8	46.5	94.0	3.421 ± 0.267	0.57	5	0.986	17.43	22.37	36.93	58.60	66.89	132.14
Sixth generation (F_6)	23.6	71.6	178.4	2.637 ± 0.273	1.38	7	0.999	21.09	26.34	58.47	87.73	130.84	243.18

SE = Standard error of regression line.
b = Axis intercept.

F = Degree of Freedom.
R = Correlation coefficient

Table 2 Resistance ratio (RR) for adults of different generations of *Tribolium castaneum* (Herbst), exposed to a fixed concentration of phosphine (28.26 µg/l) at LT_{50}, LT_{99} levels, at $26 \pm 1°C$ and $60 \pm 5\%$ r.h.

Selection Strain $(mg\ h^{-1}l^{-1})$	dose	LT_{50} (h)	(RR)	LT_{99} (h)	(RR)	Slope $\pm$ SE
Lab. strain (parent)	–	5.2	1.0	17.0	1.0	4.525 $\pm$ 0.762
Second generation (F_2)	0.223	7.9	1.52	24.2	1.42	4.744 $\pm$ 0.472
Fourth generation (F_4)	0.560	19.8	3.81	94.0	5.53	3.421 $\pm$ 0.267
Sixth generation (F_6)	0.667	23.6	4.54	178.4	10.49	2.637 $\pm$ 0.273

Table 3 Lethal times and parameters of probit regression line estimates for adults of different generations of *T. castaneum* (Herbst), exposed to a fixed concentration of phosphine (28.26 µg/L) at $6\pm1°C$ and $60\pm5\%$ r.h.

| Generation | Lethal times (h) | | | Parameters of regression line | | | | 95% confidence limits at | | | | | |
| | LT_{50} | LT_{90} | LT_{99} | Slope$\pm$SE | b | F | R | $LT_{50}(h)$ | | $LT_{90}(h)$ | | $LT_{99}(h)$ | |
								Lower	Upper	Lower	Upper	Lower	Upper
Parent (Lab. strain)	21.7	42.3	73.2	4.396+0.225	-0.88	4	0.990	18.97	24.89	33.77	53.06	51.47	104.06
2nd generation (F_2)	29.1	57.9	102.0	4.255+0.554	-1.23	4	0.999	24.04	35.19	43.89	76.46	66.12	157.34
4th generation (F_4)	48.8	136.1	316.0	2.857+0.219	0.18	5	0.905	41.87	56.81	99.73	185.71	196.88	507.23
6th generation (F_6)	58.9	171.8	413.8	2.738+0.006	0.15	5	0.999	52.02	66.02	138.04	213.74	284.95	600.84

SE = Standard error of regression line.
b = Axis intercept

F = Degree of freedom.
R = Correlation coefficient.

Table 4 Resistance ratio (RR) for adults of different generations of *Tribolium castaneum* (Herbst), exposed to a fixed concentration of phosphine (28.26 µg/l) at LT_{50}, LT_{99} levels, at $6 \pm 1°C$ and $60 \pm 5\%$ r.h.

Strain	Selection dose (mg h^{-1}l^{-1})	LT_{50} (h)	(RR)	LT_{99} (h)	(RR)	Slope $\pm$ SE
Lab. strain (parent)	–	21.7	1.00	73.2	1.00	4.396 $\pm$ 0.225
Second generation (F_2)	0.822	29.1	1.34	102.0	1.39	4.255 $\pm$ 0.554
Fourth generation (F_4)	1.379	48.8	2.25	316.0	4.32	2.857 $\pm$ 0.219
Sixth generation (F_6)	1.665	58.9	2.71	413.8	5.65	2.738 $\pm$ 0.006

Table 5 Lethal response (LC.T$_{50}$, LC.T$_{90}$, LC.T$_{99}$ and LC.T$_{99.9}$) and resistance ration (RR) at LC.T$_{99}$ and LC.T$_{99.9}$ levels for adults of parent and phosphine-selected strain of *T. castaneum* at 26$\pm$1°C and 6$\pm$1°C

Strain	At 26$\pm$1°C				RR at		At 6$\pm$1°C				RR at	
	LC.T$_{50}$	LC.T$_{90}$	LC.T$_{99}$	LC.T$_{99.9}$	LC.T$_{99}$	LC.T$_{99.9}$	LC.T$_{50}$	LC.T$_{90}$	LC.T$_{99}$	LC.T$_{99.9}$	LC.T$_{99}$	LC.T$_{99.9}$
	(mg h·l^{-1})	(mg h·l^{-1})	(mg h·l^{-1})	(mg h·l^{-1})			(mg h·l^{-1})	(mg h·l^{-1})	(mg h·l^{-1})	(mg h·l^{-1})		
Parent (Lab. strain)	0.147	0.283	0.480	0.709	1.00	1.00	0.613	1.195	2.069	3.086	1.00	1.00
Second generation (F$_2$)	0.223	0.413	0.684	0.992	1.43	1.40	0.822	1.636	2.883	4.358	1.39	1.41
Fourth generation (F$_4$)	0.560	1.314	2.656	4.442	5.53	6.27	1.379	3.846	8.930	16.532	4.32	5.36
Sixth generation (F$_6$)	0.667	2.023	5.042	9.823	10.5	13.85	1.665	4.855	11.694	22.235	5.65	7.21

Monro *et al.* (1972) found that *S. granarius* adults selected for tolerance to phosphine were found after twenty-eight selections to be able to tolerate exposure to the gas for more than three times as long as unselected insects.

Price and Dance (1983) reported that resistant insects (*Rhizopertha dominica*, *Oryzaephilus surinamensis* and *Cryptolestes ferrugineus*) absorb less phosphine than susceptibles. Chaudhry and Price (1989) demonstrated that uptake of phosphine by resistant strains of *R. dominica* and *T. castaneum* was lower than by susceptible ones.

Bond (1975) stated that both the uptake of the fumigant and the respiration rate of the insects declined steadily as the temperature was lowered from 25 to 0°C.   Kem (1977) mentioned that a strain of *T. castaneum* resistant to phosphine could be developed by exposing successive generations to the fumigant at rates calculated to give 60–80% mortality and rearing the survivors. The parental stock was derived from field-collected adults. The degree of resistance was estimated from comparison of the LD_{50s} of the selected and unselected strain. As a result of such selection pressure for ten generations, a strain of 11.95 times as resistant to phosphine as the unselected strain developed.

Saxena and Bhatia (1980) reported that a phosphine-resistant strain of *T. castaneum* was developed in the laboratory in India by selection of adults in successive generations; after sixteen generations, the adults showed 5.9-fold resistance to the fumigant as compared with the parent strain.

The results indicate that *T. castaneum* adults have the genetic potential to develop resistance to phosphine and the treatment of the successive generations of this insect with phosphine produced insects with increased tolerance to the gas at the two test temperatures.

The biological characteristics of the phosphine-resistant strain (F_4) of *T. castaneum* (Herbst)

The biology of the resistant strain of *T. castaneum* selected for four generations with phosphine was studied at $30 \pm 1°C$ and $75 \pm 5\%$ r.h. in comparison to that of the parental stock (laboratory strain).

The data obtained are illustrated in Tables 6, 7 and 8 and in Figure 1.

Results showed that there was no significant difference between the laboratory strain and the phosphine-resistant strain (F_4) of *T. castaneum* in the average pre-oviposition period, in average total duration of the larval instars, in average duration of the pupal instars, and in the mean of the total developmental period (Table 6).

Table 6 Some biological parameters for the laboratory strain and the phosphine-resistant strain (F_4) of *T. castaneum* at $30\pm1°C$ and $75\pm5\%$ r.h.

Parameters	Laboratory strain (L)	Phosphine-resistant strain (F_4)	Probability
Average pre-oviposition period (day)	8.333 ± 1.453	8.667 ± 0.333	0.8340
Average no. of eggs per female/day	3.051 ± 0.364	5.876 ± 0.545	0.0010**
Average total no. of eggs per female during 14 days	42.714 ± 5.098	82.286 ± 7.618	0.0010**
Average hatching rate (%)	90.000 ± 4.082	76.250 ± 2.394	0.0271*
Average incubation period (days)	3.500 ± 0.167	4.300 ± 0.153	0.0023**
Average total duration of larval instars (days)	24.643 ± 0.884	22.775 ± 0.736	0.1879
Average duration of pupal instar (days)	6.071 ± 0.352	6.875 ± 0.375	0.1772
Total developmental period (days)	34.143 ± 0.661	33.875 ± 1.125	0.8298
Mortality for larval instars (%)	30	60**	
Emergence rate (%)	100	100	

* = Difference is significant at 5% level.
** = Difference is significant at 1% level.

The total developmental period averaged 34.14 ± 0.66 days for the laboratory strain and 33.88 ± 1.13 days for the phosphine-resistant strain at the above-mentioned temperature and relative humidity.

It was also observed that the emergence rate of the adults was unaffected and amounted to 100% for the two strains. The average incubation period was significantly longer in the phosphine-resistant strain, it lasted 3.5 days for the laboratory strain and 4.3 days for the phosphine-resistant strain.

The mortality rate of the larval instars was significantly higher for the phosphine-resistant strain than for the laboratory strain.

During an observation period of fourteen days, the average

Table 7 Average daily number of eggs, laid per female of *T. castaneum* during fourteen days, for the laboratory strain and the phosphine-resistant strain (F_4) at $30\pm1°C$ and $75\pm5\%$ r.h.

Day	Average number of eggs laid per female/day	
	Laboratory strain (L)	Phosphine-resistant strain (F_4)
1	2.00	4.50
2	1.00	5.00
3	2.57	6.00
4	3.86	7.14
5	7.00	10.29
6	5.57	8.14
7	3.00	5.57
8	3.43	5.86
9	3.30	6.14
10	2.00	8.00
11	3.29	5.60
12	1.70	3.00
13	1.90	3.29
14	2.70	3.70
Average total number of eggs laid during fourteen days	42.71 ± 5.098	82.29 ± 7.618**

** = Difference is significant at 1% level.

number of eggs laid per female per day was significantly higher for the phosphine-resistant strain than for the laboratory strain (see also Table 7). The average hatching rate of the eggs significantly declined from about 90% for the laboratory strain to 76% per the phosphine-resistant strain.

The comparative average number of eggs laid daily per female of the laboratory strain and the phosphine-resistant strain (F_4) of *T. castaneum* during an observation period of fourteen days at $30\pm1°C$ and $75\pm5\%$ r.h. is shown in Figure 1 and Table 7.

It is obvious that the phosphine-resistant strain laid significantly higher numbers of eggs during this observation period than the parents. The average total number of eggs laid per female during fourteen days was 42.71 ± 5.10 for the laboratory strain and

Table 8   Average duration and the total period of the larval instars for the parental strain and the phosphine-resistant strain (F_4) of *T. castaneum* at $30\pm1°C$ and $75\pm5\%$ r.h.

Larval instars	Average duration of larval instars		Probability
	Laboratory strain (L)	Phosphine-resistant (F_4)	
First	1.000 ± 0.000	2.000 ± 0.000	$0.0000**$
Second	4.143 ± 0.404	4.750 ± 0.750	0.4511
Third	4.000 ± 0.756	2.750 ± 0.250	0.2598
Fourth	3.286 ± 0.360	3.000 ± 0.000	0.5717
Fifth	6.000 ± 0.951	5.250 ± 1.031	0.6264
Sixth	3.857 ± 0.930	3.500 ± 2.010	0.8569
Seventh	2.000 ± 0.893	1.375 ± 1.214	0.6857
Eight	0.214 ± 0.101	0.125 ± 0.125	0.5993
Average no. of larval instars	7.143 ± 0.340	6.750 ± 0.479	0.5126
Average total period for the larval instars (days)	24.643 ± 0.884	22.775 ± 0.736	0.1879

$**$ = Difference is significant at 1% level.

82.29 ± 7.7 for the phosphine-resistant strain and this difference was significant at the 1% level.

Table 8 indicates the average duration and total period of the larval instars for the parental and the phosphine-resistant strain (F_4) of *T. castaneum* at $30\pm1°C$ and $75\pm5\%$ r.h. Eight larval instars were recorded for the two strains. The first larval stadium lasted one and two days for the laboratory and the phosphine-resistant strains, respectively. This period, therefore, was significantly shorter for the laboratory strain. The results also showed that there was no significant difference between the laboratory strain and the phosphine-resistant strain in the duration of the second, third, fourth, fifth, sixth, seventh and eighth larval instars. The total duration of the larval instars lasted 24.6 ± 0.88 days for the laboratory strain and 22.8 ± 0.74 days for the phosphine-resistant strain at $30\pm1°C$ and $75\pm5\%$ r.h. and no significant difference was found between the two strains for this period.

In summary, the phosphine-resistant strain (F_4) of *T. castaneum* laid significantly higher numbers of eggs than the laboratory strain,

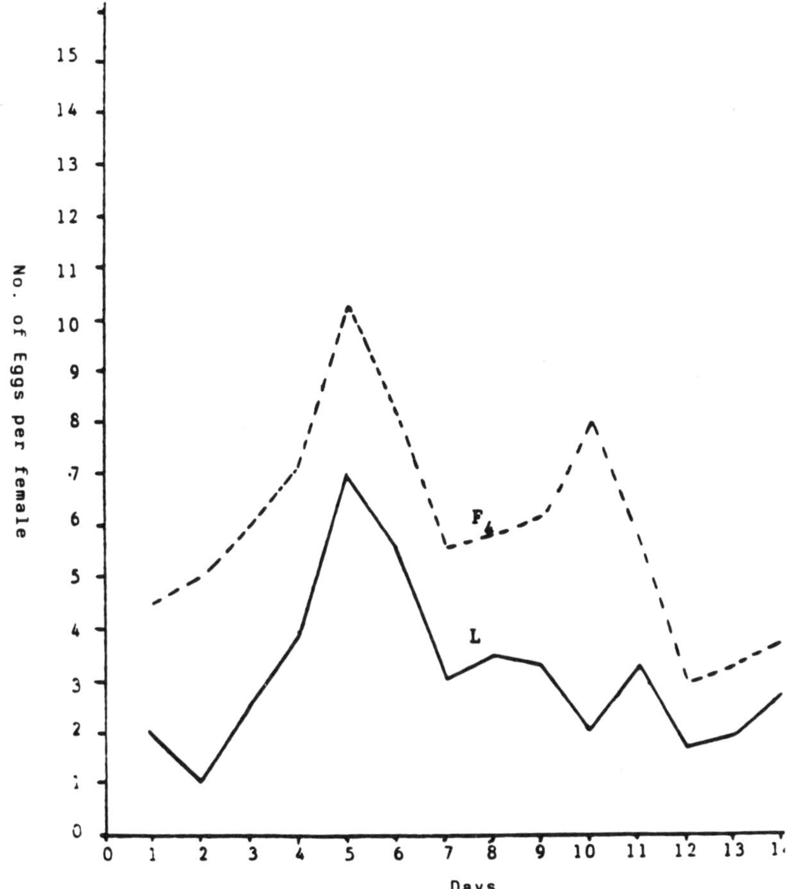

Figure 1   Comparative average number of eggs laid daily per female of laboratory (L) and phosphine-resistant (F_4) strain of *T. castaneum*.

but the hatchability of the eggs declined and the incubation period of the eggs was considerably prolonged. The larval mortality was significantly higher and the duration of the first larval instar was significantly longer for the phosphine-resistant strain. No significant differences were obtained in the two strains for the other recorded biological parameters.

Saxena and Bhatia (1980) reported that a phosphine-resistant strain of the stored product pest *T. castaneum* (Herbst) was developed in the laboratory in India by selection of adults in successive generations; after sixteen generations, the adults showed

5.9-fold resistance to the fumigant, as compared with the parent strain. The phosphine-resistant females had 24% lower fecundity than the parent strain, but no differences were observed in the duration of the various developmental stages between the two strains.

El-Nahal and El-Halafawy (1973) reported that sublethal doses of carbon disulphide increased the duration of larval and pupal stages of *T. confusum* and reduced hatchability of the eggs and the effect was proportional to the dose.

Mahgoub (1982) found that when the effects of LC_{50} and nonlethal doses of malathion, cypermethrin and decamethrin were investigated on *T. confusum* and *T. castaneum*, selected for tolerance, all insecticides used exerted deleterious effects on the biology of the parents and subsequent generations selected for tolerance. The effects on oviposition, hatchability, larval and pupal development periods and longevity of adults varied considerably with various insecticides.

Madhaven *et al.* (1974) mentioned that sublethal doses of DDT, endrin, endosulfan and parathion prolonged the larval period of *T. castaneum* in approximately direct proportion to the concentration of the insecticides.

Acknowledgment

This work is part of the Research Project No. MO PP 256 PP 06.

The authors acknowledge the contributions of the MOA (Ministry of Agriculture and Land Reclamation) and USAID to this research project.

References

ABBOTT, W.W. (1925). A method of computing the effectiveness of an insecticide. *Journal of Economic Entomology*, **18**, 265–267.

BARBARA, D.H., BELL, C.H., MILLS, K.A. and GOODSHIP, G. (1976). The toxicity of phosphine to all developmental stage of thirteen species of stored product beetles. *Journal of Stored Products Research*, **12**, 235–244.

BOND, E.J. (1973). Increased tolerance to ethylene dibromide in a field population of *Tribolium castaneum* (Herbst). *Journal of Stored Products Research*, **9**, 61–63.

BOND, E.J.(1975). Control of insects with fumigants at low temperatures: response to methyl bromide over the range 25°C to −6.7°C. *Journal of Economic Entomology*, **68**(4), 539–542.

CHAMP, B.R. and DYTE, C.E. (1976). *Report of the FAO Global Survey of Pesticide Susceptibility of Stored Grain Pests*. FAO Plant Products Protection Service, No. 5.

CHAUDHRY, M.Q. and PRICE, N.R. (1989). Biochemistry of phosphine uptake in susceptible and resistant strain of two species of stored product beetles. *Comp biochem physiol comp pharmacol toxicol*, **94**(2), 425–430.

DRESMARCHELIER, J.M. (1984). Effect of carbon dioxide on the efficacy of phosphine against different stored product insects. *Mitteilung aus der Biologischen Bundesanstalt für Land- und Forstwirtschaft-Berlin-Dahlam*, Heft 220.

DYTE, C.E. and BLACKMAN, D.G. (1970). The spread of insecticide resistance in *Tribolium castaneum* (Herbst), Col. Tenebr. *Journal of Stored Products Research*, **6**, 255–261.

EL-NAHAL, A.K.M. and EL-HALAFAWY, M.A. (1973). The effects of sublethal doses of methyl bromide on the biology of the confused flour beetle, *Tribolium confusum* (Duv.) Coleoptera Tenebrionidae. *Bulletin of the Entomological Society of Egypt. Economic Series* (1973, Publ. 1974), 201–211.

FINNEY, D.J. (1971). *Probit Analysis*, 3rd edn. Cambridge University Press.

HALLIDAY, D., HARRIS, A.H. and TAYLOR, R.W.D. (1983). *Development of Resistance to Phosphine by Insect Pests of Stored Grains*. British Crop Protection Council, Croydon, UK.

HESELTINE, H.K. and THOMPSON, (1957). Use of aluminium phosphide for the fumigation of grain. I. Mulling, **129**, 676–677.

HOWE, R.W. and HOLE, B.D. (1967). Predicting the dosage of fumigant needed to eradicate insects from stored products. *Journal of Applied Ecology*, **4**, 337–351.

KEM, T.R. (1977). Selection of a strain of *Tribolium castaneum* (Herbst) resistant to phosphine. *Journal of Entomological Research*, **1**(2), 213–217.

LINDGREN, D.L., VINCENT, L.E. and STRONG, R.G. (1958). Studies on hydrogen phosphide as a fumigant. *Journal of Economic Entomology*, **51**, 900–903.

MADHAVEN, O.T., NAIR, M.R.G.K. and CHANDRIKA, S. (1974). On the effect of sublethal doses of insecticides on the biology of *Tribolium castaneum* (Herbst). *Agricultural Research Journal of Kerala*, **7**(2), 104–106.

MAHGOUB (1982). Biological and toxicological studies on some stored product insects. M.Sc. Thesis, Faculty of Agriculture, Moshtohor, Zagazig University, Egypt.

MONRO, H.A.U., UPITIS, E. and BOND, E.J. (1972). Resistance

of a laboratory strain of *Sitophilus granarius* (L.), Col. Curcl. to phosphine. *Journal of Stored Products Research*, **8**, 199–207.

NOACK, S. and REICHMUTH, Ch. (1978). Ein rechneriches Verfahren zur Bestimmung von beliebigen Dosis-Werten eines Wirkstoffes aus empirisch ermittelten Dosis-Werkungs-Daten. *Mitteilung aus der Biologischen Bundesanstalt für Land- und Forstwirtschaft-Berlin-Dahlam* Heft 185, 1–49.

PRICE, N.R. and DANCE, S.J. (1983). Some biochemical aspects of phosphine action and resistance in three species of stored product beetles. *Comparative Biochemistry and Physiology*, C. **76**(2), 277–281.

SAXENA, J.D. and BHATIA, S.K. (1980). Laboratory selection of the red flour beetle, *Tribolium castaneum* (Herbst) for resistance to phosphine. *Entomology*, **5**(4), 301–306.

41
Anti-resistance Strategies against the Red Spider Mite, *Panonychus ulmi* (Koch) in Belgian Orchards

G. STERK

Research Station of Gorsem (IWONL), Department of Zoölogy, Brede Akker 3, B-3800 Sint Truiden, Belgium

Introduction

Panonychus ulmi (Koch), the red spider mite, is a widespread and important pest of apple and other fruit trees, including plum, cherry and pear. Eggs overwinter on the bark, mainly on the smaller branches and spurs. Hatching begins under Belgian conditions mainly in April or May. On emerging from eggs, the larvae move to the underside of the leaves and start feeding. Mites remain on the leaves while moulting from one stage to the next. Summer eggs, which look paler than winter ones, are laid on the underside of leaves. There are at least five generations each year. From August onwards, winter eggs are laid on the branches (Vanweswinkel, 1979). Immature and adult mites cause speckling of the foliage by sucking. The leaves become dull green, brownish, and finally bronze. Such foliage can drop prematurely. Attacks are most important in hot, dry summers, and sometimes more than a hundred mobile stages can be found on one leaf.

As a result, size, colouring of the fruit and bud formation are affected. During the 1960s and 1970s, *Panonychus ulmi* (Koch) became resistant against several compounds like omethoate, dimethoate and some acaricide, causing severe problems in Belgian orchards. The classical acaricides like the organotin compounds are still effective, but resistance occurs already in some countries. There are difficulties at the moment in several European countries with the

Insecticides: Mechanism of Action and Resistance

ovolarvicidal acaricides. Therefore, the Department of Zoology of the Research Station of Gorsem started to promote an anti-resistance strategy against this important pest.

Implementation of the Resistance Management Strategy for Spider Mite Control

This is based on the ideas of the Insecticide Resistance Action Committee (IRAC) of the International Group of National Associations of Agrochemical Manufacturers (GIFAP) (Lemon, 1988).

An approach to resistance management in spider mites in apple and pear orchards was proposed and promoted in 1990 by the Research Station of Gorsem through discussions with sale companies, by written warnings for the fruitgrowers and through articles in specialized magazines. Also, a special IRAC video, named 'The paradox of resistance' was shown on several occasions to fruitgrowers and salespeople. The response to this project was highly positive from companies, the government and the fruitgrowers. This means that implementation of this strategy was quite successful (Sterk, 1990).

This management is based on the following options (Lemon, 1988):

1. The use of mixtures of acaricides or coformulations is highly recommended. These should be subject to different resistance mechanisms.
2. The alternation or rotation of acaricides is promoted.
3. The use of acaricides should be moderated by introducing as much as possible biological agents, by using damage thresholds and improved scouting, and by local treatments. The thresholds are based on the number of winter eggs on the branches or on the density of mobile stages per leaf, and are linked with the abundance of predatory mites in the orchard. The predatory mite that was introduced is an organophosphate- and carbamate-resistant strain of *Typhlodromus pyri* (Oudemans).

The acaricides available were grouped according to known or expected cross-resistance patterns.

Group A: Organotins (fenbutatinoxide, azoyclotin)
Group B: Ovolarvicidal acaricides (clofentezine, hexythiazox)
Group C: Bridged diphenyl compounds (bromopropylate, dicofol)
Group D: Pyrethroids (biphentrin, fenpropathrin, fluvalinate)

Group E: Amitraz
Group F: Pyridaben

The resistance management is based on the following rules, developed by the Fruit Crop Working Group of the IRAC (Lemon, 1988):

1. Not more than one compound from any one group of acaricides should be applied to the same crop in the same season.
2. Any one compound should be used only once per season on any one crop.
3. Compounds from the same group must not be mixed.
4. Compounds should be used in such a way that detrimental effects on predatory insects and mites are minimized.
5. Use compounds only at the manufacturer's recommended rates and timings.
6. Monitoring should be conducted to detect early signs of resistance.

Implementation in Practice

Following this advice, a certain number of choices for anti-resistance management are available to the Belgian fruitgrower in apple orchards. These are given in Table 1 (Sterk, 1990).

The most important sales companies took these schemes over and published them in their annual spraying programmes. This is very important, because of their strong influence on the Belgian fruitgrowers.

Remarks

1. The classical acaricides are the organotins, amitraz and bromopropylate.
2. Fenbutatinoxide is not in the list because of its weak activity against red spider mite. This compound is only used in IPM orchards, together with predatory mites. Dicofol is only used on a very small scale and is strongly dependent on the weather conditions, therefore it is not recommended.
3. An extra problem is caused by the apple rustmite, *Aculus schlechtendali* (Nal.). This small parasite has developed resistance against synthetic pyrethroids and organophosphate pesticides. Clofentezine and hexythiazox do not show sufficient activity against this parasite.

Table 1   Choices available to the Belgian fruitgrower for anti-resistance management

Winter eggs	Start hatching of the winter eggs	50% hatching of the winter eggs	100% hatching of the winter eggs	Summer treatment
Clofentezine	–	–	–	Classical acaricide or Pyridaben
–	Hexythiazox	–	–	Classical acaricide or Pyridaben
–	–	Bromopropylate	Azocylotin	Amitraz or Pyridaben
–	–	Bromopropylate	Amitraz	Azocylotin or Pyridaben
–	–	Amitraz	Azocylotin	Bromo propylate or Pyridaben
–	–	Pyridaben	–	Classical acaricide

4. The synthetic pyrethroids with acaricidal activity are not in the list, because they are only sprayed against caterpillars, or shortly before harvest. They have also lost their activity against rustmites due to resistance.

5. Clofentezine, hexythiazox and fenbutatinoxide are harmless for predatory mites and can be used in IPM programmes in apple orchards.

6. In pear orchards the same strategies might be followed with a few exceptions. Red spider mites are not that important on pears in Belgium, so the use of selective compounds to save predatory mites is not necessary. More important here is selectivity for beneficial insects, especially anthocorids. Amitraz is the only compounds that can be used in summer against the pear sucker, *Psylla pyri* (L.). so it is not recommended as an acaricide in pear orchards to prevent resistance.

Results

It is difficult to evaluate the results after only one year. But due to the help of the industry in this matter and according to the information we have received, we are sure that a great number of fruitgrowers are now well aware of the resistance problem and are adapting their spraying schemes to avoid problems with the red spider mite. Through our warning system, the articles in our magazine for the fruitgrower, *De fruitteelt: berichten uit wetenschap en praktijk*, and several lectures we gave during the wintertime, we were able to reach a great majority of the Belgian fruitgrowers with our message. An enquiry was held in 1990 and will be repeated after a few years to follow the impact of our strategy.

Discussion

Due to commercial and fungicide problems, it seems that the introduction of predator mites in Belgian apple orchards will only take off rather slowly. Therefore, a decent anti-resistance management against the red spider mite, *Panonychus ulmi* (Koch), is certainly necessary. The interest of the industry, the government and the fruitgrowers was very encouraging in 1990. We hope to continue these efforts in the following years.

Acknowledgements

We thank the IWONL for its financial support. Thanks are also due to IRAC, the chemical industry and sales companies, the government and the fruitgrowers for their cooperation.

References

ALFORD, D.V. (1984). *A Colour Atlas of Fruit Pests, Biology and Control*, pp. 234–235. Wolfe Publishing Ltd, London.

LEMON, J. (1988). Resistance monitoring methods and strategies for resistance management in insect and mite pests of fruit crops. In *Proceedings of the British Crop Protection Conference (Brighton)*, Vol. 3, pp. 1089–1096.

SOLOMON, M. (1987). Fruit and hops. In *Integrated Pest Management*, pp. 329–360. Academic Press.

STERK, G. (1990). Resistentie of 'the parasite strikes back'. *De fruitteelt: berichten uit wetenschap en praktijk*. 3de jg, 4, pp. 19–24; 5, pp. 35–41.

STERK, G. and PEREGRINE, J. (1989a). The influence of application time on the activity of two ovolarvicidal acaricides against the fruit tree red spider mite *Panonychus ulmi* (Koch). In *Med. Fac. Landbouww. Rijksuniv. Gent 54/3b*, pp. 965–968.

STERK, G. and PEREGRINE, J. (1989b). Studies on the effects of two ovolarvicidal acaricides on beneficial insects and mites. In *Med. Fac. Landbouww. Rijksuniv. Gent 54/3b*, pp. 969–973.

VANWETSWINKEL, G. (1961). De rode spin en hat resistentie-probleem. In *De Belgische Hop*, Vol. **42**, pp. 9–16.

VANWETSWINKEL, G. (1976). De rode spin en haar bestrijding. In *Belgische Fruitrevue*, **28**(4), 88–87.

VANWETSWINKEL, G. (1979). De rode spin (*Panonychus ulmi* (Koch)) in de fruitteelt. In *Landbouwtijdschrift*, **32**(2), 457–460.

42
Multiple Resistance in Two-spotted Spider Mites (*Tetranychus urticae* Koch) from East German Glasshouses

P. RICHTER and D. OTTO

Biological Research Centre Berlin, Stahnsdorfer Damm 81, 0-1532 Kleinmachnow, Germany

Introduction

As a response to a long history of chemical treatment one of the major pests on fruit and vegetables in Europe, the two-spotted spider mite (*Tetranychus urticae* Koch) has developed a broad pattern of acaricide resistance. In East Germany difficulties in combating this species appeared in the middle of the 1960s, and a homogeneous strain highly resistant to organophosphorus compounds has been isolated by Klunker (1970). This strain exhibited the negatively correlated cross-resistance to formetanate and other carbamates described by Steinhausen (1968).

The species has now accumulated a considerable number of genes conferring resistance to all major classes of acaricides; it belongs to a number of especially critical cases in which nearly all of the affordable, previously effective insecticides have been depleted (Georghiou, 1990). In Kleinmachnow, mites have been selected with multiple resistance to dimethoate, methamidophos and formetanate (Otto, 1988). In the absence of an actual report on the distribution of multiple resistance in *T. urticae* in East Germany this induced us to carry out a representative survey among regularly treated spider mites in greenhouses.

Insecticides: Mechanism of Action and Resistance
© 1992   Intercept Ltd, P.O. Box 716, Andover, Hants SP10 1YG, UK

Material and Methods

Mite-infected cucumber leaves have been collected in greenhouses of agricultural cooperatives in Altbartelsdorf near Rostock(Mecklenburg Land), Groß Gaglow near Cottbus (Brandenburg Land) and Mittelhausen near Erfurt (Thuringia Land). The colonies were maintained in glasshouse compartments on bean plants (*Phaseolus vulgaris*) without any pesticide treatment. The chemicals tested for miticide efficacy are listed in Table 1. If available, the mortality data of a susceptible reference strain (S strain) were combined with the average of all experiments carried out in our laboratory since 1975; otherwise new experiments were started.

Table 1   Acaricides involved in the testing for resistance

Acaricide	Active ingredient	Producer
Apollo	500 g l^{-1} clofentezin	Schering AG
Bi 58 EC	380 g l^{-1} dimethoate	Chemie AG Bitterfeld
Dicarzol	50 g l^{-1} formetanate	Schering AG
Falimilbon 30 EC	300 g l^{-1} dinobuton	Fahlberg-List Magdeburg
Fentoxan	40% fenazox	Fahlberg-List Magdeburg
Filitox	565 g l^{-1} methamidophos	Chemie AG Bitterfeld
Lannate 90	90% methomyl	DuPont de Nemours International
Mesurol	50% mercaptodimethur	Bayer AG
Milbol EC	18.5% dicofol	Delicia Delitzsch
Mitac 20	20% amitraz	Schering AG
Nissuron 10 WP	10% hexythiazox	Nippon Soda Co. Ltd
Omite 57 E	600 g l^{-1} propargite	Uniroyal Chemicals Co. Ltd
Peropal	50% azocyclotin	Bayer AG
Sanmite 20 WP	20% pyridaben	Nissan Chemical Industries Ltd
Tenysan	20% tetradifon	Fahlberg-List Magdeburg
Torque	50% fenbutatin oxide	Schell International Co. Ltd

To measure the activity of adults, discs (diam 18 mm) punched out of bean leaves were dipped for 5 s into the acaricide solution prepared from tap water. Redundant drops were sucked up and the discs placed top surface down on strips of moist filter paper. About twelve to fifteen adult females were then transferred to each disc using a fine hairy brush.

Mortality was assessed after three days of exposure at $25 \pm 1°C$ and 50% r.h. Individuals were recorded as dead if they were unable to move for one body length after having been touched with a brush.

To evaluate the efficacy of ovicides, about fifteen females were allowed to oviposit on untreated leaf discs for one day. One day after the removal of the females the discs were dipped into the appropriate acaricide solution for 5 s, slightly dried and held at $25 \pm 1°C$ and 50% r.h. for eight days. The number of hatched and unhatched eggs was recorded.

Mortality data of fifteen to twenty concentrations were subjected to spline regression analyses (SRA) (Otto *et al.*, this volume, Chapter 34) to distinguish susceptible and resistant population parts.

The degree of resistance (R_i) was determined as the ratio of the LC_{50} of the resistant part of the population to the LC_{50} of the susceptible reference strain. The resistance level (resistance frequency) was calculated from data designated by SRA as a plateau between susceptible and resistant mites.

Of a number of acaricides, where resistance was not expected, only recommended concentrations have been tested. To evaluate dicofol resistance, which is commonly found to be incompletely recessive (Cranham and Helle, 1985), several hundred adult females were exposed to a discriminating concentration of 2500 µM dicofol (0.5% Milbol EC).

Results

All three glasshouse strains are resistant to the organophosphorus, carbamate and formamidine acaricides tested (Tables 2 and 3). The degrees of resistance were found to be highest to dimethoate (Figure 1) and methamidophos.

Although the reference strain held in our laboratory is not very susceptible to methomyl, carbamate resistance is clearly indicated (Figure 2). Mercaptodimethur is active on the S strain but completely inactive on the Erfurt and Groß Gaglow strains (data not given). There is no negatively correlated cross-resistance to formetanate (Figure 3).

Resistance to amitraz is more pronounced in eggs than in adults (Figure 4). The degree of resistance is difficult to measure because at least the Erfurt and the Groß Gaglow strains seem to harbour more than one group of resistant individuals. From these experiments it cannot be decided whether heterozygous and homozygous females are distinguished; however, the differences are too small to be localized by SRA in the ovicide experiments.

Table 2 Degrees of resistance of multiple-resistant *T. urticae* females

Acaricide	Erfurt	Strain Groß Gaglow	Rostock
Dimethoate	143.5	164.8	97.4
Methamidophos	58.0	121.7	61.0
Formetanate	22.9	8.2	17.8
Methomyl	13.2	20.6	10.5
Amitraz	>8.7	2.1...23.6	3.3
Pyridaben	27.0	32.9	>100

Table 3 Resistance levels of multiple-resistant *T. urticae* (% resistant females per strain)

Acaricide	Erfurt	Strain Groß Gaglow	Rostock
Dimethoate	90.3	85.3	82.6
Methamidophos	93.6	72.6	76.1
Formetanate	88.3	76.1	74.8
Methomyl	86.4	80.2	64.4
Amitraz	87.4	82.1	65.9
Pyridaben	71.6	80.8	70.2

Despite intense searching no resistance to dicofol has been detected in the Erfurt strain, but a small number of resistant females have been found in the Rostock and Groß Gaglow strains (Figure 5, Table 4). Apparently these females had to be homozygous because dicofol resistance is reported to be recessive in the laboratory (Martinson *et al.*, 1991). The strains revealed a somewhat lesser susceptibility to dicofol than the S strain, nevertheless this does not indicate that dicofol cannot be used to combat them. Evidently the mechanisms of dicofol resistance is separately inherited and not concomitantly selected with resistance to the other compounds.

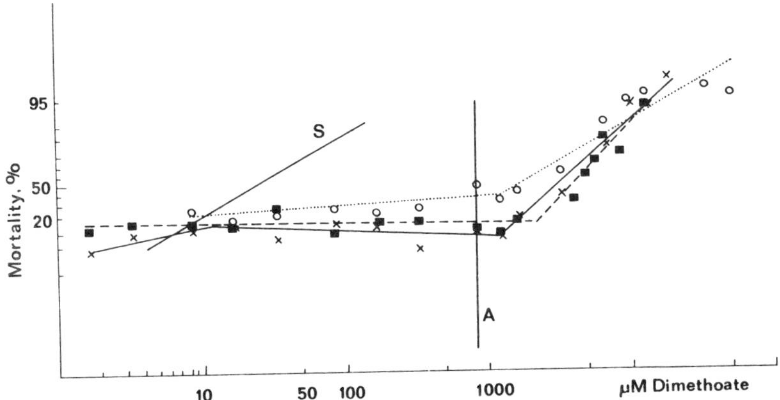

Figure 1   Toxicity of dimethoate to multiple resistant *T. urticae*: x—x, Erfurt; ■- - -■, o....o, Rostock; S, susceptible reference strain; A, recommended application concentration 0.05% Bi 58.

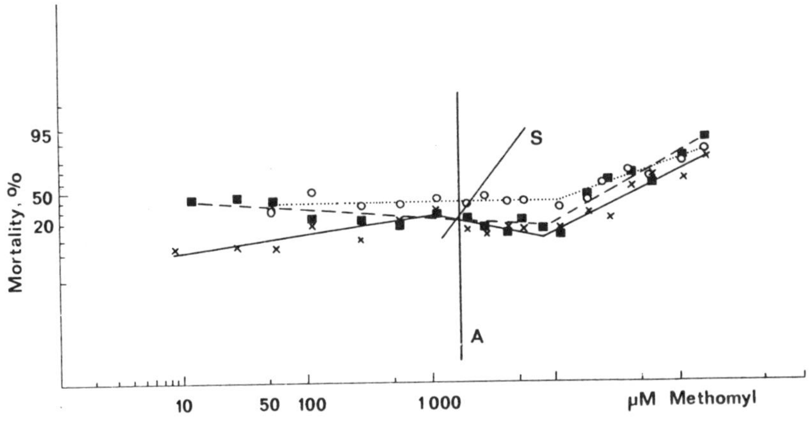

Figure 2   Toxicity of methomyl to multiple resistant *T. urticae*: x—x, Erfurt; ■- - -■, Groß Gaglow; o....o, Rostock; S, susceptible reference strain; A, recommended application concentration 0.03% Lannate 90.

A high level of resistance has also been found to occur to pyridaben, a compound that had not been used in East Germany. Sanmite is narrowly active on the S strain and completely inactive on the greenhouse strains (Figure 6).

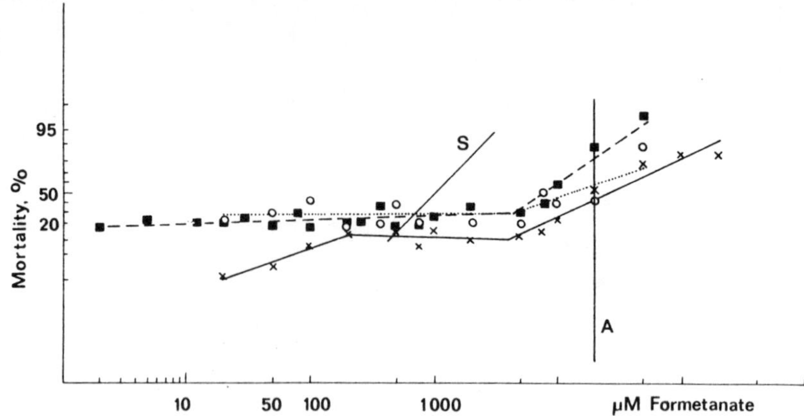

Figure 3   Toxicity of formetanate to multiple resistant *T. urticae*: x—x, Erfurt; ■- - -■, Groß Gaglow; o....o, Rostock; S, susceptible reference strain; A, recommended application concentration 0.02% Dicarzol.

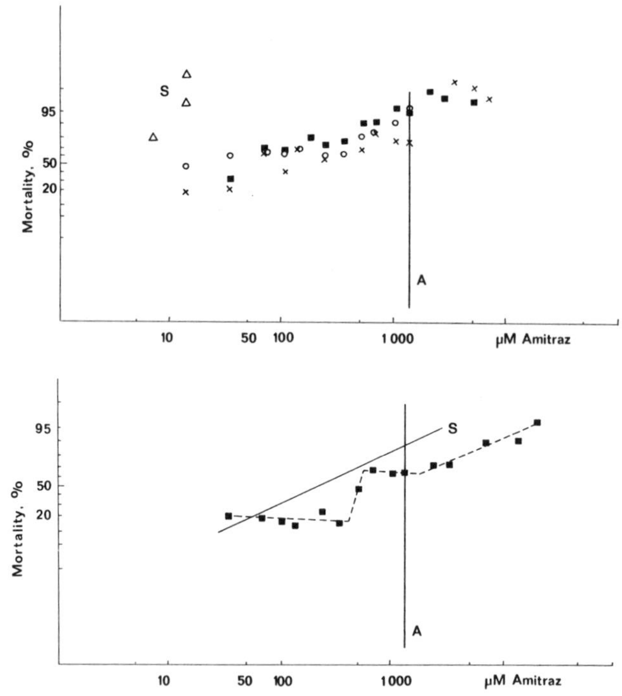

Figure 4   Toxicity of amitraz to multiple resistant *T. urticae*: x—x, Erfurt; ■- - -■, Groß Gaglow; o....o, Rostock; S, susceptible reference strain; A, recommended application concentration 0.02% Mitac 20. (a) Toxicity to adult females. (b) Ovicidal activity.

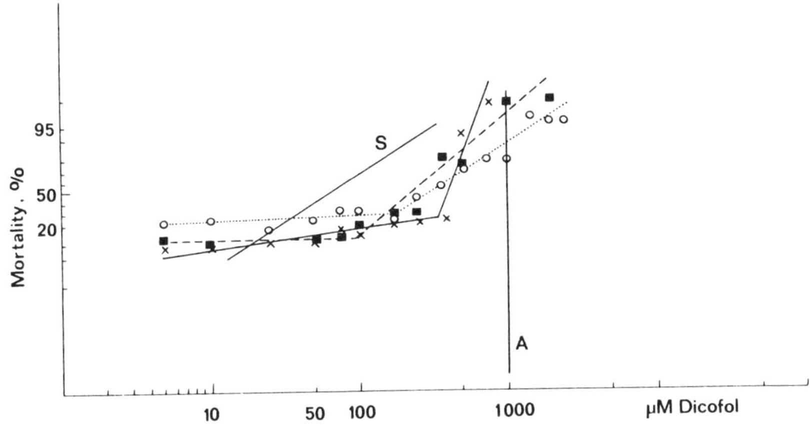

Figure 5   Toxicity of dicofol to multiple resistant *T. urticae*: x—x, Erfurt;   ■- - -■, Groß Gaglow; o....o, Rostock; S, susceptible reference strain; A, recommended application concentration 0.02% Milbol EC.

Table 4   Resistance of *T. urticae* to 2500 µM (925 ppm) dicofol

Strain	Number of tested females	Number of resistant females	%
Erfurt	1,134	0	0
Groß Gaglow	358	2	0.56
Rostock	849	19	2.24

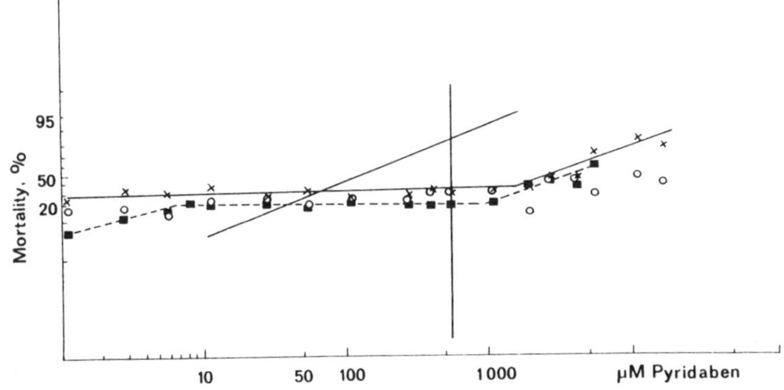

Figure 6   Toxicity of pyridaben to multiple resistant *T. urticae*: x—x, Erfurt;   ■- - -■, Groß Gaglow; o....o, Rostock; S, susceptible reference strain; A, 0.01% Sanmite.

Among ovicides, tetradifon is inactive in the Erfurt, Groß Gaglow and Rostock strains but sufficiently active on the S strain. All strains are highly susceptible to fenazox (data not shown).

The compounds listed in Table 5 have proved to be useful. Fenbutatinoxide was completely active when mortality was assessed after six days. Applied at the recommended concentration, clofentezin and hexythiazox were highly effective against the Erfurt strain.

Table 5   Susceptibility of *T. urticae* females to acaricides that are not affected by the multiple resistance system (three days' exposure)

Acaricide	Concentration tested	Mortality (%)		
		Erfurt	Groß Gaglow	Rostock
Dinobuton	0.1% Falimilbon	100	100	100
Azocyclotin	0.1% Peropal	100	100	91.4
Propargite	0.1% Omite	92.0	100	91.4
Fenbutatin oxide	0.2% Torque	76.6	82.0	46.3

Discussion

The patterns of resistance revealed in the Erfurt, Groß Gaglow and Rostock strains are very similar. The level of resistance to organophosphates, carbamates and formamidines is generally very high. After the efficacy of dimethoate had depleted, other organophosphates (e.g. dichlorvos) were used, especially against whiteflies, thus selecting the appropriate allele in *T. urticae*. The experiments do not indicate how many genes are involved to confer cross-resistance to carbamates, formamidines and pyridaben but the results strongly indicate that they are jointly inherited.

Contrary to this, dicofol resistance is low in the Rostock and Groß Gaglow strains and not found in the Erfurt strain. This is in contrast to the results obtained by Fergusson-Kolmes *et al.* (1991) who showed cross-resistance between amitraz and dicofol. In East Germany, dicofol is still a useful component in rotational schemes.

We doubt that the resistance to tetradifon is caused by the same allele as dimethoate resistance because earlier experiments carried out in our laboratory (Klunker, unpublished) did not reveal cross-resistance of organophosphate-resistant strains to this compound.

The pyridaben resistance is a clear case of cross-resistance although the compound belongs to a new class of acaricides (Hirata *et al.*, 1988) and has not been in use in this region. The populations remained susceptible to dinobuton acaricides, Omite, Peropal, Torque and Fentoxan.

Acknowledgement

This work was supported by the Chemie AG Bitterfeld. We are grateful to E. Blechschmidt and B. Himmrich for their technical assistance.

References

CRANHAM, J.E. and HELLE, W. (1985). Pesticide resistance in Tetranychidae. In *Spider Mites: Their Biology, Natural Enemies and Control* (W. Helle and M.W. Sabelis, eds), Volume 1B, pp. 405–421. *World Crop Pests.* Elsevier, Amsterdam.

FERGUSSON-KOLMES, L.A., SCOTT, J.G. and DENNEHY, T.J. (1991). Dicofol resistance in *Tetranychus urticae* (Acari: Tetranychidae): cross-resistance and pharmacokinetics. *Journal of Economic Entomology*, **84**, 41–48.

GEORGHIOU, G.P. (1990). Overview of insecticide resistance. In *Managing Resistance to Agrochemicals* (M.E. Green, H.M. LeBaron and W.K. Moberg, eds), *ACS Symposium Series*, Volume **421**, pp. 18–41.

HIRATA, K., KUDO, M., MIYAKE, T., KAWAMURA, Y. and OGURA, T. (1988). NC 129 – a new acaricide. *Proceedings of the 1988 British Crop Protection Conference – Pests and Diseases*, Volume 1, pp. 41–48.

KLUNKER, R. (1970). Untersuchungen zur Dimethoat-Resistenz der Gemeinen Spinnmilbe (*Tetranychus urticae* Koch). *Archiv für Phytopathologie und Pflanzenschutz*, **6** 17–29.

MARTINSON, T.E., DENNEHY, T.J., NYROP, J.P. and REISSIG, W.H. (1991). Field measurements of selection for two-spotted spider mite (Acari: Tetranychidae) resistance to dicofol in apple orchards. *Journal of Economic Entomology*, **84**, 7–16.

OTTO, D. (1988). Zur Resistenz der Gemeinen Spinnmilbe *Tetranychus urticae* KOCH und der Grünen Pfirsichblattlaus *Myzus persicae* SULZ. gegen Methamidophos. *Nachrichtenblatt für den Pflanzenschutz in der DDR*, **42**, 184–186.

STEINHAUSEN, W.R. (1968). Ein neues Akarizid zur Bekämpfung phosporsäureesterresistenter Spinnmilben mit durch Resistenz erzeugter höherer Empfindlichkeit. *Zeitschrift für angewandte Zoologie*, **55**, 107–114.

43
Susceptibility of Spider Mite Strains to Amitraz and Dicofol

C.S. AVEYARD,[1] P. RICHTER,[2] and D. OTTO[2]

[1]Schering AG, Pflanzenschutz, Postfach 65 03 11, W-1000 Berlin 65,
   Germany
[2]Biologische Zentralanstalt Berlin, Institut für Phytopharmakologie,
   Stahnsdorfer Damm 81, 0-1532 Kleinmachnow, Germany

Introduction

The effective management of acaricide resistance requires a good understanding of cross-resistance relationships between acaricides in order to allow for the alternation of compounds which do not select for the same resistance mechanism. Resistance in spider mites to amitraz has been reported in South Africa (Giliomee and Pringle, 1985) and in Japan (Inoue, 1984) where resistance was considered to be due to a single incompletely dominant gene.

Multiple resistance between dicofol and propargite has been previously reported in *Tetranychus* spp. (Grafton-Cardwell *et al.*, 1987). In addition, Cranham and Helle (1985) referred to dicofol resistance in spider mites being inherited as a single incomplete recessive gene.

It was therefore surprising when Fergusson-Kolmes *et al.* (1991) reported cross-resistance between dicofol and amitraz in a strain of *Tetranychus urticae* on apples from the State of New York, USA. Their use of near isogenic strains produced by backcrossing indicated that the application of either amitraz or dicofol would select for resistance to both compounds.

Consequently, further work was carried out to examine the

Insecticides: Mechanism of Action and Resistance
© 1992   Intercept Ltd, P.O. Box 716, Andover, Hants SP10 1YG, UK

possible occurrence elsewhere of cross-resistance between dicofol and amitraz.

Materials and Methods

The leaf disc dip test has been used to measure the susceptibility of adult spider mites to the test compounds. Leaf discs or sections approximately 10–15 mm diameter were cut from either bean leaves for *T. urticae* or plum leaves for *Panonychus ulmi* and dipped in the test solutions for approximately five seconds. The leaf discs were then placed on filter paper kept moist by a continuous water reservoir. After drying, 10–15 adult female mites were transferred to the leaf discs. Mite mortality was assessed after 2–3 days' exposure at room temperature.

An ovolarvicide test was also conducted on similar lines where adult female mites were allowed to lay eggs on the leaf disc for one day prior to being removed and treating the discs a day later. Mortality was assessed eight days after treatment, based on the number of eggs remaining unhatched and the survival of hatched larvae. In larvicide tests, the hatched larvae on the bean leaf discs were treated at 0–3 days after egg hatch and mortality was assessed five days after treatment.

Results and Discussion

At Kleinmachnow, the dicofol-resistant (R-Dic) strain was found to be very homogeneous. There was no significant presence of susceptible individuals when the data were subjected to spline regression analysis (Heinig *et al.*, 1989). The R-Dic strain was highly resistant to dicofol, to such a degree that the high concentrations needed to give significant mite mortality caused phytotoxicity to the leaf discs (Table 1). An LC_{50} ratio to the laboratory standard was calculated to be 156.3. In contrast, a positive concentration–mortality relationship was found with amitraz when tested on adult mites on the R-Dic strain (Table 2). There was no difference in the LC_{50} values for both the laboratory-susceptible strain and the R-Dic strain. In addition, the activity of amitraz was equally high in both strains in the ovolarvicide test (Table 3).

At Frohnau, three strains (MR, SIM 2 and DR strains) of *Tetranychus urticae* were found to show reduced susceptibility to dicofol (Tables 4 and 5) compared to a normal laboratory-susceptible strain (N strain). In contrast, amitraz was equally active against all these mite strains. Another strain (OPR) showing resistance to the organophosphate demeton-*S*-methyl was as susceptible to both amitraz and dicofol as the N strain.

Table 1   Adulticide tests with dicofol on dicofol-susceptible (S) and dicofol-resistant strains (R-Dic) of *T. urticae*, Kleinmachnow 1990

Dicofol	% Mortality (Abbott)	
(ppm a$_i$)	S strain	R-Dic
0	(1.4)	(2.8)
9.6	10.8	0
24	15.0	1.1
48	28.9	1.2
168	94.8	2.4
360	100.0	0
960	–	5.1
9600	–	64.8

Table 2   Adulticide tests with amitraz on dicofol-susceptible (S) and dicofol-resistant strains (R-Dic) of *T. urticae*, Kleinmachnow 1990

Amitraz	% Mortality (Abbott)	
(ppm a$_i$)	S strain	R-Dic
0	(2.4)	(2.6)
10	17.7	16.4
20	57.8	55.4
30	–	83.7
40	81.9	97.1
70	–	97.4
100	100.0	100.0
400	100.0	100.0

Laboratory bioassays carried out on *Panonychus ulmi* adult females obtained from one out of two pear sites in France showed reduced activity with amitraz but not with dicofol (Table 6). Conversely, *P. ulmi* obtained from two apple sites in Italy was equally susceptible to amitraz but not to dicofol (Table 7).

Table 3   Ovolarvicide tests with dicofol-susceptible (S) and dicofol-resistant strains (R-Dic) of *T. urticae*, Kleinmachnow 1990

Amitraz	% Mortality (Abbott)	
(ppm a_i)	S strain	R-Dic
0	(8.3)	(5.9)
2	–	78.2
4	99.7	96.7
10	100.0	100.0
20	100.0	100.0

Table 4   Laboratory tests on adult females from three strains of *Tetranychus urticae*, Berlin 1990

ppm a_i		% Control (Abbott)		
		N strain	SIM 2	MR
Amitraz	25.0	100	99	99
Dicofol	40.0	100	0	0
Clofentezine	200.0	–	0	73
	8.0	100	–	–
	1.6	50	–	–

N strain, Normal laboratory-susceptible strain.
SIM 2, Apples, clofentezine-resistant – Australia.
MR, Roses Australia, clofentezine-resistant

In conclusion, laboratory tests on various life stages of twelve strains of spider mites, including *Tetranychus urticae* and *Panonychus ulmi*, failed to reveal any indication of cross-resistance between amitraz and dicofol.Several authors (Pree, 1987; Grafton-Cardwell *et al.*, 1987) have reported the occurrence of multiple resistance between dicofol and other acaricides, namely cyhexatin and propargite. The occurrence of multiple resistance between amitraz and dicofol cannot be excluded but should be regarded as a rare occurrence. Consequently, spider mites should continue to be regarded as having different mechanisms of resistance for amitraz and dicofol and that both these compounds can be used in resistance management strategies.

Table 5 Susceptibility of larvae of three strains of *T. urticae*, Berlin, Germany 1988

		% Control (Abbott)		
Compound	ppm a_i	N strain	OPR strain	DR strain
Amitraz	64	100	100	100
	10	96	100	98
Dicofol	160	100	100	22
	25	76	65	0
Demeton-	160	100	15	99
S-methyl	25	88	1	91
Untreated	–	(7)	(3)	(5)

N strain, Normal susceptible strain.
OPR strain, Organophosphate-resistant.
DR strain, Dicofol-resistant.

Table 6 Susceptibility of adult *P. ulmi* to amitraz and dicofol in pears, France 1990

	% Control (Abbott)			
	Amitraz		Dicofol	
ppm a_i	Site 1	Site 2	Site 1	Site 2
0	(25)	(20)	(25)	(20)
50	87	9	37	53
250	43	16	83	84
500	83	44	100	94
1500	100	34	97	100

Table 7 Susceptibility of adult *P. ulmi* to amitraz and dicofol in apples, Italy 1990

ppm a_i	% Mortality (Abbott)			
	Amitraz		Dicofol	
	Site 1	Site 2	Site 1	Site 2
0	(9)	(5)	(9)	(5)
25	85	83	60	8
250	–	–	75	38
300	96	95	–	–

Acknowledgements

We acknowledge all Schering colleagues, particularly Mr Patrice Lagouarde, Dr Luigi Groppi and Dr Dietrich Baumert who have contributed to the data reported in this chapter.

References

CRANHAM, J.E. and HELLE, W. (1985). Pesticide resistance in Tetranychidae. In *Spider Mites: Their Biology, Natural Enemies and Control* (W. Helle and M.W. Sabelis, eds), Volume 1B, *World Crop Pests*. Elsevier, Amsterdam.

FERGUSSON-KOLMES, L.A., SCOTT, J.G. and DENNEHY, T.J. (1991). Dicofol resistance in *Tetranychus urticae* (Acari: Tetranychidae): Cross-resistance and pharmacokinetics. *Journal of Economic Entomology*, **84**(1), 41–48.

GILIOMEE, J.H. and PRINGLE, K.L. (1985). The sensitivity of six colonies of the red spider mite, *Tetranychus cinnabarinus* (Boisd.) to amitraz. *Journal of the Entomological Society of Southern Africa*, **48** (2), 325–330.

GRAFTON-CARDWELL, E.E., GRANETT, J. and LEIGH, T.F. (1987). Spider mite species (Acari: Tetranychidae) response to propargite: Basis for an acaricide resistance management programme. *Journal of Economic Entomology*, **80**, 579–587.

HEINIG, P., OTTO, D. and MOLL, E. (1989). Spline regression as a tool for analysis of population genetics in resistant populations. *Tag.-Ber., Acad. Landwirtsch.-Wiss. DDR, Berlin*, **274**, 215–218.

INOUE, K. (1984). Resistance to amitraz in the citrus red mite, *Panonychus citri* (McGregor) in relation to population genetics.

Japanese Journal Applied Entomology and Zoology, **28**, 260–268.
PREE, D.J. (1987). Inheritance and management of cyhexatin and dicofol resistance in the European red mite (Acari: Tetranychidae). *Journal of Economic Entomology*, **80**(6), 1106–1112.

44
Metabolic Resistance Factors of *Tetranychus urticae* (Koch)

U. SCHOKNECHT and D. OTTO

Biological Research Centre Berlin, Stahnsdorfer Damm 81, 0-1532 Kleinmachnow, Germany

Spider mites (*Tetranychus urticae*) are among those species with a particularly high tendency to develop resistance to acaricides. Publications already exist reporting resistance in mite populations to the organophosphorus compounds (Otto, 1988), dicofol (Kono *et al.*, 1981), amitraz (Richter and Schulz, 1990), cyhexatin (James *et al.*, 1988) and propargite (Keena and Granett, 1990).

The occurrence of resistance is dependent upon the population dynamics, selection pressure and the genetic capability of the species to express resistance factors. In addition to changes in target proteins, detoxification of pesticides is one of the major causes of resistance. Esterases, mixed function oxidases and glutathione *S*-transferases are involved in the degradation of xenobiotics. The most important esterases in the degradation of acaricides are phosphatases and carboxylesterases. Cytochrome-P-450 and flavine-dependent oxidases degrade foreign compounds in mammals, whereas in arthropods only cytochrome-P-450-dependent oxidases have so far been described. The enzyme reactions generally increase the water solubility of lipophilic xenobiotics, and in this way accelerate excretion (for detailed information see Schoknecht and Otto, 1989).

To investigate detoxifying enzymes, several strains of *T. urticae* with different resistance levels to acaricides were used (Table 1). All data were related to the susceptible strains – S.

Insecticides: Mechanism of Action and Resistance
© 1992   Intercept Ltd, P.O. Box 716, Andover, Hants SP10 1YG, UK

Table 1 Resistance of *Tetranychus urticae* populations to several acaricides

Strain		Dimethoate (Bi 58)	Methamidophos (Filitox)	Dicofol (Milbol)	Cyhexatin (Plictran)
S	LC_{50}*	0.0015%	0.0011%	0.014%	0.0024%
R DIM	R_i**	2800	20.0	1.4	1.5
R METH	R_i	very high	1479.5	1.0	0.9
R DIC	R_i	0.7	6.2	n.d.	4.2
P GG	R_i	164.8	121.7	3.8	n.d.
P E	R_i	134.5	58.0	5.9	n.d.
F AKEN	R_i	0.1	n.d.	n.d.	n.d.

* = Lethal concentrations 50. ** = Resistance index. n.d. = Not detected.

The strains R DIM, R DIC, and R METH were selected in the laboratory with dimethoate, dicofol and methamidophos, respectively. Whereas R DIM is only slightly resistant to methamidophos, R METH is cross-resistant to methamidophos and dimethoate. P GG and P E are multiresistant strains from glasshouses in Groß Gaglow (near Cottbus) and Erfurt (both in East Germany). The strain F AKEN was collected from a nature reserve at the river Elbe. Its susceptibility to dimethoate is 10-fold higher than that of the strain S.

The activity of detoxifying enzymes with model substrates differed between the examined strains (Table 2). The activity of glutathione$S$-transferases with CDNB (1-chloro-2,4-dinitrobenzene) was highest of the multiresistant strain P E. This activity was also increased in the strains R DIC and P GG, but low in the other strains. The glutathione S-transferase activity in the strain from the nature reserve (not included in Table 1) amounted to 55% of the activity of the susceptible strain. Esterase activity with α-naphthylacetate was highest in the glasshouse strain from Groß Gaglow. The laboratory strain selected with dicofol additionally possesses higher p-nitroanisole O-demethylase activity than the other strains.

The existence of cytochrome P 450 in *T. urticae* and in *Myzus persicae* was demonstrated. Peaks at 450 nm in carbon monoxide difference spectra are caused by this haem protein. For comparison, corresponding spectra were recorded with microsomes from *Musca domestica* and *Drosophila melanogaster*, since higher amounts of cytochrome P 450 could be obtained from these species (Figure 1).

Table 2   Activity of detoxifying enzymes from *Tetranychus urticae* populations with model substrates

Strain	Glutathione *S*-transferase (CDNB) (nmol min^{-1} mg^{-1})	Esterase (α-naphthylacetate) (nmol ♀$^{-1}$ min^{-1})	Oxidase (*p*-nitroanisole) (nmol min^{-1} mg^{-1})
S	97 ± 21	64 ± 26	4.6
R DIM	81 ± 17	66 ± 25	5.8
R DIC	155 ± 25**	53 ± 21***	7.8
R METH	77 ± 28	61 ± 21	n.d.
P GG	122 ± 8*	104 ± 36***	n.d.
P E	240 ± 70*	59 ± 27	6.4

* $\alpha < 0.1$.
** $\alpha < 0.01$.
*** $\alpha < 0.005$.

Assay conditions
Conjunction of 1-chloro-2,4-dinitrobenzene (CDNB) with glutathione (GSH), according to Habig *et al.* (1974): 0.4 mM CDNB, 4 mM GSH, 200 g protein (105 000 g supernatant) in 1 ml 0.05 Tris-HCL buffer pH 7.5.

α-naphthylacetate (α-NA) esterase activity; according to Gomori (1953): equivalent of 1/2 adult female (homogenate) and 0.025 mM α-NA in 500 l Na phosphate buffer pH 7.0.

O-Demethylation of *p*-nitroanisole; according to Shigematsu (see Nedelkina *et al.*, 1988): 600 l microsomal fraction (0.3–1.5 mg protein), 6.25 mM NADPH, 0.05 mM *p*-nitroanisole (dissolved in acetone), 1.25 mM MgCl$_2$ in 2 ml 0.1 M K phosphate buffer pH 7.4.

The sensitivity of the esterase assay with α-naphthylacetate allowed measurements with single adult females on microtitre plates. In this way the distribution of esterase activity within the strains could be examined. Equivalent experiments were carried out with the combined homogenate of a hundred individuals from the R DIM strain to compare the deviation of the resultant values. The diversity of the enzyme activity resembles a Gauss curve in all populations (Figure 2). There is no evidence that the populations are divided into subpopulations with different esterase activities as is the case with *M. persicae* (see Devonshire and Moores, 1982). However, a separation into classes might depend on the substrate used.

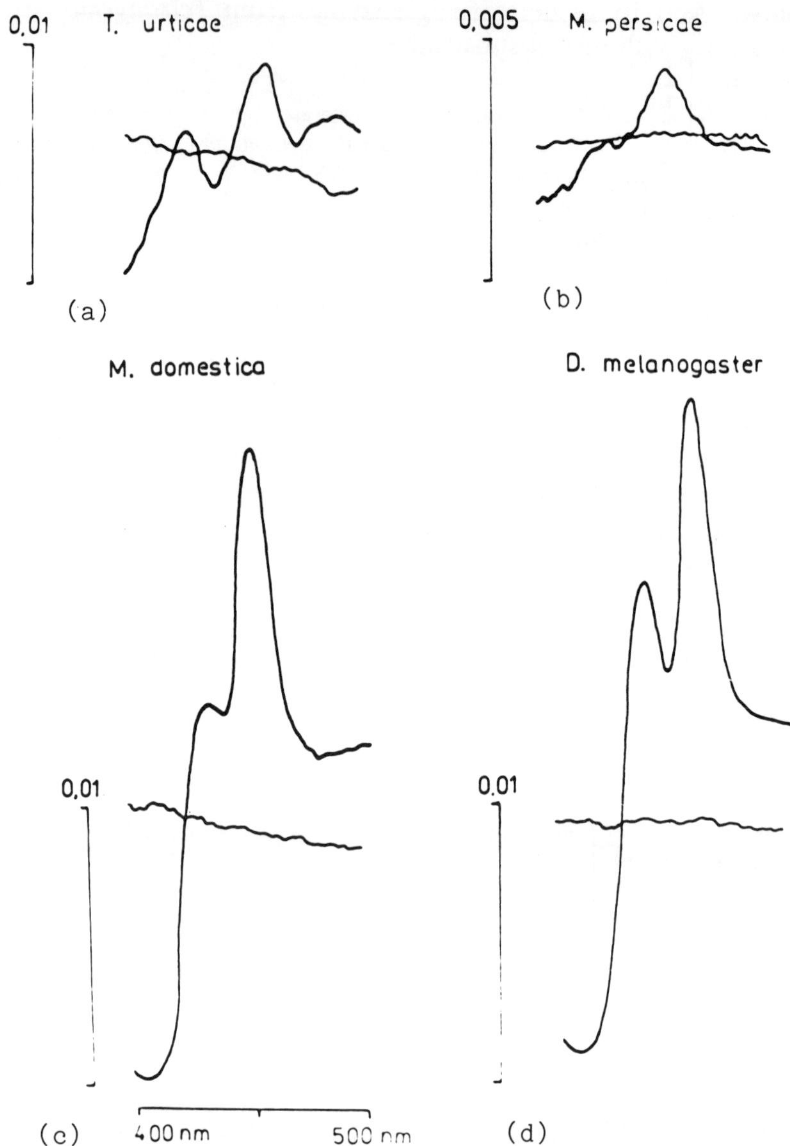

Figure 1   Carbon monoxide difference spectra of cytochrome P 450. (Recorded according to Omura and Sato (1964): (a) microsomal fractions of *Tetranychus urticae*; (b) *Myzus persicae*; (c) *Musca domestica*; (d) *Drosophila melanogaster* in 0.1M K phosphate buffer pH 7.5 with 0.1 mM EDTA and o.1 mM dithiothreitol containing 20% glycerol.)

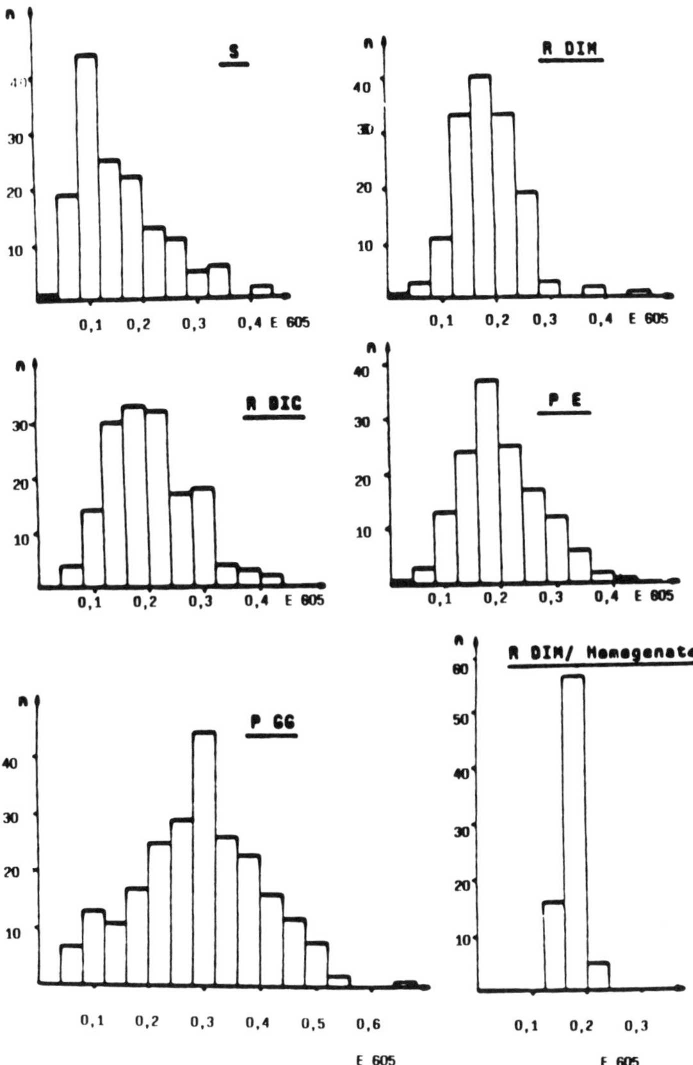

Figure 2 Distribution of the activity of α-naphthylacetate esterases within populations of *Tetranychus urticae*. (*n* = number of animals that belong to the classes of distinctive esterase activity, S, R DIM, R METH, R DIC, P E, P GG: strains of *Tetranychus urticae*, see Table 1.)

More detailed investigations concentrated on glutathione *S*-transferase (GST). Several methods were tried to prepare these enzymes. Anion exchange chromatography resulted in up to a 10-

fold enrichment of the specific activity with a recovery of 40–70% of the original activity. Affinity chromatography with glutathione-marked agarose yielded 19-fold specific activity and 29% recovery. The purity of the product was highest after affinity chromatography as demonstrated by HPLC with silica gel 300 (Figure 3). The main protein peak at 10.3 minutes represents GST activity.

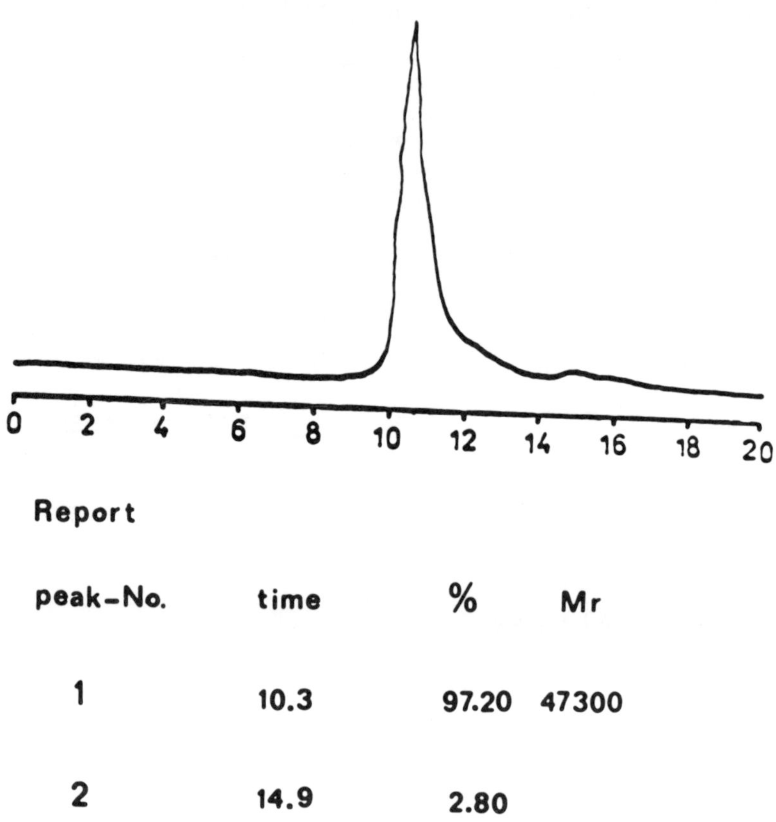

Figure 3   HPLC elution profile from silica gel 300 of glutathione *S*-transferases prepared by affinity chromatography of GSH agarose. (Elution buffer: 0.05 M Na phosphate pH 7.0, the curve shows the protein content after the elution, glutathione *S*-transferases occur after 10.8 minutes.)

Since the metabolism of acaricides may be catalysed by definite GST isoenzymes, the occurrence of isoenzymes in the strains examined was of special interest. The occurrence of isoenzymes was previously demonstrated by Šula (1989). Using a similar method, we found 4–7 isoenzymes with isoelectric points between 4 and 7. The strains possess typical isoenzyme patterns (Figure 4). With respect to the report of Clark *et al.* (1986) that isoenzymes with isoelectric points higher than 6.5 conjugate diazinone, lindane and methylparathione in houseflies, it should be pointed out that isoenzymes with isoelectric points higher than 6.5 were found only in the resistant strains. The substrate specificity of the isoenzymes is still unknown.

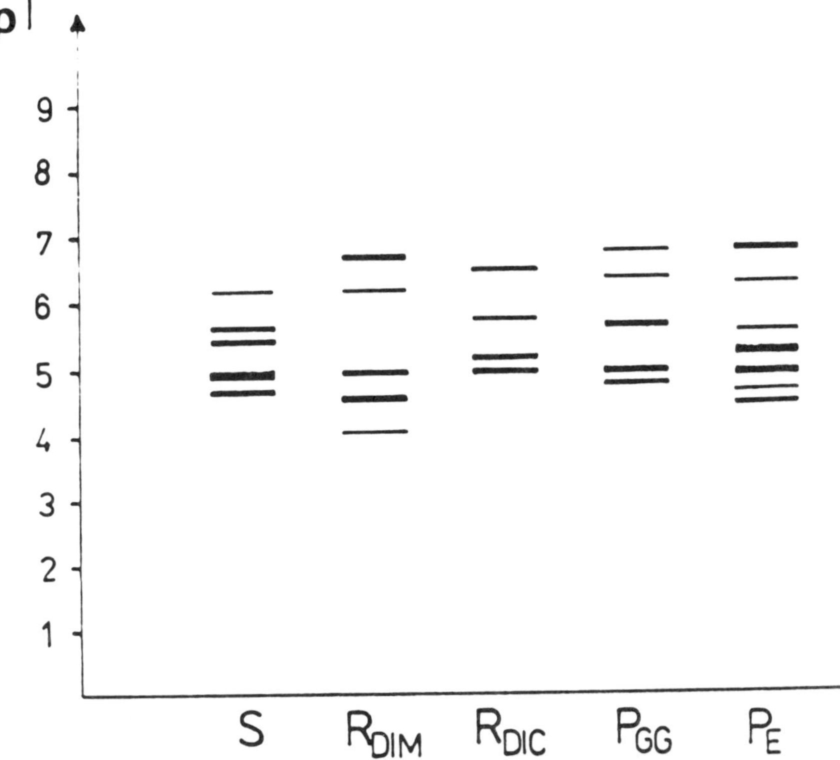

Figure 4   Isoenzymes of glutathione *S*-transferases in several strains of *Tetranychus urticae*. (105.000 g supernatants were separated by isoelectrofocusing (pH range 4–9), activity of glutathione *S*-transferases was visualized by the method of Kenney and Boyer (1981) with CDNB as the second substrate.)

Several compounds were tested as inhibitors of glutathione *S*-transferase (Figure 5). The mode of action of the herbicide synergist tridiphane is based on the delay of the degradation of the active compound atrazine (Lamoureux and Rusness, 1987). Organic tin compounds – generally known as inhibitors of ATPases – were also described as inhibitors of glutathione *S*-transferase. Acaricides such as cyhexatin and fenbutatinoxide have not yet, however, been examined for their ability to inhibit GSTs. The fungicide fentinacetate was included for comparison. The imidazoline derivatives RBr 1 and RBr 2 have been described as inhibitors of esterases, oxidases and glutathione *S*-transferases of *M. domestica* by Nedelkina *et al.* (1985).

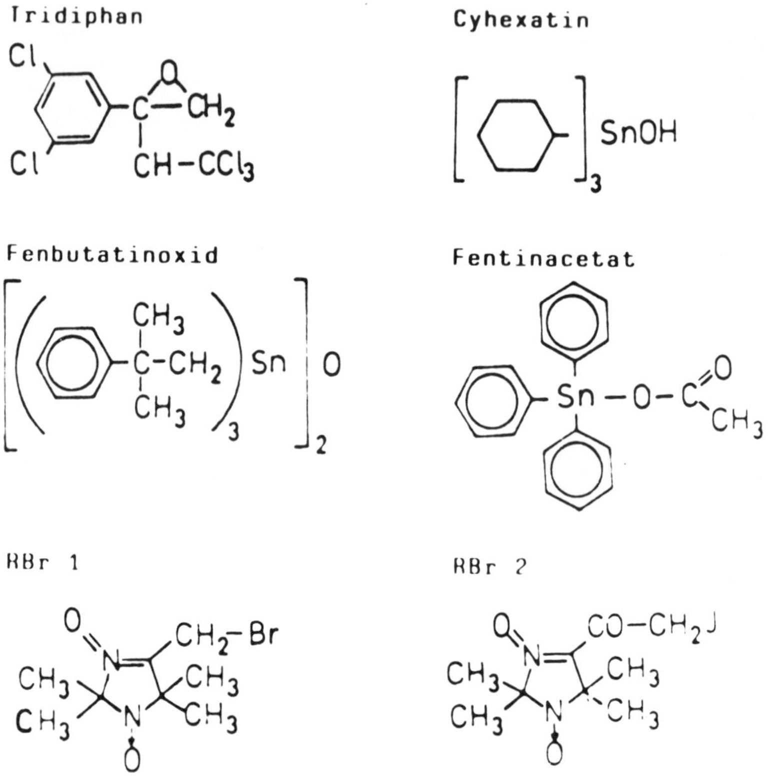

Figure 5 Structures of the compounds tested as possible inhibitors of glutathione *S*-transferases. (RBr 1 = 4-Bromomethyl-2,2,5,5-tetramethyl-3-imidazoline-3-oxide-1-oxyl; BRr 2 = 4-iodoacetyl-2,2,5,5-tetramethyl-3-imidazoline-3-oxide-1-oxyl.)

The compounds were added at a final concentration of 10^{-4} M to *in vitro* assays with CDNB. Results are expressed as a percentage of the control activity (Table 3). Data for spider mites are the mean values of measurements from five populations. The range of values is also shown. The organic tin compounds cyhexatin and fentinacetate were most effective in the experiments with spider mites. In this species the imidazoline derivatives inhibited about half of the glutathione *S*-transferase activity. Fenbutatinoxide was less effective. Tridiphane did not affect the enzyme activity. There were significant differences between the sensitivity of glutathione *S*-transferase in spider mites and houseflies. This could be related to different isoenzymes occurring in these species.

Table 3   Inhibition of glutathione *S*-transferases from *Tetranychus urticae* and *Musca domestica*. (Results are expressed as a percentage of the control activity with CDNB in an *in vitro* assay; for conditions see Table 2)

Compound (10^{-4} M)	Remaining activity (%)	
	Tetranychus urticae (5 populations) Min ... x̄ ... Max	*Musca domestica*
Tridiphane	87 ... *95* ... 99	65
Cyhexatin	10 ... *20* ... 42 (P GG)	88
Fenbutatinoxide	59 ... *77* ... 94	0
Fentinacetate	21 (P GG)	10
RBr 1	28 ... *46* ... 75 (R DIC)	n.d.
RBr 2	30 ... *47* ... 60	n.d.

In general, the properties of glutathione *S*-transferases in spider mites resembles those of other arthropod species as reviewed by Clark (1989) (Table 4).

Summarizing all results it can be pointed out that *T. urticae* possesses several xenobiotic-metabolizing enzymes that can cause resistance although nothing is known yet about the activity of these enzymes to acaricides.

Table 4 Properties of glutathione *S*-transferases from *Tetranychus urticae* in comparison to other arthropod species

Character	*Tetranychus urticae*	Arthropods
Compartment	Cytosol (97%)	Cytosol
pH optimum	Weakly basic	neutral–weak basic
Activity with		
CDNB (mol min^{-1} mg^{-1})	0.097–0.243	0.004–214
Km (CDNB)	0.12–0.26 mM	0.23–0.71 mM
Molecular weight	47 kDa	36–50 kDa
Occurrence of isoenzymes	4–7	1–5
Isoelectric points	pH 4.0–6.7	pH 5–10
Induction	Tridiphane	Tridiphane and other compounds

Acknowledgement

The authors thank Dr S.V. Nedelkina from the Institute of Cytology and Genetics Novosibirsk, USSR, for cooperating in the investigation of cytochrome-P-450-dependent oxidases and Dr J Šula for instructions on the technique of the demonstration of GST isoenzymes.

References

CLARK, A.G. (1989). The comparative enzymology of the glutathione S-transferases from non-vertebrate organisms. *Comparative Biochemistry and Physiology*, **92B**(3), 419–446.

CLARK, A.G., SHAMAAN, N.A., SINCLAIR, M.D. and DAUTERMAN, W.C. (1986). Insecticide metabolism by multiple glutathione S-transferases in two strains of the housefly, *Musca domestica* (L.). *Pesticide Biochemistry and Physiology*, **25**, 169–175

DEVONSHIRE, A.L. and MOORES, G.D. (1982). A carboxylesterase with broad substrate specificity causes organophosphorus, carbamate and pyrethroid resistance in peach-potato-aphids *M. persicae* (Sulz.). *Pesticide Biochemistry and Physiology*, **182**, 235–246.

GOMORI, G. (1953). Human esterases. *Journal of Laboratory and Clinical Medicine*, **42**, 445–453.

HABIG, W.H., PABST, M.J. and JAKOBY, W.B. (1974). Glutathione S-transferases: The first enzymatic step in mercapturic acid formation. *Journal of Biological Chemistry*, **249**, 7130–7139.

JAMES, D.G., EDGE, V.E. and ROPHAIL, J. (1988). Influence of temperature on cyhexatin resistance in *Tetranychus urticae* (Acari: Tetranychidae). *Experimental and Applied Acarology*, **5**(1/2), 15–22.

KEENA, M.A. and GRANETT, J. (1990). Genetic analysis of propargite resistance in pacific spider mites and two-spotted spider mites (Acari: Tetranychidae). *Journal of Economic Entomology*, **83**(3), 655–661.

KENNEY, W.C. and BOYER, T.D. (1981). Detection of glutathione S-transferase isoenzymes after electrofocussing in polyacrylamide gels. *Analytical Biochemistry*, **116**, 344–348.

KONO, S., SAITO, T. and MIYATA, T. (1981). Mechanism of resistance to dicofol in the twospotted spider mite *Tetranychus urticae* KOCH (Acari: Tetranychidae). *Japan Journal of Applied Entomology and Zoology*, **25**, 101–107.

LAMOUREUX, G.L. and RUSNESS, D.G. (1987). Synergism of diazinon toxicity and inhibition of diazinon metabolism in the housefly by tridiphane: Inhibition of glutathione S-transferase activity. *Pesticide Biochemistry and Physiology*, **27**, 318–329.

NEDELKINA, S.V., LEONOVA, I.N., SALGANIK, R.I., ZENK-OVA, T.J.,VAINER, L.M. and POPOVA, V.I. (1985). Inhibition of insecticide detoxication enzymes in insects by the cytochrome P 450 substrate alkylating analog. *Biokhimiya*, **50**(3), 1284–1289.

NEDELKINA, S.V., SOLOMENNIKOVA, I.V., LUKASHIN, N.S., LEONOVA, I.N., RAUSCHENBACH, I.J. and SALGA-NIK, R.I. (1988). Study of insecticide detoxication enzymatic systems in the permethrin-resistant population of the colorado potato beetle. *Biokhimiya*, **53**(1), 11–18.

OMURA, I. and SATO, R. (1964). The carbon-monoxide binding pigment of liver microsomes. II. Solubilization, purification, and properties. *Journal of Biological Chemistry*, **239**, 2370–2378.

OTTO, D. (1988). Zur Resistenz der Gemeinen Spinnmilbe *Tetranychus urticae* KOCH und der Grünen Pfirsichblattlaus *Myzus persicae* SULZ. gegen Methamidophos. *Nachrichtenblatt für den Pflanzenschutz in der DDR*, **42**, 184–186.

RICHTER, P. and SCHULZ, H.-U. (1990). Untersuchungen zum Resistenzspektrum einer Gewächshauspopulation der Gemeinen Spinnmilbe *Tetranychus urticae* Koch, aus Groß Gaglow (Kreis Cottbus). *Archiv für Phytopathologie und Pflanzenschutz, Berlin*, **26**(6), 551–556.

SCHOKNECHT, U. and OTTO, D. (1989). Enzymes involved in the metabolism of organophosphorus, carbamate and pyrethroid insecticides. *Chemistry of Plant Protection*, **2**, 117–156, Springer Verlag.

ŠULA, J. (1989). Glutathione S-transferases in several strains of *Tetranychus urticae*. In *Insecticides – Mechanisms of Action and Resistance, Proceedings* (Academy of Agricultural Sciences of the GDR, Vol. 274), pp. 281–286. Berlin.

45
Considerations on the Resistance of *Phytoseiulus persimilis* Athias-Henriot by Means of Genetic Model Approaches*

A.Y.M. EL-LAITHY

National Research Centre, Department Pests and Plant Protection, Cairo 12311 Dokki, Egypt

Introduction

Resistance of pests to pesticides is a vital problem which has been focused by several researchers. Hoskins and Gordon (1956) defined characteristics of resistance, some genetic aspects and mechanisms. A comprehensive report about the evolution of the phenomenon of resistance of different pest species was discussed by Georghiou (1972) who explains its correlation with biological, physiological and behavioural factors. An analytical study of Crow (1957) is concerned with the genetic origin of resistance either as a pre- or post-adaptive mechanism. He concluded a pre-adaptive origin. In order to get a precise interpretation of the interaction between factors, including resistance, studies on resistance as an evolutionary phenomenon, using mathematical computer models, were performed by Georghiou and Taylor (1977a,b), Plapp *et al.* (1979) and Chin (1987).

The latest study was carried out by Tabashnik (1990) who elucidated the different types of resistance models based on the

---

* This chapter was presented at the Symposium under the title 'Prediction of Acaricidal Resistance of the Predatory Mite *Phytoseiulus persimilis* Athias-Henriot Based on Simulation of a Genetic Model'.

Insecticides: Mechanism of Action and Resistance
© 1992   Intercept Ltd, P.O. Box 716, Andover, Hants SP10 1YG, UK

following criteria: assumptions about the mode of inheritance of resistance; model approach, which may be analytical or empirical or a simulation or optimization; a modern definition of factors affecting resistance which include biological, operational and ecological factors. Further problems are the evolution of resistance in one pest species to one pesticide or to several pesticides used in mixtures and the interaction between species resulting from natural pest enemies.

Although resistance of pests to pesticides is considered the dilemma which attends pesticide application, it shows simultaneously a beneficial phase when it develops in populations of biocontrol agents such as predators and parasites. Since 1970, the latter concept has been widely investigated, but it is relatively restricted to predacious mites which are widely used in biological control programmes against the two-spotted spider mite *Tetranychus urticae* Koch in both orchards and greenhouses.

Croft and Strickler (1983) reported that among the phytoseiid predators resistance was recorded either as natural in orchards resulting from a frequent application of pesticides or selected in the laboratory (Hoy, 1985). At least seven resistant species are known, the most widespread exploited of which were *Metaseiulus occidentalis, Amblyseius fallacis, Typhlodromus pyri* and *Phytoseiulus persimilis*. The report about resistance of arthropod natural enemies by Hoy (1990) mentioned the effectiveness of resistant predatory mites in biological control, whereas the insect predators and parasites showed generally a poor resistance level against pesticides in the field which was interpreted as a result of lacking variability in these insect populations.

The predatory mite *Phytoseiulus persimilis* Athias-Henriot is widely used for the biological control of two-spotted spider mites in greenhouses (McMurtry, 1981). Therefore, many studies deal with the development of a resistant breeding line of this species against different chemical groups of pesticides (Schulten and Klashorst, 1974; Schulten *et al.*, 1976; Konig and Kassan, 1986; El-Laithy, 1986).

The present investigation, as the first trial, attempts to incorporate the given data about inheritance of resistance in a population of *P. persimilis* in a genetic model approach to study the evolution of resistance at different selection pressures along with the calculation of the necessary selection cycles to obtain a homozygous resistant strain under the given assumptions.

Assumptions on Resistance

The development of resistance in terms of change in the frequency of resistance genes was simulated on an IBM PC. Assumptions of this model approach are based on recent knowledge derived from experimental data published by several authors to achieve practical deductions for the concept of resistance evaluation in the predatory mite *P. persimilis*. The data presented in Table 1 reflect the modes of inheritance of resistance against organophosphorus insecticides of phytoseiid predators reviewed by Hoy (1985).

Table 1 OP resistance genes in phytoseiid predatory mites

Species	Genetic description	Compound	Authors
Amblyseius fallacis	Partially dominant single gene	Azinphos-methyl	Croft *et al.* (1976)
Phytoseiulus persimilis	Single dominant gene	Ethion and Carbofos	Silberminsz and Petrushov (1980)
P. persimilis	Single dominant factor	Parathion and demeton-*S*-methyl	Schulten and Klashorst (1974)
Metaseiulus occidentalis	Major dominant gene	Azinphos-methyl	Hoy (1985)
Typhlodromus pyri	Semi-dominant major gene	Parathion	Overmeer and van Zon (1983)

Experimental Background

A dimethoate-resistant strain of the predatory mite *P. persimilis* (El-Laithy, 1986) was reselected over twelve cycles until the LC_{50} increased from 630 to 1000 ppm dimethoate. The assumptions about resistance depend on surveyed experimental results from several authors (Table 1). According to these results resistance to organophosphorus compounds is attributed to a single dominant gene in *P. persimilis*. Furthermore, findings of El-Laithy (1986) from tests with one susceptible and three different resistant strains indicated that male *P. persimilis* were twice as resistant to dimethoate as females. This situation could be interpreted in view of the findings of Lyon's theory (Lyon, 1968). In this sense the maternal karyotype includes a diploid set where only one X-set is active, while the other is inactive (mostly considered of hetero-chromatin). On the other hand, the male's karyotype consists only of an X-set of chromosomes. In this view the male paternal group shows a high degree of resistance to mutagenic euchromatin agents and an inherent high degree of resistance in sons (males) whereas females are mostly less resistant and exhibit a high degree of mortality. The resistance gene (R), therefore, contributes mostly to the new generation of males since their X-set has less heterochro-matin (genetically active). In the light of this interpretation it may be the reason for the lower effect of dominant R gene on the female set of chromosomes.

A concentration of 1000 ppm of dimethoate was considered a discriminatory dose to differentiate the proportions of resistant and susceptible genotypes in the population as the LD–P lines of the three genotypes RR, Rr and rr were not overlapping. McKenzie *et al.* (1982) estimated the frequency of the different genotypes of blowfly using respectively 0.5 µl, 0.01 µl and 0.025 µl of diazinon as discriminatory doses.

Figure 1 contrasts the different genotypes of the models in which RR is the reselected strain until LC_{50} increased from 630 to 1000 ppm, rr is the susceptible reference strain with a LC_{50} of 33 ppm dimethoate whereas the Rr genotype was assumed to have three different values according to the proposed grade of dominance (similar to RR a complete dominance is considered or between RR and rr a partial dominance.)

Description of the Model Approach and Parameters

As stated previously only a single gene in the predatory mite *P. persimilis* can be involved in OP resistance. Although the known data about the dominance grade presented in Table 1 showed generally the complete dominance of OP resistance genes, these results depend on the toxicological data for the genotype of parents as well as from backcrosses, which are not reliable for an absolute determination of dominance. A similar limitation was also mentioned by Plapp *et al.* (1979). Therefore, the present invest-igation includes the simulation of resistance development under the assumption that the resistance gene is either dominant or semi-dominant.

Two iterative sets of equations derived from Hardy–Weinberg (Sperlich, 1973) were used to calculate the change of gene frequency under different selection coefficients and different grades of dominance.

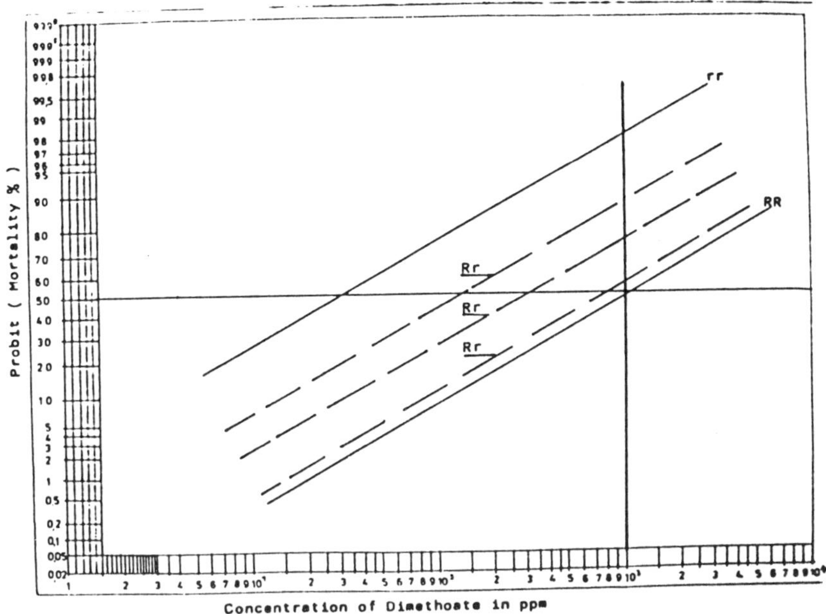

Figure 1   LD-P lines of the model strains (*RR, rr* and *Rr*).

Complete dominance:

$$q_1 = \frac{q_0(1 - sq_0)}{1 - sq(2o)}$$

Incomplete dominance:

$$q_1 = \frac{q_0(1 - hs + q_0hs - sq_0)}{1 - 2q_0hs(1 - q_0) - sq(2o)}$$

where q_0 = initial gene frequency;
q_1 = gene frequency after one cycle of selection;
s = selection coefficient;
h = measure of dominance grade;

where

$$h = \frac{1 - WRr}{1 - Wrr}$$

with *WRr* and *Wrr* denoting the survival rates (fitness) of the corresponding genotype.

Complete dominance of R allele to r allele

The proposition of resistant genotypes RR and Rr as shown in Figure 1 is 0.5. This equals a frequency of R gene $= \sqrt{0.5} = 0.07$. Applying the Hardy–Weinberg equation $p^2 + 2pq + q^2 = 1$ to these genotypes, P denotes r alleles and q denotes the frequency of r alleles. Proceeding on $p + q = 1$, the equation is

$$0.7 + q = 1$$
$$q = 0.3$$

Incomplete dominance

The initial frequencies of R allele are supposed to be 0.7, 0.5 and 0.3, i.e. the frequencies of the corresponding q are 0.3, 0.5 and 0.7. Factor h, measuring the degree of dominance, is used with increasing values, which reflect a decreasing degree of dominance: 0.3, 0.5 and 0.7. The selection coefficient (SC) for complete and incomplete dominance was determined as 0.25, 0.5, 0.75 which showed different selection pressures.

Recovery of a selected population

The last parameter under investigation is the time internal for the population to reach the required size after each selection cycle under the given climatic conditions. It was assumed that neither migration nor immigration occurs, besides the absence of refugia, which means that all individuals were exposed to the treatments. The time for a population to double can be calculated by the formula used by Boyking and Campbell (1982) as follows:

$$\text{Time for the population to double} = \frac{\text{Log}_e 2}{rm}$$

This equation is a derivation from the basic formula of Andrewartha and Birch (1954)

$$T = \frac{\log_e R_o}{rm}$$

where
R_o = net reproduction rate;
rm = intrinsic rate of natural increase;
T = mean generation time.

The latter equation could be applied to estimate different multiplication rates as follows:

$$\text{Time for the population to multiply } n \text{ times} = \frac{\log_e n}{rm}$$

Discussion of the Computer Simulation Results on Gene Frequency

Dominance

Figure 1 shows the change in the frequency of the susceptible allele r (i.e. its elimination rate) when the resistant allele R was completely dominant. The rate of change is slow. Figures 2a–c show the change of r allele frequency under ascending degree of incomplete dominance, i.e. $h = 0.3$, 0.5 and 0.7. The elimination rate of r allele increases also in an ascending manner. We can conclude that when the degree of incomplete dominance of R allele increases, the elimination rate of r allele also increases. Similar deduction was mentioned by Georghiou and Taylor (1977a).

Initial frequency of r allele

In the start population shown in Figure 1 the initial frequency was 0.3; t was then assumed to have higher values, i.e. 0.5 and 0.7 (Figures 2e–k). The number of selection cycles needed to reduce these initial frequencies to 0.1, for example, were obviously decreasing. In Figures 2e, f, g the cycles to reduce the initial frequency from 0.5 to 0.1 average 17, 14 and 11 (each value represents a degree of incomplete dominance in an ascending order). In Figures 2h–k about twenty-four, twenty and seventeen cycles were necessary to reduce the initial frequency from 0.7 to 0.1. Comparing these cycles with those in Figures 2a–c, we can conclude that the elimination rate of the r allele slightly increased as its initial frequency increased.

It is also remarkable that at each initial frequency the change of r allele occurred during fewer selection cycles as the degree of incomplete dominance increased. These results are compared with the study of Georghiou and Taylor (1977a) who implied a very low initial frequency and, therefore, mentioned no influence on the rate of change of the same allele.

Selection coefficient

The figures show that the change of frequency of r allele was steeper as the SC (selection pressure LC_{25}, LC_{50}, LC_{75}) increased.

This effect was relatively strong in the early cycles of selections whereas it gradually weakened in the later cycles.

Population size after selection

Provided the population was reared under room conditions, the *rm*

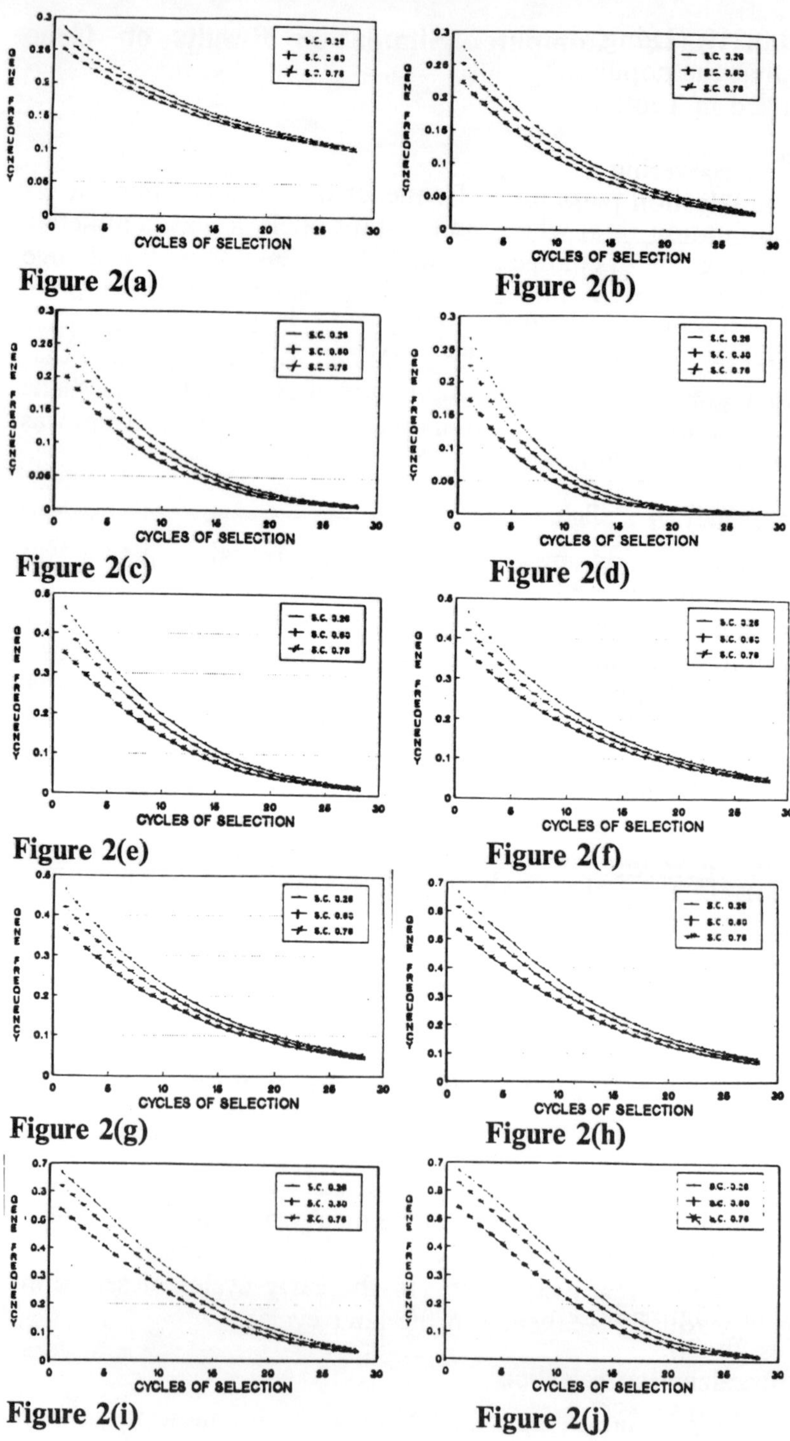

Figure 2(a)

Figure 2(b)

Figure 2(c)

Figure 2(d)

Figure 2(e)

Figure 2(f)

Figure 2(g)

Figure 2(h)

Figure 2(i)

Figure 2(j)

value is 0.219 (Laing, 1968). To estimate the population size, we assume that the population will be exposed to selection pressures as summarized in Table 2.

Table 2　Recovering of predatory mite populations after exposure to various selection pressure

Selection pressure Parameter	LC_{50}	LC_{60}	LC_{80}	LC_{90}
Survival rate	50	40	20	10
Multiplication rate (n) to recover the normal population size	2	2.5	5	10
Time interval (days)	3.16	4.18	7.34	10.5

Finally, one question arises which is to be clarified by further investigations: What will be the behaviour of the population of the predatory mite-resistant strain with an LC_{50} of 1000 ppm dimethoate if it is selected for about thirty cycles? In other words: Are the cycles obtained in the present study sufficient to obtain a homozygous resistant strain of the predatory mite or must another development be assumed?

Acknowledgement

The author is deeply in debt to Dr A.S. El-Balall, Department of Genetics, National Research Centre in Dokki/Egypt, for his valuable advice throughout the course of this study and for criticizing the manuscript.

Figures 2a–j　Computer simulations showing changes in frequency of r gene under different selection coefficients when R allele is dominant (Figure 2a) and when R allele is in incomplete dominance (Figures 2b–2j).
Initial frequencies: 2a–d ($r = 0.3$; $R = 0.7$); 2e–g ($r = 0.5$; $R = 0.5$); 2h–j ($r = 0.7$; $R = 0.3$). (SC: selection coefficient)

References

ANDREWARTHA, H.G. and BIRCH, L.C. (1954). *The Distribution and Abundance of Animals*. University of Chicago Press, Chicago.

BOYKING, L.S. and CAMPBELL, W.V. (1982). Rate of population increase of the two spotted spider mite (Acari: Tetranychidae) on peanut leaves treated with pesticides. *Journal of Economic Entomology*, **75**, 966–971.

CHIN, K.M. (1987). A simple model of selection for fungicide resistance in plant pathogen populations. *Phytopathology*, **77**, 666–669.

CROFT, B.A. and STRICKLER, K. (1983). Natural enemy resistance to pesticides: documentation, characterization theory and application. In *Pest Resistance to Pesticides* (G.P. Georghiou and T. Saito, eds), pp. 669–702. Plenum Publishing Corp., New York.

CROFT, B.A., BROWN, A.W. and HOYING, S.A. (1976). Organophosphorus resistance in the predacious mite *Amblyseius fallacis. Journal of Economic Entomology*, **69**, 64–68.

CROW, J.F. (1957). Genetics of insect resistance to chemicals. *Annual Review of Entomology*, **2**, 227–246.

EL-LAITHY, A.Y.M. (1986). Zur Resistenz der Raubmilbe *Phytoseiulus persimilis* A.-H. gegenüber Insektoakariziden. *Ph.D. Thesis*, Humboldt-Universität zu Berlin.

GEORGHIOU, P.G. (1972). The evolution of resistance to pesticides. *Annual Review of Ecological Systems*, **3**, 133–168.

GEORGHIOU, P.G. and TAYLOR, C.E. (1977a). Genetic and biological influences in the evolution of insecticide resistance. *Journal of Economic Entomology*, **70**, 319–323.

GEORGHIOU, P.G. and TAYLOR, C.E. (1977b). Operational influences in the evolution of insecticide resistance. *Journal of Economic Entomology*, **70**, 653–658.

HOSKINS, W.M. and GORDON, H.T. (1956). Arthropod resistance to chemicals. *Annual Review of Entomology*, **1**, 89–122.

HOY, M.A. (1985). Recent advances in genetics and genetics improvement of the phytoseiidae. *Annual Review of Entomology*, **30**, 345–370.

HOY, M.A. (1990). Pesticide resistance in arthropod natural enemies: variability and selection responses. In *Pesticide Resistance in Arthropods* (R.T. Roshad and B.E. Tabashnik, eds), pp. 203–236. Chapman and Hall, New York and London.

KONIG, K. and KASSAN, S.A. (1986). Resistenz und Kreuzresistenz der Raubmilben *Phytoseiulus persimilis* A.-H. gegenüber organischen Phosphorsäureestern. *Zeitschrift für angewandte Entomologie*, **101**, 206–215.

LAING, J.E. (1968). Life history and life table of *Phytoseiulus persimilis* A.-H. *Acarologia*, **10**, 578–588.

LYON, M.F. (1968). Chromosomal and subchromosomal inactivation. *Annual Review of Genetics*, **2**, 31–52.

McKENZIE, LA., WHITTEN, M.J. and ADENA, M.A. (1982). The effect of genetic background on the fitness of Diazinon-resistant genotypes of the Australian sheep blowfly *Lucilia cuprina*. *Heredity*, **49**, 1–9.

McMURTRY, J.A. (1982). The use of phytoseiids for biological control: progress and future prospects. In *Recent Advances in Knowledge of the Phytoseiidae* (M.A. Hoy, ed.), pp. 23–48.

OVERMEER, W.P.J. and ZON, A.Q.A van (1983). Resistance to parathion in the predacious mite *Typhlodromus pyri* (Acarina: Phytoseiidae). *Med. Fac. Lnadboum. Rijskuniv.*, **48**, 247–251.

PLAPP, J.R., BROWNING, C.R. and SHARPE, P.J.H. (1979). Analysis of rate of development of insecticide resistance based on simulation of a genetic model. *Environmental Entomology*, **8**, 494–500.

SCHULTEN, G.G.M. and KLASHORST, G. van de (1974). Genetics of resistance to parathion and demethon-s-methyl in *Phytoseiulus persimilis* A.-H. (Acari: Phytoseiidae). *Proceedings of the Fourth International Congress on Acarology, Saalfelden*, pp. 519–524.

SCHULTEN, G.G.M., KLASHORST, G. van de and RUSSELL, V.M. (1976). Resistance of *Phytoseiulus persimilis* A.-H. (Acari: Phytoseiidae) to some insecticides. *Zeitschrift für angewandte Entomologie*, **80**, 337–341.

SILBERMINSZ, I.V. and PETRUSHOV, A.Z. (1980). *Phytoseiulus persimilis* A.-H. Dominant resistance to phosphororganic compounds and its practical use. *Zelshkokhozyajstvennaya biologiaya*, **15**,882–886.

SPERLICH, D. (1973). Populationsgenetik. Grundlagen und experimentelle Ergebnisse. In *Grundlagen der modernen Genetik*, Volume 12, 197pp. Gustav Fischer Verlag, Stuttgart.

TABASHNIK, B.E. (1990). Modelling and evaluation of resistance management tactics. In *Pesticide Resistance in Arthropods* (R.T. Roshad and B.E. Tabashnik, eds), pp. 153–182. Chapman and Hall, New York and London.

46
Analysis of Organophosphorus Resistance Mechanisms in the Predacious Mite *Typhlodromus pyri* from Moravian Vineyards

J. ŠULA and M. ZACHARDA

Czechoslovak Academy of Sciences, Institute of Entomology, Branišovská 31, České Budějovice, ČSFR

Introduction

Previous selection with insecticides in the field can influence the level of resistance to new chemical pesticides through cross-resistance. This resistance can provide benefits when present in key natural predator species (Croft and Strickler, 1983). Pests of vine in South Moravia have been exposed extensively to organophosphorus (OP) insecticide and miticide spraying. *Typhlodromus pyri* Scheuten, an important predator of spider and eriophyid mites in orchards and vineyards, has developed resistance to OP and carbamate insecticides (Kapetanakis and Cranham, 1983; van de Baan *et al.*, 1985; Šula and Zacharda, 1991). Resistance to azinphos-methyl was found to be relatively low while higher resistance has been found to parathion and carbaryl (Kapetanakis and Cranham, 1983) and azinphos-ethyl (Šula and Zacharda, 1991). The biochemical basis of pesticide resistance has been studied in several species of phytoseiid mites (Motoyama *et al.*, 1971; Mullin *et al.*, 1982, Roush and Plapp, 1982; Scott *et al.*, 1983; van de Baan *et al.*, 1985; Fournier *et al.*, 1987; Anber and Overmeer, 1988; Anber and Oppenoorth, 1989; Mueller *et al.*, 1989). The resistance based on higher hydrolytic, glutathione

Insecticides: Mechanism of Action and Resistance
© 1992   Intercept Ltd, P.O. Box 716, Andover, Hants SP10 1YG, UK

S-transferase and/or mixed function oxidase activity or reduced sensitivity of a target enzyme, acetylcholinesterase, was noticed in resistant phytoseiid mites (Mullin *et al.*, 1982; Scott *et al.*, 1983; Chang and Whalon, 1986).

Resistance of predators is generally considered to be important in integrated pest management (IPM) practices when insecticides are still indispensable and predators should survive their use. Information on the biochemical basis of this resistance can help to make decisions regarding what pesticides are compatible with these predators. In this chapter azinphos-ethyl resistance detected in a population of *T. pyri* from South Moravian vineyards was studied. Since the resistant populations of *T. pyri* are crucial agents for controlling spider and eriophyid mites in vineyards and apple orchards, we tried to establish the biochemical basis of the resistance in one of them.

Material and Methods

Samples of susceptible (S) population of *Typhlodromus pyri* Scheuten were collected on a horse chestnut (*Aesculus hippocastaneum*) in a cemetery near České Budějovice where no pesticides have been sprayed. A population sample of *T. pyri* resistant (R) to OP pesticides was collected in a regularly sprayed vineyard in South Moravia near Mikulov. Mite cultures were reared under laboratory conditions (20–27 °C, 90% r.h. and eighteen hours daylight as described by McMurtey and Scriven, 1965). Mites were fed with all stages of *Tetranychus urticae*. Slightly modified residual leaf-dip testing (Hoy and Knop, 1979) was used to assess the levels of susceptibility of these mites to Gusathion E 40. Results of toxicological tests were treated statistically using a probit analysis to obtain values of LC_{50} after twenty-four and forty-eight hours. Long-term residual toxicities of OP compounds Gusathion E 40 (azinphos-ethyl), Metation E 50 (fenitrothion), Decemtion EK 20 (phosmet), and Zolone EC 35 (phosalone) were determined using the same bioassay technique. Mortalities were assessed each twenty-four hours within seven days. The long-term residual effects of the pesticides were expressed as cumulative percentage mortalities.

Mites were homogenized in a respective buffer and centrifuged at $10\,000 \times g$ at 0 °C. Supernatants were then used for the measurements of enzyme activities.

The esterase activity was measured spectrophotometrically with 0.75 mM 4-nitrophenylacetate (4-NFA) and 1-naphthylacetate (1-NA) as substrates in 0.07 M phosphate buffer pH 7.0 according to

Huggins and Lapides (1947) and Johnston and Ashford (1980), respectively.

Glutathione *S*-transferase activity has been determined according to Habig *et al.* (1974) with 5 mM reduced glutathione and 1 mM 1-chloro-2,4-dinitrobenzene (CDNB) as substrates. Product formation was calculated using extinction coefficient 9.6 mM^{-1} cm^{-1}.

Acetylcholinesterase activity was determined according to Ellman *et al.* (1961). Acetylthiocholine iodine was used as a substrate in a concentration range of 6–100 μM. Kinetic constants K_m and V were determined according to Wilkinson (1961).

Acetylcholinesterase inhibition was performed according to Hart and O'Brien (1974) using azinphos-ethyl as the inhibitor in a concentration range of 1×10^{-6} –5×10^{-5} M at 25 °C. The inhibitor, dissolved in ethanol, was added to a reaction mixture. Final substrate (acetylthiocholine iodine) concentrations were 10 to 100 μM.

Blanks without mites were used to correct for non-enzymatic activity in all enzyme activity determinations.

Results

Toxicological bioassays

Residual toxicity leaf-dip tests discovered some levels of resistance to OP insecticides in the R strain of *T. pyri*. The R strain was up to 227-fold resistant to azinphos-ethyl (LC$_{50}$ value for susceptibility to Gusathion E 40 was 0.4425% and 0.0020% of active ingredient for the R and S strains, respectively).The tests aimed at determinations of long-term residual toxicities of OP pesticides at their highest recommended concentrations for spraying in vineyards also corroborated these results. Approximately 10 to 25% of samples of the field populations inhabiting sprayed vineyards survived and recovered from one-week exposure to the pesticides.

Esterase activity

As shown in Table 1 the enzyme activity of the R strain was significantly higher (2.6-fold to 5-fold, $P < 0.05$) than in the S strain when 4-NFA or 1-NA was used as a substrate, respectively. Affinity to 1-NA was in both strains higher than to 4-NFA, the difference being greater in the R strain.

Glutathione *S*-transferases

Activity of glutathione *S*-transferases was measured with CDNB as a substrate and reduced glutathione as a co-substrate. The R strain possessed approximately 1.6 times higher activity than the S strain (Table 1).

Table 1 Esterase and glutathione S-transferase in *T. pyri* females*

| Strain | Esterase | | Glutathione |
	1-NA	4-NFA	S-transferase
S (n)	1.84 (3)	1.5 (4)	7.88 (6)
	± 0.27	± 0.60	± 3.06
R (n)	9.48 (3)	3.04 (4)	12.56 (5)
	± 0.73	± 0.92	5.98

* Activity expressed as nmol per mite per h.

The results of the measurements of acetylcholinesterase activity are shown in Table 2. The activity in the R strain is significantly lower (P < 0.005) than in the S strain – roughly 9-fold. The K_m values for splitting the acetylthiocholine iodine differ in the R and S strains, being approximately 1.22×10^{-5}M and 2.74×10^{-5}M, respectively. The results of the inhibition experiments of strains S and R with azinphos-ethyl are shown in Table 2. The rate of inhibition, expressed as k_i, bimolecular phosphorylation rate constant, in the R strain is roughly 7-fold slower than that of the S strain. Both the equilibrium dissociation constant K_d and the phosphorylation constant k_2 are smaller in the R strain than in the S strain.

Table 2 Acetylcholinesterase activity and inhibition in *T. pyri* females

Strain	K_m (M^{-1}) x 10^{-5}	AChE activity V (nmol per mite per h)	K_d (M^{-1}) x 10^{-5}	Inhibition constants k_2 (min^{-1})	k_i (M^{-1} min^{-1}) x 10^4
S (n)	2.74 (4)	2.19	8.55	6.06	7.08
	± 1.43	± 0.69			
R (n)	1.22 (4)	0.27	17.39	1.70	0.98
	± 0.96	± 0.16			

Discussion

Our results indicate an important role of esterases in *T. pyri* resistance to the azinphos-ethyl. In Phytoseiidae several authors found non-specific esterase activity as a mechanism of resistance to OP and pyrethroid insecticides. In *A. fallacis* similarly enhanced esterase activity was observed in the multiresistant strain (Mullin *et al.*, 1982) and in the permethrin-resistant strain (Scott *et al.*, 1983; Chang and Whalon, 1986) using 1-NA as a substrate. Scott and co-authors (1983) considered the esterases to be the main resistance mechanism. The role of esterase activity in azinphos-ethyl detoxification in our resistant population of *T. pyri* thus seems to be important judging from the difference between activities in the S and R strains. Besides higher esterase activity raised glutathione *S*-transferase activity was noticed, though this increase was not as high as in the case of esterases. Similar elevation of transferase activity was found by Mullin *et al.* (1982) in the multiresistant strain of *A. fallacis*. Glutathione *S*-transferases are supposed to be active mainly against methyl-analogues of phosphorodithioates (Croft and Strickler, 1983) and their importance in resistance to OPs was ascertained also in *P. persimilis* in resistance to methidathion (Fournier *et al.*, 1987, 1988). The finding of a 7-fold slower rate of inhibition by azinphos-ethyl in the acetylcholinesterase of the R strain of *T. pyri* shows another cause of resistance. A similar mechanism was found in spider mites, ticks, rice leafhoppers, mosquitoes, houseflies and also in phytoseiid mites. The first evidence was published by van de Baan *et al.* (1985) in *T. pyri*, resistant to parathion and propoxur. The activity of the acetylcholinesterase in the R strain was significantly lower and the bimolecular rate constant differed in comparison with the S strain 36-fold and 14-fold for paraoxon and propoxur, respectively. A similar case was observed by Anber and Overmeer (1988) in *A. potentillae* in the parathion-resistant strain. The activity of the modified enzyme was 2-fold lower than that of the susceptible strain and the k_i value was 19-fold and 311-fold lower for the inhibition by azinphos-methyl and paraoxon, respectively. Mueller *et al.* (1989) described an approximately 2-fold decrease in k_i for the inhibition of acetylcholinesterase by omethoate in the dimethoate-resistant strain of *P. persimilis*.

It has been shown that in most cases only one or two mechanisms of resistance were found in analysed mites and only in *A. fallacis* several detoxification enzyme systems were revealed (Mullin *et al.*, 1982). In our *T.pyri* field strain metabolically active detoxification systems are present together with altered acetylcholinesterase. This insensitive enzyme, enables some of the *T. pyri* mites to survive pesticide treatments in the field as it favours the effect of detoxication. It results in selection and stabilization of phytoseiid mite field population consisting of one species – *T. pyri*, compatible with all routine IPM practices which have been performed in chemically sprayed commercial vineyards in South Moravia for many years.

References

ANBER, H.A.I. and OPPENOORTH, F.J. (1989). A mutant esterase degrading organophosphates in a resistant strain of the predacious mite *Amblyseius potentillae* (Garman). *Pesticide Biochemistry and Physiology*, **33**, 283–297

ANBER, H.A.I. and OVERMEER, W.P.J. (1988). Resistance to organophosphates and carbamates in the predacious mite *Amblyseius potentillae* (Garman) due to insensitive acetylcholinesterase. *Pesticide Biochemistry and Physiology*, **31**, 91–98.

CHANG, C.K. and WHALON, M.E. (1986). Hydrolysis of permethrin by pyrethroid esterases from resistant and susceptible strains of *Amblyseius fallacis* (Acari: Phytoseiidae). *Pesticide Biochemistry and Physiology*, **25**, 446–452.

CROFT, B.A. and STRICKLER, K. (1983). Natural enemy resistance to pesticides: documentation, characterization theory and application. In *Pest Resistance to Pesticides* (G.P. Georghiou and T. Saito, eds), pp. 669–702. Plenum Publishing Corp., New York.

ELLMAN, G.L., COURTNEY, K.D., ANDERS, V. and FEATHERSTONE, R.M. (1961). A new and rapid colorimetric determination of acetylcholinesterase activity. *Biochemical Pharmacology*, **7**, 88–95.

FOURNIER, D., CUANY, A., PRALAVORIO, M., BRIDE, J.M. and BERGE, J.B. (1987). Analysis of methidathion resistance mechanisms in *Phytoseiulus persimilis* A.-H. *Pesticide Biochemistry and Physiology*, **28**, 271–278.

FOURNIER, D., PRALAVORIO, M., CUANY, A. and BERGE, J.B. (1988). Genetic analysis of methidathion resistance in *Phytoseiulus persimilis* (Acari: Phytoseiidae). *Journal of Economic Entomology*, **81**, 1008–1013.

HABIG, W.H., PABST, M.J. and JAKOBY, W.B. (1974). Glutathione S-transferases: The first enzymatic step in mercapturic acid formation. *Journal of Biological Chemistry*, **249**, 7130–7139.

HART, G.J. and O'BRIEN, R.D. (1974). Stopped-flow studies of the inhibition of acetylcholinesterase by organophosphates in the presence of substrate. *Pesticide Biochemistry and Physiology,***4,** 239–244.

HOY, M.A. and KNOP, N.F. (1979). Studies on pesticide resistance in the phytoseiid *Metaseiulus occidentalis* in California. In *Recent Advances of Acarology* (J.G. Rodrigues, ed.), Vol. I, pp. 89–94. Academic Press, New York.

HUGGINS, C. and LAPIDES, J. (1947). Chromogenic substrates: IV. Acylesters of *p*-nitrophenol as substrates for the colorimetric determination of esterase. *Journal of Biological Chemistry*, **170,** 467–482.

JOHNSTON, K.J. and ASHFORD, A.E. (1980). A simultaneous-coupling azo dye method for the quantitative assay of esterase using α-naphthyl acetate as substrate. *Histochemical Journal*, **12,** 221–234.

KAPETANAKIS, E.G. and CRANHAM, J.E. (1983). Laboratory evaluation of resistance to pesticides in the phytoseiid predator *Typhlodromus pyri* from English Apple orchards. *Annals of Applied Biology*, **103,** 389–400.

McMURTEY, J.A. and SCRIVEN, G.T. (1965). Insectary production of phytoseiid mites. *Journal of Economic Entomology*, **58,** 282–284.

MOTOYAMA, N., ROCK, G.C., and DAUTERMAN, W.C. (1971). Studies on the mechanism of azinphosmethyl resistance in the predacious mite, *Neoseiulus (T) fallacis* (family: Phytoseiidae). *Pesticide Biochemistry and Physiology*, **1,** 205–215.

MUELLER, A., ADAM, H. and EL-LAITHY, A.Y.M. (1989). Erste Befunde zum Resistenzmechanismus einer gegen Dimethoat resistenten Zuchtlinie der Raubmilbe *Phytoseiulus persimilis* A.-H. *Archiv für Phytopathologie und Pflanzenschutz, Berlin*, **25,** 73–79.

MULLIN, C.A., CROFT, B.A, STRICKLER, K., MATSU-MURA, F. and MILLER, J.R. (1982). Detoxification enzyme differences between a herbivorous and predatory mite. *Science*, **217,** 1270–1272.

ROUSH, R.T. and PLAPP, JR, F.W. (1982). Biochemical genetics of resistance to aryl carbamate insecticides in the predacious mite, *Metaseiulus occidentalis. Journal of Economic Entomology*, **75,** 304–307.

SCOTT, J.G., CROFT, B.A. and WAGNER, S.W. (1983). Studies on the mechanism of permethrin resistance in *Amblyseius fallacis* (Acarina: Phytoseiidae) relative to previous insecticide use on apple. *Journal of Economic Entomology*, **76,** 6–10.

ŠULA, J. and ZACHARDA, M. (1991). Differential potential of three predacious mites (Acari: Phytoseiidae) to develop resistance to azinphos-ethyl. In *Modern Acarology*, Vol. 2 (F. Dusbábek and V. Bukva, eds). Co-published by SPB Academic Publishing bv, The Hague, and Academia Prague, in press.

VAN DE BAAN, H.E., KUIJPERS, L.A.M. OVERMEER, W.P.J. and OPPENOORTH, F.J. (1985). Organophosphorus and carbamate resistance in the predacious mite *Typhlodromus pyri* due to insensitive acetylcholinesterase. *Experimental and Applied Acarology*, **1**, 3–10.

WILKINSON, G.N. (1961). Statistical estimations in enzyme kinetics. *Biochemical Journal*, **80**, 324–332.

47
Propargite Metabolism in the Bulb Mite *Rhizoglyphus echinopus* (Fumouze and Robin) (Acari: Acaridae)

CHARLES O. KNOWLES* and MOHAMED S. HAMED

Department of Entomology, University of Missouri, Columbia, Missouri 65211

The specific acaricide propargite has been used for about two decades to control a variety of pest mites on numerous crops. There have been several reports of the development of resistance by spider mites to propargite (Mansour and Plaut, 1979; Chapman and Penman, 1984; Keena and Granett, 1985, 1987; Dennehy *et al.*, 1987; Zil'bermints and Zhuravleva, 1988). We recently found that a strain of bulb mite, *Rhizoglyphus echinopus* (Fumouze and Robin), was tolerant to this acaricide (Knowles *et al.*, 1988). Apparently there have been no studies of the fate of propargite in any acarine species. Therefore, we examined the metabolism of radioactive propargite in bulb mites.

Materials and Methods

Mites
Bulb mites were reared at high humidity in the dark at $26 \pm 1°C$ in Petri dishes containing a wheat-germ-based medium designed for acarid mites (Bot and Meyer, 1967).

---

*To whom correspondence should be addressed

Insecticides: Mechanism of Action and Resistance
© 1992   Intercept Ltd, P.O. Box 716, Andover, Hants SP10 1YG, UK

Compounds

Propargite labelled with radiocarbon in the phenyl moiety (sp. act. 30 mCi/mmol) in benzene solution containing 2% propylene oxide stabilizer was provided by Uniroyal Chemical Co., Inc. (Naugatuck, Conn.). Using thin-layer chromatography (TLC) (see below) it was found to be >98% radiochemically pure. Also provided by Uniroyal were non-radioactive samples of propargite and two potential metabolites *t*-butylphenol and 2-(4-*t*-butylphenoxy)cyclohexanol (glycol ether).

Metabolism studies

Glass vials (5 ml) were coated with 0.2 ml of a propargite solution containing about 100 000 dpm, and the solvent was allowed to evaporate. To each vial were added 50 mg (about 250) of bulb mites. Mites were left in the vials for 1 h; during the initial 2 min, the vials were rotated on a mini-mill at 60 rpm to facilitate contact of the mites with the acaricide. After the 1 h exposure, mites were transferred to holding vials. The radiocarbon remaining in the exposure vials were estimated by liquid scintillation counting (Beckman LS 7500, Beckman Instruments, Fullerton, Calif.) as described by Hamed and Knowles (1988).

Treated mites were kept in holding vials for 6, 12, 24 and 48 h after exposure, and the analysis of metabolism was conducted as described previously (Hamed and Knowles, 1988) with the following exception. Because some of the mites usually adhered to the sides of the holding vial and thus were difficult to remove, we previously used methylene chloride to assist with their transfer. This procedure resulted in the radioactive material in the holding vial as a consequence of 'rub-off' and excretion or both and the radioactive material in the external mite rinse being together in one fraction. In the present study we carefully removed the mites from the holding vial and then rinsed the vial with methylene chloride. Radioassay of an aliquot of the methylene chloride yielded the total radiocarbon in the holding vial rinse. Other fractions analysed for total radiocarbon included the external mite rinse, internal mite extract, and the mite residue (Hamed and Knowles, 1988).

The nature and concentration of the radiocarbon in the holding vial rinse and that in the internal mite extract were determined by TLC on glass plates coated with silica gel GF_{254}. The solvent system was cyclohexane–ethyl acetate (4:1) average R_f values were 0.71 for propargite, 0.54 for *t*-butylphenol, and 0.46 for glycol ether. The radioactivity in the silica gel zones was quantitated (Hamed and

Knowles, 1988).

To examine *in vitro* degradation of propargite, bulb mites were homogenized in 0.2 M potassium phosphate buffer (pH 7.5) at a concentration of 60 mg per ml. The homogenate was centrifuged at 12 000 x *g* for 15 min at 4°C to yield supernatant and pellet fractions. The pellet was resuspended in buffer at the original volume. [^{14}C]Propargite was dissolved in 95% ethyl alcohol, and distilled water was added to give an acaricide concentration of 10 nmol per ml.

The incubation system consisted of 0.2 ml of whole homogenate, supernatant, or pellet equivalent to 12 mg of mites (wet weight); 0.1 ml of propargite solution equivalent to 1 nmol; and 0.1 ml of buffer or of a buffer solution containing 0.106 mg of NADPH (Sigma Chemical Co., St Louis, Missouri) or 0.122 mg of glutathione (GSH) (Sigma) or both. These mixtures were incubated with shaking in a water bath at 30°C for 1 h. Controls in which buffer was substituted for mite preparation and for NADPH or GSH or both were incubated to estimate non-enzymatic degradation of propargite. Reactions were stopped with 0.2 ml of 1 M hydrochloric acid and extracted with ethyl ether (3 × 3 ml). Radioactivity in the ether extract and aqueous fraction was measured (Hamed and Knowles, 1988), and the ether extract was subjected to TLC as mentioned above.

Results

At the end of the 1 h exposure period, $88 \pm 4\%$ ($n = 12$) of the radiocarbon remained in the holding vial and $12 \pm 4\%$ was transferred to the mites. Thus the mite dose was about 1.27 µg/g. The total recovery of applied radiocarbon for exposure and holding phases was $102.6 \pm 3.3\%$ ($n = 12$). TLC of rinses of propargite-treated vials incubated for 1 h in the absence of bulb mites revealed only the presence of the parent compound. Therefore, the propargite was stable during the exposure period.

The distribution, nature and amount of radiocarbon following exposure of bulb mites to propargite are given in Table 1. The total external radiocarbon, which consisted of the sum of that in the holding vial and external mite rinses, was relatively constant during the 48 h period with a mean of $84 \pm 1.3\%$ ($n = 4$). However, there were differences in the relative levels of radiocarbon in the holding vial rinse and in the external mite rinse ($56.6 \pm 5.9\%$, $n = 4$) was about twice that in the holding vial rinse ($27.5 \pm 6.0\%$, $n = 4$). At 6, 12 and 48 h post-exposure, propargite was the major compound in the holding vial rinses and comprised >50% (84.9%, 76.3%, 51.0%) of the radioactivity in that fraction. However, at 48 h propargite had decreased to 7.8% of the total recovered radioactivity and 32.6% of the radioactivity in the holding vial rinse. Low levels of the metabolites *t*-butylphenol and glycol ether were present in the holding vial rinses at most analysis times, but much of the non-parent radiocarbon in the holding vial rinse was unidentified (remainder and origin).

Table 1 Distribution, nature and relative concentration of radiocarbon following exposure of bulb mites to propargite

Distribution and nature of radiocarbon	x̄ (SD) % Radiocarbon at indicated time (h)*			
	6	12	24	48
Holding vial rinse	21.9(3.3)	35.5(5.3)	28.6(1.7)	23.9(4.6)
Propargite	18.6(0.2)	27.1(0.1)	14.6(0.1)	7.8(0.1)
t-Butylphenol	0.4(<0.1)	<0.1(<0.1)	0.2(<0.1)	<0.1(<0.1)
Glycol ether	1.0(<0.1)	0.5(<0.1)	0.7(<0.1)	0.4(<0.1)
Remainder	1.4(0.1)	5.6(0.1)	9.5(0.1)	12.5(0.1)
Origin	0.5(<0.1)	2.3(<0.1)	3.6(0.1)	3.2(0.1)
External mite rinse	63.1(0.4)	49.7(4.9)	54.0(2.6)	59.1(4.5)
Internal mite extract	9.6(2.5)	8.0(0.5)	8.9(1.0)	8.4(0.3)
Propargite	5.1(0.1)	3.9(<0.1)	3.2(<0.1)	2.5(<0.1)
t-Butylphenol	0.3(<0.1)	0.2(<0.1)	0.1(<0.1)	0.1(<0.1)
Glycol ether	0.5(<0.1)	0.3(<0.1)	0.3(<0.1)	0.3(<0.1)
Remainder	1.6(<0.1)	2.3(<0.1)	3.6(<0.1)	4.4(<0.1)
Origin	1.3(0.1)	1.3(<0.1)	1.7(<0.1)	1.1(<0.1)
Mite residue	5.4(1.2)	6.8(1.0)	8.5(0.7)	8.6(0.2)

* $n = 3$.

The total internal radiocarbon, which consisted of the sum of that in the internal mite extract and mite residue, also was relatively constant during the 48 h period with a mean of 16.1% ± 1.3% (n = 4) (Table 1). Initially levels of radiocarbon in the mite extract were higher than those in the mite residues. However, radiocarbon levels in the mite extract subsequently decreased with a concomitant increase in radioactivity in the residue with the result that radioactivity levels in the two fractions were about equal at 48 h. Propargite levels in the internal extract decreased from 5.1% at 6 h to 2.5% of the recovered radiocarbon at 48 h. Low levels of *t*-

butylphenol and glycol ether were present as was unidentified radioactive material (remainder and origin).

Table 2 gives the degradation of propargite by bulb mite homogenates. Low activity was observed with the whole homogenate and the 12 000 × *g* pellet; however, the 12 000 × *g* supernatant was quite active. Fortification of the 12 000 × *g* supernatant with NADPH or GSH greatly enhanced propargite metabolism. Propargite was stable in buffer alone or in the presence of buffer and NADPH, but GSH in buffer degraded about 15% of the propargite to *t*-butylphenol during the 1 h incubation. Thus not all of the degradation in the presence of GSH was catalysed by the mite supernatant. Although *t*-butylphenol and glycol ether were detected as propargite degradation products, most of the radioactive material was not extractable with ethyl ether and remained in the aqueous fraction.

Discussion

The probable paths for propargite metabolism in bulb mites are presented in Figure 1. Theoretically, propargite can undergo initial cleavage at two sites, the ester linkage between the sulphur-containing side chain and cyclohexyl ring and the ether linkage connecting the cyclohexyl and phenyl moieties. Cleavage of the former would yield the glycol ether and propargyl alcohol, whereas cleavage of the latter would yield *t*-butylphenol and a compound containing the cyclohexyl moiety with the intact sulphur-containing side chain. The *t*-butylphenol also could arise from cleavage of the ether linkage of glycol ether; the other product of this reaction would by the cyclohexane diol. Since the radiolabel in our sample of propargite was associated with the substituted phenyl moiety, we could not detect compounds such as propargyl alcohol and cyclohexane diol that would assist in establishing product–precursor relationships. We suspect that the *t*-butylphenol was from glycol ether as shown. However, the fact that only *t*-butylphenol was detected from propargite in the presence of GSH and buffer indicates that some cleavage at the ether linkage of propargite itself may have occurred, at least non-enzymatically.

Propargite was toxic to two-spotted spider mites, *Tetranychus urticae* Koch, in laboratory studies. The 48 h LC_{50} for spider mites dipped for 5 s in propargite dissolved in acetone–water was 128 ppm (Knowles *et al.*, 1988). In contrast, however, only 35% mortality was observed at 72 h when bulb mites were dipped for 5 s in an emulsion of propargite at 1000 ppm (Knowles *et al.*, 1988). Thus, it was

Table 2   Degradation of propargite by bulb mite homogenates*

ng per h per 100 mg mite tissue (wet weight)

Product	Whole homogenate	Pellet	Supernatant	Supernatant + NADPH	Supernatant + GSH	Supernatant + NADPH + GSH
Organic extract						
t-Butylphenol	2.9	1.5	3.9	14.6	61.3	42.4
Glycol ether	<1.0	<1.0	<1.0	16.1	144.2	188.4
Remainder	1.5	<1.0	10.3	1.5	27.8	109.3
Origin	<1.0	<1.0	<1.0	1.5	17.5	17.5
Aqueous extract	61.3	62.8	261.4	365.0	438.0	502.2
Total degraded	65.7	64.3	275.6	398.7	688.9	859.8

* Whole homogenate was centrifuged at 12 000 × g to yield supernatant and pellet fractions. A 0.2 ml aliquot equivalent to 12–mg of mite tissue (wet weight) was incubated for 1 h with 1 nmol of propargite. Values are means of duplicate analyses.

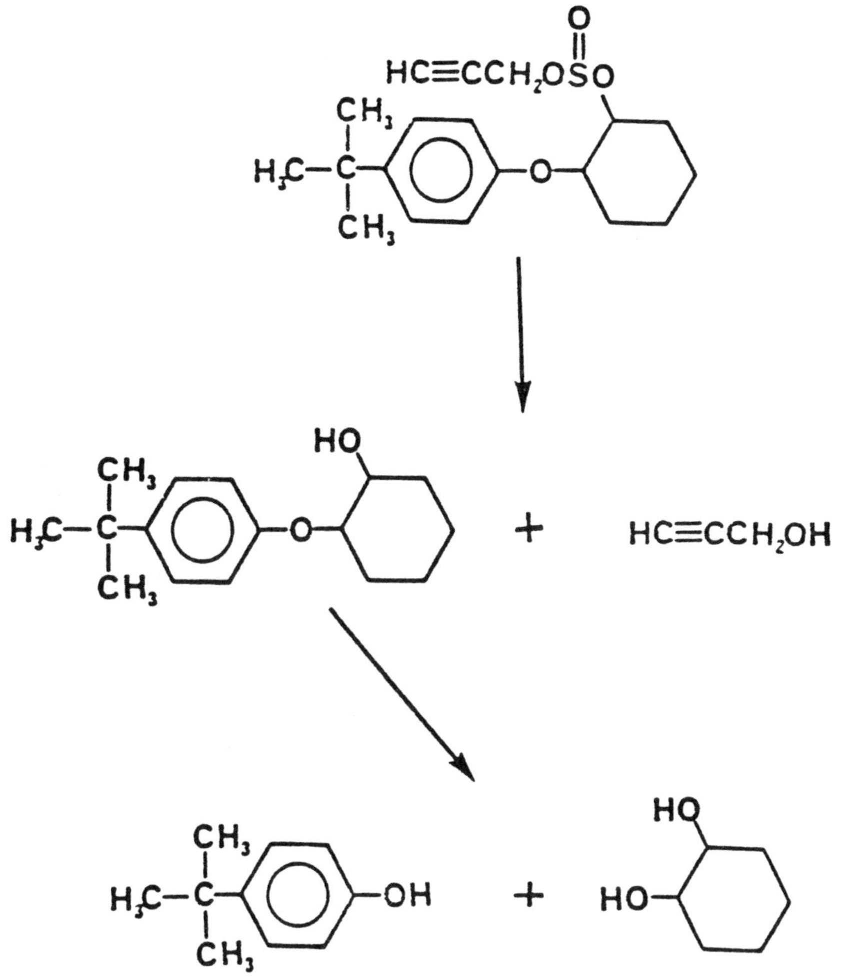

Figure 1   Probable paths for propargite metabolism in bulb mites.

concluded that propargite was not toxic to this particular strain of bulb mites. The results of the pharmacokinetic studies described herein reveal that propargite penetrated into bulb mites and was metabolized and excreted, although the rates of these processes did not appear to be very fast. Further, some of the propargite or metabolite(s) or both were bound to mite tissue. Internal levels of unbound propargite in the mite extract ranged from 5.1% of the recovered radiocarbon at 6 h to 2.5% at 48 h. Assuming that propargite itself is the actual toxicant it would appear that these levels might have been sufficiently high for successful attack on a sensitive target. However, until the target of this acaricide in bulb mite is elucidated definitive conclusions cannot be made.

Acknowledgement

We thank Habsah A. Kadir for technical assistance.

This research is a contribution from the Missouri Agricultural Experimental Station, Columbia. Journal Series No. 10753.

References

BOT, J. and MEYER, M.K.P. (1967). An artificial rearing medium for acarid mites. *Journal of Entomological Society of South Africa*, **29**, 199.

CHAPMAN, R.B. and PENMAN, D.R. (1984). Resistance to propargite by European red mite and two-spotted mite. *New Zealand Journal of Agricultural Research*, **27**, 103–105.

DENNEHY, T.J., GRANETT, J., LEIGH, T.F. and COLVIN, A. (1987). Laboratory and field investigations of spider mite (Acari: Tetranychidae) resistance to the selective acaricide propargite. *Journal of Economic Entomology*, **80**, 565–574.

KENNA, M.A. and GRANETT, J. (1985). Variability in toxicity of propargite to spider mites (Acari: Tetranychidae) from California almonds. *Journal of Economic Entomology*, **78**, 1212–1216

KENNA, M.A. and GRANETT, J. (1987). Cyhexatin and propargite resistance in populations of spider mites (Acari: Tetranychidae) from California almonds. *Journal of Economic Entomology*, **80**, 560–564.

KNOWLES, C.O., ERRAMPALLI, D.D. and EL-SAYED, G.N. (1988). Comparative toxicities of selected pesticides to bulb mite (Acari: Acaridae) and twospotted spider mite (Acari: Tetranychidae). *Journal of Economic Entomology*, **81**, 1586–1591.

MANSOUR, F.A. and PLAUT, H.N. (1979). The effectiveness of various acaricides against resistant and susceptible carmine spider mites. *Phytoparasitica*, **7**, 185–193.

ZIL'BERMINTS, I.V. and ZHURAVLEVA, L.M. (1988). Resistance of red spider mite to recently developed acaricides. *Agrokhimiya* (5), 111–116 (in Russian). *Chemical Abstracts*, **109**, 33832v, 1988.

Author Index